Lecture Notes in Economics and Mathematical Systems

Managing Editors: M. Beckmann and H. P. Künzi

Mathematical Programming

199

Evaluating Mathematical Programming Techniques

Proceedings of a Conference
Held at the National Bureau of Standards
Boulder, Colorado
January 5–6, 1981

Edited by John M. Mulvey

Springer-Verlag
Berlin Heidelberg New York 1982

Managing Editors
Prof. Dr. M. Beckmann
Brown University
Providence, RI 02912, USA

Prof. Dr. H.P. Künzi
Universität Zürich
CH-8090 Zürich, Schweiz

Editor
Prof. John M. Mulvey
School of Engineering/Applied Science
Princeton University
Princeton, NJ 08544, USA

Sponsored by:
Committee on Algorithms, Mathematical Programming Society
Center for Applied Mathematics, National Bureau of Standards
National Science Foundation (MSC-8007250)
Department of Energy

ISBN 3-540-11495-5 Springer-Verlag Berlin Heidelberg New York
ISBN 0-387-11495-5 Springer-Verlag New York Heidelberg Berlin

Library of Congress Cataloging in Publication Data
Main entry under title: Evaluating mathematical programming techniques. (Lecture notes in economics and mathematical systems; 199) Sponsored by Committee on Algorithms, Mathematical Programming Society and others. Bibliography: p. Includes index.
1. Programming (Mathematics)--Congresses. I. Mulvey, J.M. (John M.) II. Mathematical Programming Society (U.S.). Committee on Algorithms. III. Series. QA402.5.E94 519.7
82-5603 AACR2
ISBN 0-387-11495-5 (U.S.)

Printing and binding: Beltz Offsetdruck, Hemsbach/Bergstr.
2142/3140-543210

ABSTRACT

This book summarizes a two day conference that was held in Boulder, Colorado (January 5-6, 1981) to consider how mathematical programming techniques ought to be evaluated. Approximately fifty academic researchers, software developers and users participated in this first meeting devoted exclusively to computational testing of mathematical programs.

Following the introductory addresses, the remainder of the report is organized around nine topics:

1. Design and use of problem generators and hand selected test cases
2. Non-linear optimization codes and empirical tests
3. Integer programming and combinatorial optimization
4. Comparing three computational studies
5. Testing methodologies
6. Approaches to software evaluation from other disciplines
7. Special topics
8. Advances in network programs
9. On establishing a group for testing mathematical programs

The Committee on Algorithms (COAL) of the Mathematical Programming Society (MPS) sponsored two panel discussions -- one dealing with a comparative evaluation of three computational studies, and one involving the establishment of a group (or center) for testing mathematical programming techniques. Summaries of these sessions are contained in sections 4 and 9, respectively.

WELCOMING REMARKS

by

John M. Mulvey

Welcome to sunny Boulder, Colorado. This first conference on "Testing and Evaluating Mathematical Programming Algorithms and Software" has many sponsors, including the Committee on Algorithms (COAL) of the Mathematical Programming Society, the National Bureau of Standards, the National Science Foundation, and the Department of Energy. For their financial and moral support, we are most grateful.

Computational testing, or what Peter Denning in the October, 1980 issue of CACM called experimental computer science, has been greatly scrutinized over the past several years. The Feldman report stimulated extensive discussions about the role of experimentation in computer science. Also, the NSF has begun to emphasize these activities in their funding appropriations.

In the field of mathematical programming, researchers who carry out computational testing are confronted with a variety of issues for which there is little unanimity. Some examples are the following: What type of test cases should be used -- randomly generated or hand picked? If random problems are used how can the structure of the real-world be introduced so that the conclusions are meaningful? If hand picked, how do we know that they are representative? What evaluation criteria (performance measures) should be employed? How can testing be conducted so that an algorithm's effectiveness can be inferred from small or moderate size examples, without testing large-scale examples? Is this feasible? The basis for the conference lies in confronting these difficult issues.

As a matter of history, computational tests have been conducted at least since 1953, when Alan Hoffman et al. compared the simplex method, the fictitious play method of George Brown, and the relaxation method of T.S. Motskin. We have witnessed a rapid increase in this activity since 1967. Three computational studies were reported at the 1967 MPS International Symposium in Princeton, New Jersey. By 1976 at the Budapest Symposium, the number of computational studies had grown to 12. And in Montreal (1979), 32 computational analyses were presented. Clearly, this field is growing and becoming a respectable topic for research.

Despite these encouraging signs, a series of storm clouds lie on the horizon.

1. Often, more questions are raised during an empirical investigation than are answered. A consensus among researchers is rarely reached.
2. Computational experiments are extremely expensive undertakings. The results may not generalize outside the immediate set of test problems solved.
3. There are few maxims to reduce the computational burden or to maximize the benefits of empirical testing. Statistical experimental design principles may not always apply.

These problems lead to frustration and occassional withdrawal by prominent researchers. Frankly, it is often easier to theorize than to experiment.

Yet it appears that real progress has been made over the past few years. The purpose of this conference is to document past accomplishments and to develop a plan of action for future activities. If this area of study is to become a recognized discipline, we must draw attention to solid contributions -- as well as pinpoint substandard scholarship.

Due to a serious illness in his family, Dr. Phillip Wolfe is unable to attend this conference and give the keynote address. He asked me to present the following statement.

"I am very sorry that family problems have prevented me from attending this meeting, to which I had been looking forward a long time. My talk was going to consist of two parts: the first, some anecdotes and horror stories from the early days of competition between various algorithms and proposals for solving linear and nonlinear programming problems; the second, praise for the very significant work of the Committee on Algorithms of the Mathematical Programming Society in helping get the whole business sorted out.

Now it is much more fun to relate anecdotes than it is to write them down for someone else to read, and I suspect it is even more fun to write them down than to have to listen to them, so you will be spared the first part. However, I do not want to skip the second part, because I think that supporting the Committee on Algorithms has been about the most valuable service that the Society has performed for the mathematical programming community. I need not belabor that point; the fact that you are here indicates that you are likely to share that view, and I am sure that most members of the Society do, too. Instead, it might be of interest to point out the influence the Committee itself has had on the Society's activities in general.

When it was formed in 1971, the Mathematical Programming Society had only two functions: support of the journal, Mathematical Programming, and ensuring the continuation of the International Symposia on Mathematical Programming. But for the efforts of a few people, that might still be the case today. However, at its first council meeting, in 1973, Michael Powell volunteered to head a Committee on Activities to investigate what other things the Society might do, and early in 1974 a report from the Committee proposed guidelines for a "Working Committee on Algorithms."

For a while the "Working Committee" did not live up to its name, but under Harvey Greenberg's leadership in 1975 it began to take off. Its first public sessions, at the Ninth Mathematical Programming Symposium in Budapest in 1976, were very well received by the public; however, they had a further effect on the rather small group that had been nurturing mathematical programming's public role since 1971 and before. They showed us an important activity the Society could support in pursuit of its aims to help mathematical programming in general, and there was a new group of enthusiastic people willing to carry it out on their own. For me, at least, this was inspiring; it was evidence that the Society could really amount to something in the technical world, and that further steps should be taken to make the Society a solid base for professional activity. We thus proceeded to reorganize the financing of the Journal to bring more of its cash flow into the Society; revised the Constitution, broadening the scope of the Society's envisaged activities; began a program of support and sponsorship of other meetings in mathematical programming; recently, founded the Society's Newsletter OPTIMA, following the lead of the COAL Newsletter; and, feeling that we really had something to offer people, started a membership drive. John Mulvey, following Dick Cottle, is making this a success, as he did the Committee on Algorithms during his Chairmanship.

I hope these few remarks indicate some of the importance the work of this Committee has had for mathematical programming beyond the scope of its charter, and the grounds the Society has for thanking Harvey Greenberg, John Mulvey, Ric Jackson, Karla Hoffman, the other members of the Committee, and all of you, for what you have done and are continuing to do so well. I know you will have an interesting and successful meeting, and hope you have many more of them."

Special thanks are given to Linda Taylor and Sally Blount for assisting in the development of these proceedings.

TABLE OF CONTENTS

Opening Address (Darwin Klingman) 1

1. Design and Use of Problem Generators and Hand Selected Test Cases

   Test problems for computational experiments -- issues and techniques (Ronald L. Rardin and Benjamin W. Lin) 8

   NETGEN-II: A system for generating structured network-based mathematical programming test problems (Joyce J. Elam and Darwin Klingman) 16

   The definition and generation of geometrically random linear constraint sets (Jerrold H. May and Robert L. Smith) 24

   Construction of nonlinear programming test problems with known solution characteristics (Gideon Lidor) 35

   A comparison of real-world linear programs and their randomly generated analogs (Richard O'Neill) 44

2. Nonlinear Optimization Codes and Empirical Tests

   Evidence of fundamental difficulties in nonlinear optimization code comparisons (Ernie Eason) 60

   A statistical review of the Sandgren-Ragsdell comparative study (Eric Sandgren) 72

   A methodological approach to testing of NLP-software (Jacques C.P. Bus) 91

3. Integer Programming and Combinatorial Optimization

   A computational comparison of five heuristic algorithms for the Euclidean traveling salesman problem (William R. Stewart, Jr.) 104

   Implementing an algorithm: performance considerations and a case study (Uwe Suhl) 117

   Which options provide the quickest solutions (William J. Riley and Robert L. Sielken, Jr.) 135

   An integer programming test problem generator (Michael Chang and Fred Shepardson) 146

4. Comparative Computational Studies in Mathematical Programming

Introduction
(Ron S. Dembo) 161

Remarks on the evaluation of nonlinear programming algorithms
(David M. Himmelblau) 163

Comments on evaluating algorithms and codes for mathematical programming
(Robert B. Schnabel) 166

Some comments on recent computational testing in mathematical programming
(Jacques C.P. Bus) 170

Remarks on the comparative experiments of Miele, Sandgren and Schittkowski
(Ken M. Ragsdell) 174

5. Testing Methodologies

In pursuit of a methodology for testing mathematical programming software
(Karla L. Hoffman and Richard H.F. Jackson) 177

Nonlinear programming methods with linear least squares subproblems
(Klaus Schittkowski) 200

An outline for comparison testing of mathematical software -- illustrated by comparison testings of software which solves systems of nonlinear equations
(Kathie L. Hiebert) 214

A portable package for testing minimization algorithms
(A. Buckley) 226

6. Approaches to Software Testing from Other Disciplines

Transportable test procedures for elementary function software
(William J. Cody) 236

Testing and evaluation of statistical software
(James E. Gentle) 248

TOOLPACK -- An integrated system of tools for mathematical software development
(Leon J. Osterweil) 258

Overview of testing numerical software
(Lloyd D. Fosdick) 268

The application of Halstead's software science difficulty measure to a set of programming projects
(Charles P. Smith) 277

7. Special Topics

Mathematical programming algorithms in APL
(Harlan Crowder) 290

8. Advances in Networks

Solution strategies and algorithm behavior in large-scale network codes (Richard S. Barr) 305

Recursive piecewise-linear approximation methods for nonlinear networks (Robert R. Meyer) 315

Computational testing of assignment algorithms (Michael Engquist) 323

9. On Establishing a Group for Testing Mathematical Programs

Introduction (John M. Mulvey) 329

Panel Discussion

10. Appendix

Conference program 337

List of participants 340

A model for the performance evaluation in comparative studies (Klaus Schittkowski) 343

Remarks on the comparative evaluation of algorithms for mathematical programming problems (Angelo Miele) 350

Comments on a testing center (Angelo Miele) 353

Systematic approach for comparing the computational speed of unconstrained minimization algorithms (Salvador Gonzalez) 355

The evaluation of optimization software for engineering design (Ken Ragsdell) 358

OPENING ADDRESS

by

Darwin Klingman

I. Computer Implementation Technology

During the past decade, algorithmic research has focused on the design and development of efficient digital computer optimization software. From this research, an important interface between mathematics and computer science, called computer implementation technology, has evolved. While the origins of this technology date back to 1952 and the implementation of the stepping-stone method on the National Bureau of Standards' Eastern Automatic Computer (Hoffman [4]), it has only been recently recognized as a major discipline in its own right.

Computer implementation technology seeks to determine efficient special procedures for carrying out subalgorithms of a general method on a digital computer by investigating: (1) what kinds of information are most effective to generate and maintain for executing operations, (2) which data structures are best to record, access and update this information, and (3) what methods are most suitable for processing these data structures to make the desired information available when it is needed.

Garnering this knowledge requires experimentation that artfully blends the best elements of mathematics and computer science. The computer is used both to evaluate the efficiency of resulting algorithmic processes (embodied in executable programs) and to provide statistics about the operation of key components under varying test conditions. Properly designed and utilized, these statistics allow researchers to gain valuable insights on how to improve the design of various components. This iterative modification, integration and evaluation of key processes is directly analogous to the laboratory research of other disciplines and leads to the view that computer implementation technology is the laboratory research of mathematics and computer science.

The application of computer implementation technology to network optimization has brought about unprecedented advances in solution efficiency. The remarkable gains of the early to mid 1970's for solving transportation, transshipment, and generalized transshipment problems are widely known, thus enabling network codes to out-perform LP codes by two orders of magnitude. The pioneering study by Gilsinn and Witzgall [2] demonstrated that effective use of computer implementation technology could reduce solution times for shortest path problems from one minute to slightly more than one second by using the same general shortest path algorithm, computer, and compiler.

The momentum launched by these studies continues into the present. New advances in all areas of optimization have superseded the procedures previously found to be best. Current computer implementations clearly outstrip the best codes of the recent past; our understanding of the important relation between algorithmic design and implementation continues to grow.

Computer implementation technology has uncovered several interesting points. For instance, while the literature contains many shortest path algorithms, there are only a few general methods for solving shortest path problems. Each general algorithm has within it subalgorithms. As a result, there are special subproblems, or sets of operations that must be handled in order to execute the general algorithm, e.g., finding the minimum of a set, breaking a loop, reconnecting subtrees, carrying out computations over the nodes and arcs of subtrees. The literature contains descriptions of a number of methods for handling these subproblems; unfortunately, these alternatives are referenced in the literature as different algorithms rather than as variants of a small class of general algorithms.

Historically, these 'algorithms' were developed and published because researchers devised ingenious ways of handling one or more of the subproblems in a mathematically efficient manner, i.e., the developer was able to show that his algorithm would require, in the worst-case, fewer addition or comparison operations than another algorithm.

The use of digital computers has shown, however, that algorithms with excellent worst-case bounds are not necessarily the most efficient (in terms of computer time) for solving real-world problems. This is due in part to the unique features of real-world problems, e.g., only a fraction of the total number of possible arcs, special network or grid structures, and small distance coefficient values. These features are often not reflected in the worst-case bounds, but more importantly, many of the 'good' (polynomially bounded) algorithms assume that information is available or updated after each iteration at no computational expense. When using a digital computer to execute the algorithm, however, the maintenance of such information requires non-trivial computer storage, retrieval, and comparison operations. Mathematically efficient algorithms do not necessarily result in efficient computer solution procedures. I feel strongly that computer implementation technology is an essential and historically neglected component of study. It is, in fact, a major practical tool for dealing with the ubiquitous issue of computational complexity, since no analysis of computational complexity can be truly meaningful without reference to the technology by which solution systems are implemented.

In the past, due to the lack of attention given to developing systematized principles and concepts, variations in a general algorithm were attributed to the skill (art) of the computer programmer. Recently, an awareness has developed within many science disciplines, and particularly within operations research, acknowledging that the design of efficient computer programs for solving mathematical problems is subject to the enunciation of key methodological and analytical principles and is therefore primarily a science rather than an art.

II. Experimental Design and Testing

Testing Before 1970

Early testing took place on the most advanced computers of the day, which tended to be large mainframe systems. The very early testing on the National Bureau of Standards' Eastern Automatic Computer employed in-core out-of-core software. Memory consisted of approximately enough words to store a few instructions and several parameters. As soon as sufficient main memory became available, most of the recorded testing changed to in-core solution codes. As people begin to conduct testing on micro- and mini- computers in the 80's, we are going to have to address this important perspective.

Early testing often consisted of hand comparisons of algorithms without the aid of a calculator or computer. All testing, including the use of computers, took place with very small problems by today's standards and a limited set of problems. Early papers contained statements like "so and so tested a problem that was a hundred by two hundred and got this result. We tested a problem of approximately the same size using a different computer, different problem and got this result. Therefore, we feel that this algorithm is superior." Today, such testing would be considered crude.

Testing in the 70's

Testing in the 70's has improved upon the procedures used to evaluate algorithms. Harlan Crowder, Ron Dembo, and John Mulvey [1] after studying the computational literature extensively have provided important suggestions for improving current testing procedures. Their paper clearly establishes that perfection was not attained in the 1970's and that development of better and more scientific evaluation procedures is needed.

Nevertheless there are a number of conclusions that can be drawn from the last decade of testing. First, when comparing algorithmic procedures, the experiments must be performed on the same computer, the same compiler, the same operating system, and the same problems.

Second, conclusions drawn on small test problems do not generalize to large-scale problems. Algorithmic procedures have to be changed for large-scale problems.

Third, patching existing codes to test new procedures can bias the conclusions against a new concept. It is very easy to say "I've got this code, and it can be used to test this new idea by simply making these changes and adding a new subroutine or two." I have done this more than once and come away with the feeling that an idea was not useful, but after careful consideration realized that the data structures in the code were not conducive to the current idea. Thus, I strongly recommend that everyone periodically start afresh in code development. Although additional effort is required, it is amazing how this process will change the computational results and conclusions.

For a long, long time there's been a myth about developing a modular LP code which experimenters can use to test new ideas. While there is merit in using a modular LP code for experimentation, it should be realized that "if and only if" conclusions cannot be drawn. If an idea works well, then it is probably good, but if it does not work well, it is not necessarily a bad idea.

Operating systems and billing routines in computers bias our conclusions -- often incorrectly. In developing good experimental design procedures, statistics must be used carefully.

The idea of counting functional evaluations is sound since it eliminates the billing system and operating system environment of the computer. For example, in performing extensive testing on in-core out-of-core codes for solving shortest path problems on a CDC 6600 and on a DEC 10, dissimilar algorithmic principles seemed to be superior based on the criterion of minimizing the total dollar cost for solving the problem. After looking into the operating and billing systems, it was found that each disk access on a DEC machine incurred a minimum charge of 128 words of data transfer. In contrast, the CDC machine incurred a charge of a minimum of 512 words of data transfer. The data transfers being made in these codes were both very small; thus this minimum fixed charge was added to the bill for each read access. These results affected the relative performance of the basic algorithmic concepts. Thus, one must be careful about drawing conclusions without studying characteristics of the operating and billing systems; the conclusions being drawn may be based on the characteristics of the operating systems and not on algorithmic characteristics.

When reporting standards, authors should be required to specify what operating system was used as well as what language and compiler. Authors should specify their problem input data format. Some formats assume a more structured problem file than others. For instance, several network codes' input format require that the nodes are numbered in increasing order where the supply nodes are numbered first and the demand nodes are numbered last. The code then checks these numbers and uses different algorithmic procedures depending on the type of node being processed. These requirements can increase the speed of a code, but it is invalid to compare a code that assumes no problem file structure to one that assumes a very rigorous structure.

The foregoing discussion points out an important and difficult problem -- how does one set up comprehensive standards for conducting and reporting computational analysis while insuring valid conclusions?

This problem occurs in all disciplines where conclusions are based on experimental testing. While chemists have different experimental design problems than sociologists, all are faced with the proverbial question, "Are the conclusions correct?" There are a number of algorithmic conclusions which appear in the literature that are incorrect. Many of these erroneous conclusions arise because the authors did not fully consider the effect of the data structures used in the implementation of the algorithm. During the 80's a more scientific set of experimental design guidelines must be developed. Experimenters should be required to fully document their experimental design -- including the data structures used in an implementation of an algorithm.

The Future

Finally, I would like to examine future testing. The challenges of computational testing in the 80's are more robust than they have been at any time. If one looks at the hardware currently available -- the micro's, the mini's, the macro's, IBM's new H computer which has eight megabites of central memory, and the array processors -- the increasing need for computational analvses becomes evident.

Testing in the 80's will show that many of the best implementation procedures of the 70's are inferior for the new hardware, operating systems, billing systems, and languages. For instance, it is possible that a full tableau LP code might turn out to be the fastest on some of these machines. Since this implementation approach was disregarded years ago, the biggest problem in discovering such a result might be simply to consider testing such an implementation. Thus, I encourage each of you not to rely solely on the past to direct future testing.

In the future, our computational testing should extend beyond randomly-generated unstructured problems and to consider randomly-generated structured problems, e.g., a minimum cost flow network which has a large number of demand nodes or some other topological property. Additionally, our analysis should include application-specific models, e.g., logistics problems, human resource planning problems, and cash management problems.

Future testing must extend beyond algorithms to also consider application systems. Data processing requirements that the optimizing software places on the computer system and the user -- input requirements, editors, report writers, and solution analyzers -- must be tested.

While a number of optimizing compilers have been developed which are very efficient, they require extensive pre- and post- processing of inputs and outputs. Consequently, the total amount of computer time required to solve a problem is larger for these optimizers than for the ones they supposedly surpassed.

In the future, modeling should also play a major role. Computational tests should be conducted to determine when an equivalent model type is more efficient for a given problem. For example, any linear program is equivalent to a non-linear generalized network, but which is better to solve? Should the problem be solved using a general linear programming system, a nonlinear generalized network system, or a hybrid system which uses a nonlinear generalized network algorithm until convergence slows and then switches to a general linear programming system?

Similarly, any 0-1 integer problem is equivalent to a 0-1 mixed integer generalized network (Glover and Mulvey [3]). It has been seen in a variety of applications that the 0-1 mixed integer generalized network model is computationally more efficient to solve when solution systems are designed to exploit the insights provided by viewing the problem in a graphical setting.

In general, the best computational models are achieved when the model formulator understands his or her problem from a structural framework. A good structural framework is often from a network viewpoint. This concept has helped many practitioners to formulate (or re-formulate) models to suit existing software. For example, it was used to reduce an empty box car scheduling model from 25 million 0-1 variables to 12,000 network variables (Klingman and Mote [5]). It has also been extremely helpful in designing special purpose algorithms for a specific application area. LP systems are too general for many practitioners, giving the user too much freedom to create computationally inefficient models.

References

[1] Crowder, H., Dembo, R., and Mulvey, J., "Reporting Computational Experiments in Mathematical Programming," Mathematical Programming, 5, (1978), pp. 316-319.

[2] Gilsinn, J., and Witzgall, C., "A Performance Comparison of Labeling Algorithms for Calculating Shortest Path Trees," Technical Note 772 (National Bureau of Standards, Washington, DC, May 1973).

[3] Glover, F., and Mulvey, J., "Equivalence of the 0-1 Integer Pro gramming Problem to Discrete Generalized and Pure Networks," Operations Research, 28, 3, (May-June 1980) pp. 829-836.

[4] Hoffman, A., Mannos, M., Sokolowsky, D., and Wiegmann, N., "Computational Experience in Solving Linear Programs," J. Siam, 1, (1953), pp. 17-33.

[5] Klingman, D. and Monte, J., "A Multi-Period Distribution and Inventory Planning Model," CCS 364, Center for Cybernetics Study, University of Texas, Austin, (Dec. 1979), pp. 1-42.

TEST PROBLEMS FOR COMPUTATIONAL EXPERIMENTS -- ISSUES AND TECHNIQUES

Ronald L. Rardin[1] and Benjamin W. Lin[2]

ABSTRACT

This paper compares and contrasts real world and random generated test problems as sources of standard tests for mathematical programming algorithms. Real problems are viewed as realizations from a test population of interest, and random problems are treated as models for the population. Methodological advantages and difficulties inherent in the alternatives are highlighted, and methods for dealing with the limitations discussed.

1. Introduction

The range of possible goals for computational studies of mathematical programming algorithms is at least as wide as the range of algorithms themselves. Some investigators seek to "tune" specific algorithms for use on a particular set of problems in a known application setting. Others are algorithmic verification studies, desiring only to prove a given code meets its design specifications. Different still are a host of algorithm development efforts seeking to assess the usefulness of procedures for solving families of problems arising in similar, but not completely known settings.

This span of goals can be viewed along a continuum of claims the investigator hopes of be able to make at the conclusion of the analysis. Algorithm tuners wish only to be able to conclude that their algorithms are effective for the limited set of problems arising in the application and site for which the procedures are being tuned. An algorithm developer hopes to justify a broader claim--that the algorithm is effective for all or nearly all commonly arising examples within the family or problems to which it is addressed. Algorithm verifiers are still more ambitious. They hope to be able to conclude that their code works as planned on every example that might confront it.

---

[1] Associate Professor, School of Industrial and Systems Engineering, Georgia Institute of Technology, Atlanta, Georgia 30332.

[2] Assistant Professor, Department of Mechanical, Industrial and Aerospace Engineering, Rutgers University, Box 909, Piscataway, New Jersey 08854.

Clearly what lies behind this continuum of claims are differences in populations of problem instances. All these analyses may be dealing, for example, with set covering problems. Still, as goals change from those of an algorithm tuner to those of an algorithm developer, to those of an algorithm verifier, the subject problem population becomes more and more diverse.

In recent years there has emerged a mathematical theory of computation that classifies algorithms according to the time and storage space they require to solve problems drawn from a specified population (see, for example, Garey and Johnson [2]). This theory seems to have real promise for determining the ultimate boundaries of the class of problems for which computer algorithms are likely to be effective. However, its categories of algorithmic efficiency are so broad that the theory gives almost no useful information for any of the computational analysis situations described above. Among algorithm tuners, developers, and verifiers, only developers might hope to benefit from such theoretical study of their proposed procedures.

Lacking much helpful theory, investigators confronting any detailed computational analysis must rely on experimentation. Algorithms being investigated must be applied to specific problems and inferences about the efficacy of the algorithms drawn by generalizing from results for such a sample of the whole population. Certainly, procedures for selection of problems in such tests directly affect the accuracy of inferences drawn from the experiment about the performance of algorithms on the entire population. Still, any experiment designer must trade the need for the sample of actually solved examples to be representative of the population against the cost of obtaining test problems and experimentally applying algorithms to them. The remainder of this paper draws on related work in [1, 3, 4, 5] to define the range of alternative sources of test problems from which experimenters may choose, and to describe their relative advantages and limitations in terms of this conflict between representativeness and convenience.

2. Sources of Test Problems

Test problems for computational experiments come from three sources published test examples conceived by researchers, documented data sets taken from "real world" applications, and randomly generated problems created artificially via a sequence of pseudo-random numbers. Illustrative examples in research papers are usually quite small in size and designed to demonstrate particular, sometimes pathelogical, algorithmic behavior. Also, there are very few such problems in existenc

Thus, they do not provide a very useful source of test data.

The remaining possibilities--real data and random generation--actually offer a continuum of sources with greater and greater degrees of abstraction from the subject problem population. At one extreme are truly real world problems drawn directly from the population of interest. Next in line are related problems available from prior research and experimentation. The connection of such examples to the population of interest in the present experiment may be somewhat obscure, but at least they match the population in mathematical form. Still further along the continuum of abstraction are random variants of real world data--problems that are artificially created by randomly perturbing some of the constants in data sets collected from actual practice. The new problems can no longer be described as actual applications data, but they presumably retain much of the inherent structure of the applied problems from which they were generated.

Most abstract of all are completely synthetic test problems. Such problems are produced by a random generation code that constructs problem data to meet user-specified problem parameter values. Such parameters may include the number of variables, the number of constraints, the magnitudes and signs of coefficients, and numerous more subtle properties such as the size of the problem's duality gap. The mathematics of the construction guarantee that resulting test problems meet the desired parameter specifications, but there is no assurance that problems produced are similar to applied problems except that they match in specified parameter values.

It is useful to think of this continuum of increasingly more abstract sources of test problems as a sequence of models of problems in the population of real interest. Examples taken directly from the population are an actual sample. As problems are further and further abstracted from the actual population, the experimenter is relying more and more on claims that the test examples actually experimented with model real problems because they match in enough meaningful ways to make inferences drawn from the test population valid for the population of actual interest.

3. Conveniences of Synthetic Problems

Anyone who has never undertaken a computational experiment might reasonably ask "Why test on anything but an actual sample of the subject problem population when use of any other test problem source raises extra doubts about conclusions drawn from the experiment?" The answer is very much the same one that causes medical researchers to experiment on rats rather than human beings and airframe designers to work with wind tunnel models instead of full size aircraft. Experiments on laboratory models are less hampered by physical and ethical limits, and require less time and cost to produce results. The following are a few of the specific ways in which the laboratory models of computational experimentation--randomly generated problems--offer such conveniences in conducting and analyzing experimental results:

- Quantity. Often when the same algorithm is applied to two quite similar problems it performs rather differently. To protect against being "fooled" by peculiar experience with a few test examples, computational experimenters need a large quantity of test problems. Any random generation scheme for obtaining test problems offers an almost infinite supply. Each change of the seeds for the pseudo-random number sequence used in the generator yields a new test problem.

- Variety. Any computational experimenter wishes to test against problems that adequately span the entire problem population in which he or she is interested. Such variety in test cases increases confidence in results and permits insights about problem characteristics that seem to be particularly easy or difficult for the algorithm. Although an algorithm tuner might find an adequate range of problems among a hand selected set of examples from the application he or she is facing, most other researchers cannot find satisfactory variety among real world, or even randomly-perturbed real world problems. On the other hand, synthetic problem generators offer almost unlimited variety. In our own work with integer programming problem generators [3,5], we have come to believe that if enough ingenuity is used in designing the generator code, almost any combination of problem properties can be placed under the control of the code user. To obtain variety the user needs only to change parameter settings.

- Measurability. Computational experiments usually measure as a response variable either solution time (perhaps counted in iterations), or solution accuracy, or some combination of the two. Both require knowledge of the test problem's solution to be accurately measured. If test problems are taken from practical experience, neither an optimal solution nor the optimal solution value may be known with much accuracy. In tests on large routing heuristics, for example, exact optimal solutions have never been obtained for even some of the most famous test examples. Researchers must settle for comparing their results to the best known solution. Similar difficulties arise in nonlinear programming with evaluating the accuracy of a solution when an exact one is unknown.

Carefully designed random problem generators can avoid these difficulties by constructing the problems so that optimal solutions are "built in", i.e. known at the completion of problem generation. All that is required is a sufficient condition for optimality in the problem class. The generation constructs problems so that the claimed optimal solution meets the specified condition.

• Portability and Reportability. The advancement of knowledge in any experimental science depends on researchers being able to cheaply and concisely report their results to their colleagues. Hand selected problems are again at a disadvantage. Nontrivial examples usually involve very large and bulky data sets. Since they are too large to be published in standard journals, researchers can obtain them only by depending on the cooperation of persons who have previously tested against the same problems. Even if all the researchers involved are willing to cooperate, proprietary considerations of their employers/ clients often preclude their supplying the data.

Random generation codes may also be fairly long (sometimes several thousand lines of FORTRAN). However, they need to be obtained only once. As soon as a researcher has a given code, he or she can undertake and report a wide range of experiments. Test problems can be summarized in a single table of generator parameter settings and random number seeds. If the generation code has a degree of machine independence, any other researcher with the same code should be able to reconstruct the problem.

4. Dealing with Validity

We have seen that as test problems become more and more abstracted from the actual problem population, conclusions drawn from experimentation become at the same time more convenient to obtain and more suspect because test data may not be representative of actual applications. Traditionally, computational researchers have dealt with this dilema by testing on whatever data they could readily obtain and hoping the data were representative. We believe serious attention to formal validation of test problem sources could add a great deal of confidence to results without sacrificing the convenience of experimentation on laboratory models.

One level of validation is probably done informally by almost every researcher. Test problems can be screened for "face validity", i.e., checked to see that they match real data in at least important parameters such as coefficient size and density.

The deeper validation question is "Are inferences about algorithmic performance drawn from experiments on the test problems valid for the real problem population of interest?" This question could be formally investigated by comparing algorithmic performance on a sample of real problems to that on a sample of the proposed test problems. If the performances matched, we could conclude the proposed problems are satisfactory models. However, some research needs to be directed to exactly how "algorithmic performance" might be measured. Possibilities include the following:

- robustness - do solution times and accuracies vary more dramatically with proposed test problems than real data?
- convergence rates - do rates and patterns of algorithmic convergence on proposed test problems mirror those observed on real data?
- step utilization - is the distribution of activity among various steps and phases of the algorithm sufficiently like that experienced with real data?
- numerical stability - do controls on numerical accuracy (e.g. basis reinversion) need to be invoked more or less often in solving proposed test problems than with real data?
- storage utilization - do patterns of in core and out of core memory utilization with proposed test problems match those with actual data?

5. Scaling

All the above conveniences of Section 3 derive from using abstract models of problems, but one form of simplification yields convenience with any type of test problem. Virtually all computational experiments would be easier to conduct if tests could be run on artificially small examples.

In order to know that results from small-scale tests are indicative of the experience that could be expected on problems of realistic size, the issue of validation arises again. Any of the schemes outlined above for comparing experience on model and real problems might be tried to establish validity.

There is, however, an added complexity in scaled experiments. Even if one could be convinced that a ranking of algorithms derived from tests on artificially small examples would extend to actual problems, it is not at all clear how much time or storage the algorithms might require on large problems. When model and real problems are about the same size one might speculate that solution times would differ only by a constant. When they are of dramatically different sizes, however, long experience with mathematical programming shows that time is likely to grow at least polynomially, and perhaps exponentially or worse. Thus in order to draw useful information from scaled experiments, both validation issues and performance forecasting concerns must be faced.

If scaled experiments are to be practical tools, much more research needs to be focused on empirical forecasting of the performance of algorithms as a function of problem size.

6. Conclusion

In the above discussion we have briefly described the dilemma that computational experimenters must face in choosing test problem sources. As in most other experimental sciences, one is confronted by a tradeoff between convenience in conducting the experiment and confidence that experimental results are meaningful for the population of problems in which the researchers is really interested. Development of convenient techniques for validating problem generators and for forecasting the impact of problem size would reduce the degree to which researchers must sacrifice either convenience or validity to obtain the other. However, we suspect that some tradeoff will always have to be made.

We recommend that experimenters consider such a tradeoff in terms of the particular problem population with which they are concerned. Algorithm tuners probably need to err in the direction of validity--using mostly actual test problems selected from the setting in which they are working. Convenience is likely to be paramount in the decisions of algorithm verifiers. Any problem of the required mathematical form will provide a test, but it is important to know the optimal solution and to have a large variety of problems at hand. Algorithm developers, who fall between these two extremes, must probably think in terms of development stages. At early phases of algorithm development it is only necessary to test whether the concept behind the algorithm is viable. Experimental convenience is very important and validity is a less serious concern, so random problems are likely to be quite satisfactory. As the algorithm evolves into a sophisticated code (perhaps offered for sale), validity becomes far more important. At such later stages the developer would probably be forced to test against real applications data.

7. References

1. Dembo, R. S. and Mulvey, J. M., "On the Analysis and Comparison of Mathematical Programming Algorithms and Software," report HBS 71-19. Graduate School of Business Administration, Harvard University, December 1976.

2. Garey, Michael R. and David S. Johnson. Computers and Intractability; A Guide to the Theory of NP-Completeness, W. H. Freeman and Company, San Francisco, 1979.

3. Lin, Benjamin W. and Ronald R. Rardin, "Development of a Parametric Generating Procedure for Integer Programming Test Problems," Journal of the ACM, 24, 465-472, 1977.

4. Rardin, Ronald L. and Benjamin W. Lin, "Controlled Experimental Design for Statistical Comparison of Integer Programming Algorithms," Management Science, 12, 1258-1271, 1980.

5. Rardin, Ronald L. and Benjamin W. Lin, "The RIP Random Integer Programming Test Problem Generator," School of Industrial and Systems Engineering, Georgia Institute of Technology, June 1980.

NETGEN-II: A SYSTEM FOR GENERATING STRUCTURED NETWORK-BASED MATHEMATICAL PROGRAMMING TEST PROBLEMS

Joyce J. Elam
Darwin Klingman
College of Business Administration
University of Texas
Austin, Texas 78712

ABSTRACT

The increased importance of designing and implementing algorithms to solve particular management problems has created the need for more robust test problem generators that can match the overall structure and parameter values of these problems. Of particular interest are management problems that can be modeled using a network structure. This paper discusses the design of a system for generating network-based mathematical programming test problems that conform to user-supplied structural and parameter characteristics.

1. Introduction

Recent years have seen an increase in the development of efficient algorithms for solving various classes of mathematical programming problems. This development has been motivated by a desire to reduce solution time and computer costs in solving current problems and/or to solve problems that are computationally infeasible using existing methods. The efficiency of an algorithm is based upon several criteria including its effectiveness with respect to different problem classes, its speed, capacity, and accuracy. Since existing theory alone cannot provide measurements for these criteria, empirical computational testing must be employed. A necessary prerequisite for such testing is the ability to construct and/or obtain test problems with known optimal solutions. The literature contains several sets of randomly generated test problems that have been used for this purpose [3,11,12,13].

One class of mathematical programming problems that has received extensive interest in recent years can be broadly defined as network and network-related problems. Pure network problems represent a special class of linear programming problems and embody a group of distinct model types: shortest path, assignment, transportation, and transshipment. Generalized network problems represent a broader classification of linear network-related problems. Other network-related problems include linear programming problems that have a network substructure such as multi-commodity networks, a pure or generalized network with extra constraints, or even a linear programming problem with GUB constraints.

The development of efficient solution methods and new modeling techniques for expressing problems in a pictorial network formulation [1,2,6,7,8,9,10] has led to the increased use of network-based models in government and business. These models range from rather straightforward network applications such as production planning and distribution to less obvious applications involving the refueling of nuclear reactors and optimal lot sizing and machine loading for multiple products. As network model-based systems are designed and implemented to handle larger and more diverse types of network and network-related problems, it is highly desirable to have the capability to generate test problems that match the overall structure and parameter value ranges of the models the systems are being developed to solve.

NETGEN [12], currently the most widely used generator for network test problems, can only generate pure network problems. In addition, NETGEN is limited in its ability to capture the characteristics of real-world problems in the network problems it can generate. For example, NETGEN cannot generate multi-period transportation or transshipment problems. The increased emphasis on modeling and the development of computer-based decision support systems built around network models and employing network algorithms have created a need for a more robust and powerful test problem generator that is driven by user-supplied problem characteristics. This paper describes the design of NETGEN-II, a system developed in response to this need.

2. NETGEN-II OVERVIEW

A distinguishing feature of NETGEN-II is the use of a model specification language for describing the structural characteristics of a model and the parameters values to be used in generating a problem from this model structure. A user can create a model specification with this language either through the interactive builder component of NETGEN-II or directly through a system-supplied editor. In addition, NETGEN-II provides access to a library of "standard" network model structures that can be used as a basis for creating a model specification. Once a model specification has been created, NETGEN-II provides the capability to randomly generate a family of problems from this specification, where each problem has the same underlying structure and parameter value ranges. NETGEN-II represents each generated prob-blem in MPSX input format. The overall architecture of NETGEN-II is shown in Figure 1.

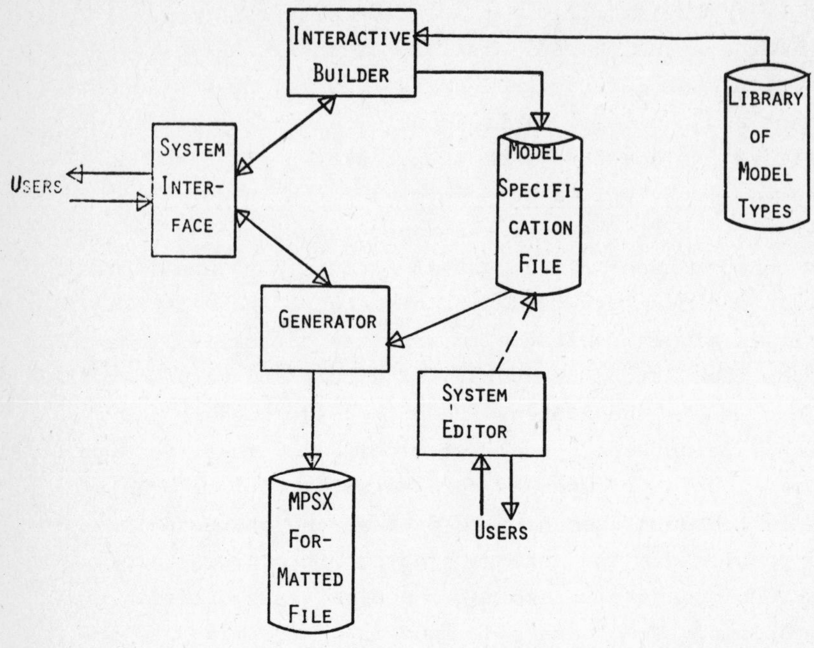

FIGURE 1. Architecture of NETGEN-II System

The remainder of this paper presents a brief overview of the model specification language. A more complete description of the language and the NETGEN-II is contained in [4,5].

2. Model Specification Language

The design of the model specification language is built around the concept of a "template" as the vehicle for generating the network structure of a mathematical programming test problem. An example template definition using the model specification language and one possible template that could be generated from this definition is shown in Figure 2. A template is defined through four types of statements: NODE CLASS, SIZE, RELATIONSHIP, and CONNECT. The NODE CLASS AND RELATIONSHIP statements define the types of nodes (called classes) and the types of linkages between these node classes (called relationships)

TEMPLATE DEFINITION

```
NODE CLASSES plant, warehs, cust

SIZE IS 2 FOR plant

SIZE IS 3 FOR warehs

SIZE IS 5 FOR cust

RELATIONSHIP ship (plant,warehs)

RELATIONSHIP sell (warehs,cust)

CONNECT IN RELATIONSHIP ship FROM all plant  TO all warehs

CONNECT IN RELATIONSHIP sell FROM ordered set 1-2 of warehs
                                  TO random set of 3 cust

CONNECT IN RELATIONSHIP sell FROM warehs 3 TO random set of 2 cust
```

TEMPLATE INSTANCE

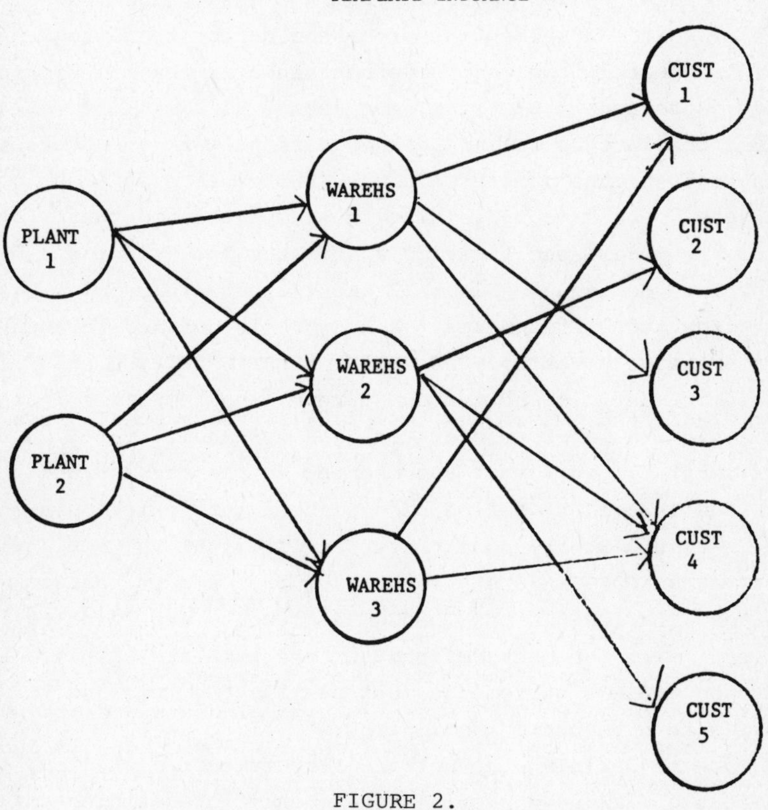

FIGURE 2.

that are to be represented in the basic graph structure of the template. A relationship always involves two node classes and is directed from the first node class to the second node class. The SIZE and CONNECT statements specify how the template is to be created using the node classes and relationships. The SIZE statements define the number of nodes in each class, where the nodes in a class are assumed to form an ordered set that is numbered consecutively beginning at 1. The CONNECT statements specify how to generate the arcs that form each relationship. For a given relationship, a CONNECT statement identifies a subset of nodes in the from node class and in the to node class of the relationship that are to be connected. The syntax defined for the CONNECT statement allows several different ways of identifying a subset. The nodes in a subset can be explicitly defined--i.e., all nodes in a class, (all plant) or all nodes between the mth and nth node in a class (ordered set 1-2 of warehs); or the identification of nodes to be contained in a subset can be deferred until the template is generated--i.e., a given number of nodes chosen randomly from a class (random set of 3 cust). When a CONNECT statement is processed during template generation, an arc is created from _each_ node contained in the to node subset to _every_ node contained in the from node subset.

After the template is defined, the entire network structure is defined by specifying the number of times the template is to be repeated and the linkages to be used in joining the templates together. Figure 3 illustrates the model specification statements for defining a network using the template defined in Figure 2 and the network problem that would result from these statements. A network is defined through four types of statements: TEMPLATE, NODE CLASS, RELATIONSHIP, CONNECT. The TEMPLATE statement defines the number of repeating templates in the network and assigns a label to each template instance. The NODE CLASS statement is identical in format and meaning to the one defined for the template and is used to define node classes that exist outside the template. (In Figure 3, all node classes exist within a template and thus, there are no NODE CLASS statements required for the network definition.) The RELATIONSHIP and CONNECT statements are identical in format and meaning to the ones defined for the template with the exception that the CONNECT statements must qualify the from and to node class subsets with a template identifier.

In addition to defining the network structure of a problem, the model specification language provides statements for defining any additional constraints that cannot be represented directly in the network structure. For example, the condition that total inventory at the

NETWORK DEFINITION

```
TEMPLATES period1, period2

RELATIONSHIP inventory (warehs,warehs)

CONNECT IN inventory FROM all warehs in period1 TO corresponding
        warehs in period2
```

NETWORK INSTANCE

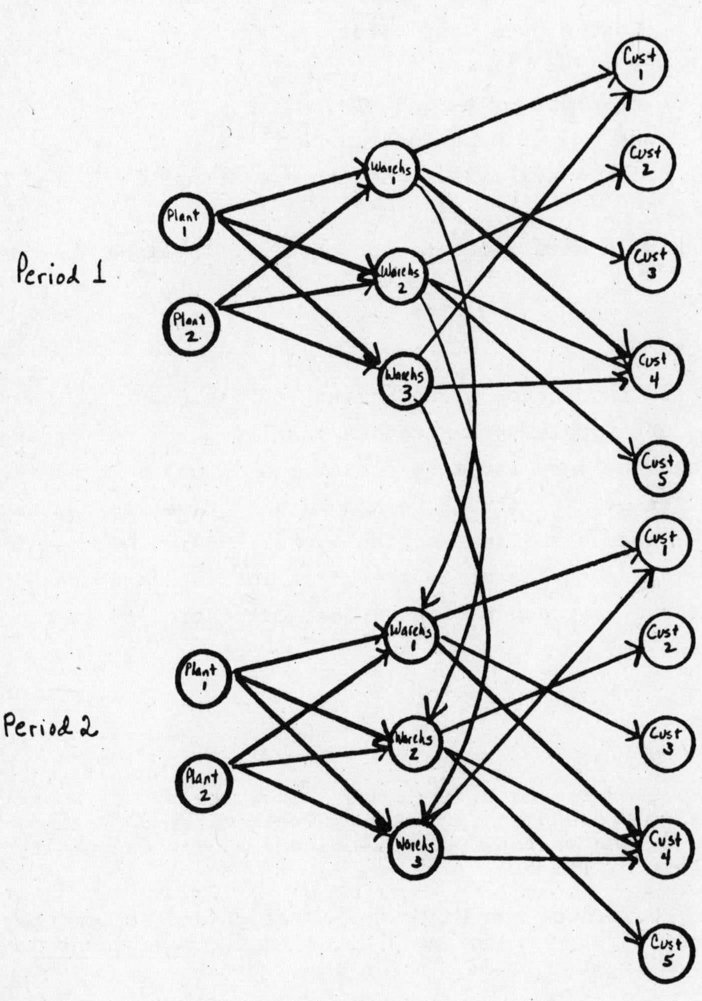

FIGURE 3.

end of the first period must not exceed 10,000 in the problem defined in Figures 2 and 3 could be expressed as follows:

SUM OF FLOWS IN inventory BETWEEN period1 AND period2 IS LESS THAN 10000.

The remaining requirement of the model specification language is to define the parameter values that are to be associated with the network structure of the problem. These parameters include supply and demand values for elements of node classes or subclasses and costs, bounds, and multipliers for elements in relationships. Parameter values can be expressed as constants, a range of values, and/or in terms of previously defined parameters. Some example model specification statements for assigning parameter values to the problem defined in Figures 2 and 3 are given below:

SUPPLY for each plant in period1 IS 500

SUPPLY FOR each plant in period2 IS PREVIOUS SUPPLY * 1.1

DEMAND FOR ordered set 1-3 of cust in each template IS random 50,150

COST FOR ship in each template for each arc IS linear random 5,15.

4. Conclusions

The increased importance of designing and implementing algorithms to solve management problems that can be modeled using network structures has created the need for more robust test problem generators that can match the overall structure and parameter values of these problems. This paper has discussed the overall design of a system that meets this need and the model specification language that is used to define the structural and parameter characteristics to be represented in a test problem.

5. References

1. R. Barr, F. Glover, and D. Klingman, "Enhancements of Spanning Tree Labeling Procedures for Network Optimization," Research Report CCS 262, Center for Cybernetic Studies, The University of Texas at Austin, (1976).

2. G. Bradley, G. Brown, and G. Graves, "Design and Implementation of Large-Scale Primal Transshipment Algorithms," Management Science 24, 1 (1977) 1-34.

3. J. C. P. Bus, "A Proposal for the Classification and Documentation of Test Problems in the Field of Nonlinear Programming," Report No. NN 9/77, Stichting Mathematisch Centrum 2 e Boerhaavestraat 49 Amsterdam 1005, HOLAND, (1977).

4. J. Burruss, J. Elam, and D. Klingman, "NETGEN-II: User's Manual", Research Report, Center for Cybernetic Studies, The University of Texas at Austin, (1980).

5. J. Burruss, J. Elam, and D. Klingman, "The Design of a Generator for Structured Network-Based Problems," Research Report, Center for Cybernetic Studies, The University of Texas at Austin, (1980).

6. A. Geoffrion, "Comments on Mathematical Programming Project Panel on Futures," SHARE XIV Meeting, Los Angeles, (1975).

7. F. Glover, D. Karney, and D. Klingman, "The Augmented Predecessor Index Method for Locating Stepping Stone Paths and Assigning Dual Prices in Distribution Problems," Transportation Science 6,2 (1972) 171-179.

8. F. Glover, D. Karney, and D. Klingman, "Implementation and Computational Study on Start Procedures and Basis Change Criteria for a Primal Network Code," Networks 4,3 (1974) 191-212.

9. F. Glover and D. Klingman, "The Simplex SON Algorithm for LP/Embedded Network Problems," Research Report CCS 317, Center for Cybernetic Studies, The University of Texas at Austin (1977).

10. F. Glover and J. Mulvey, "Equivalence of the 0-1 Integer Programming Problem to Discrete Generalized and Pure Networks," Operations Research, 28, 3 (1980), 829-836.

11. J. Haldi, "25 Integer Programming Test Problems," Working Paper No. 43, Graduate School of Business, Stanford University.

12. D. Klingman, A. Napier, and J. Stutz, "NETGEN: A Program for Generating Large-Scale Capacitated Assignment, Transportation, and Minimum Cost Flow Network Problems," Management Science 20, 5 (1974) 814-821.

13. J. Rosen and S. Suzuki, "Construction of Nonlinear Programming Test Problems," Comm. ACM 8, 2 (1965) 113.

THE DEFINITION AND GENERATION OF GEOMETRICALLY RANDOM LINEAR CONSTRAINT SETS

by
Jerrold H. May*
and
Robert L. Smith**

Abstract

The conventional procedure for generating random linear constraint sets independently and uniformly selects the constraint coefficients. Structure is usually imposed through some kind of rejection technique. Recent work of Van Dam and Telgen indicates that this type of generator tends to produce geometrically atypical polytopes. We define and construct a generator that produces geometrically random constraint sets; that is, their probability measure is invariant to the choice of coordinate system used. Moreover, an extremely efficient technique is presented for making an unbiased selection of a feasible polytope. Conventional approaches often guarantee feasibility by implicitly selecting that randomly generated polytope covering the origin. Such a procedure biases the selection in favor of large polytopes.

1. Introduction

We consider in this paper the problem of generating and characterizing random feasible regions of linearly constrained mathematical programs. In particular, a large class of such random feasible regions (or convex polytopes) is operationally defined, and efficient Monte Carlo procedures for their generation are discussed.

With the increasing proliferation of new and competing mathematical programming algorithms, sound procedures for their comparative testing have taken on new importance. Many of these evaluation schemes rely on the results of applying the algorithms to a series of randomly generated test problems. There is concern, however, about biases that may be inadvertently created by an uncritical introduction of randomness [12]. We introduce here extremely efficient procedures that generate geometrically unbiased feasible regions.

---

*Graduate School of Business, University of Pittsburgh, Pittsburgh, PA 15260

**Department of Industrial and Operations Engineering, University of Michigan, Ann Arbor, MI 48109

The general class of random polytopes considered in this paper is defined in Section 2. We make it as general as possible, with any special characteristics being motivated by geometric notions of randomness. Section 3 discusses an efficient Monte Carlo procedure for the generation of random polytopes of the type defined. In Section 4, we propose a special class of random polytopes that are shown to be those whose random properties are preserved under Euclidean transformations. Their generation leads to a choice of distribution that is geometrically random.

2. A Definition for Random Polytopes

An operational definition of random polytope is employed in this section to indirectly define the population of random polytopes we consider. More precisely, we restrict the class of allowable distributions in the Monte Carlo generation of random polytopes.

As mentioned in the Introduction, a general convex polytope, hereafter referred to as a polytope*, is the intersection of a finite number of closed half-spaces. Algebraically, $P \subseteq R^n$ is a polytope, if and only if it can be represented by $P = \{x \in R^n : Ax \leq b,\ A \in R^{m\times n}\}$. Note that P may be unbounded.

In order to define the notion of a random polytope, we need to parameterize the class of all polytopes. This could be done via (A_i, b_i), $i = 1, 2, \ldots, m$, where A_i is the ith row of A. However this parameterization is not one-to-one, since (kA_i, kb_i), for all $k > 0$ and $i = 1, 2, \ldots, m$, refers to the same polytope. We adopt the one-to-one parameterization $s = \{(p^1, i_1), (p^2, i_2), \ldots, (p^m, i_m)\} \in S = (R^n \times \{0,1\})^m$ corresponding to $P = \{x \in R^n \mid (p^j)^T x \gtrless \| p^j \|^2,\ p^j \in R^n,\ j = 1, 2, \ldots, m\}$ where the sense of the jth inequality is $\leq$ for $i_j = 0$ and $\geq$ for $i_j = 1$. Geometrically, p^j is the foot of the perpendicular from the origin to the jth constraining hyperplane (Figure 1).

The class of random polytopes being considered is now defined by imposing a multivariate probability distribution F over the parameter space S. It is clear that some $s \in S$ correspond to empty polytopes. To rule out these cases, we define F as the conditional probability distribution over S, given that the polytope generated by the probability distribution F_o over S is non-empty.

---

*What we have referred to as a polytope is more conventionally termed a polyhedral set [3]. The term polytope is then reserved for bounded polyhedral sets.

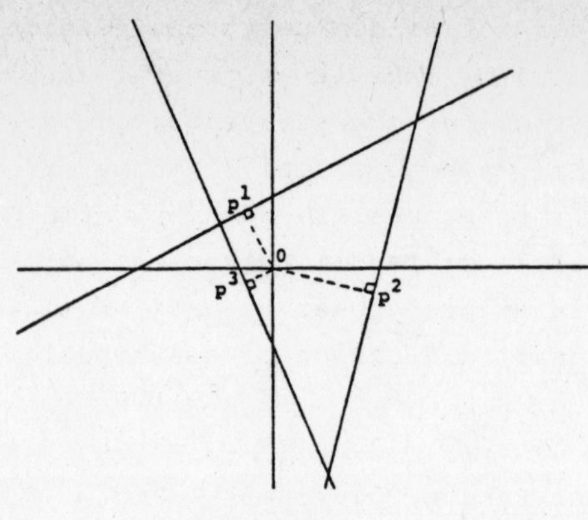

Figure 1

The properties we assume for F_o, and by implication for F, are stated below.*

Assumptions on the Generating Probability Distribution F_o

(i) The multivariate distribution over $p = (p^1, p^2, \ldots, p^m) \in R^{mn}$ is absolutely continuous with respect to Lebesgue measure.

(ii) The multivariate distribution over $(i_1, i_2, \ldots, i_m) \in \{0,1\}^m$ is given by $P(i_j = 0) = P(i_j = 1) = 1/2$ independently† for all $j = 1, 2, \ldots, m$ and independently of p.

Assumption (i) is motivated by the need to rule out degeneracy and other exceptional alignments of constraining hyperplanes. The polytopes generated are, in fact, with probability one, so-called simple polytopes, i.e., polytopes whose vertices are formed by exactly n facets. The effect of this assumption is to also, with probability one, rule out sparsity and integer constraint coefficients. Assumption (ii) in the presence of Assumption (i) is equivalent to Assumption (ii').

*Letting $\bar{S}$ be the set of all $s \in S$ corresponding to feasible polytopes, F assigns zero measure to any set in $S - \bar{S}$. Restricting F to $\bar{S}$ and considering any event $E \subseteq \bar{S}$, we get $F(E) = kF_o$ where $k = 1/F_o(\bar{S})$.

†$i_1, i_2, \ldots, i_m$, however, are dependent under the induced distribution F, since, under F, the polytope generated is necessarily non-empty.

(ii') Each point in R^n has the same likelihood of lying in the random polytope generated.

Hence, to deny (ii) in the presence of (i) necessarily leads to a procedure for selecting regions that is biased towards (or against) those regions covering exceptional points (e.g., the origin). Procedures for generating random test problems often force feasibility of the origin to assure non-emptiness of the region generated. For the class of simple product mix problems this is justified and, in fact, necessary on conceptual grounds. On the other hand, for the class of diet problems, for example, the origin should, with probability one, be excluded. We view the class of random problems defined by Assumptions (i) and (ii) as an aggregate class subsuming these and other special subclasses. There is, accordingly, no reason to favor or disfavor some points over others.

The equivalence of (ii) and (ii') is seen by conditioning on a realized configuration on the hyperplanes corresponding to p^1, p^2, ..., p^m. Let x_o be any point. Then we want the probability that x_o lies in the region selected by the choice of i_1, i_2, ..., i_m, given that the region is non-empty. But this is simply $k(1/2)^m$ by Assumption (ii), where k is the reciprocal of the probability that the region generated by F_o is non-empty. The reverse implication is argued similarly.

An important corollary of the above analysis is that *for a particular realized configuration of hyperplanes, each polytope so created is equally likely to be selected.* This follows from choosing a representative interior point of each polytope and noting that each point is equally likely to lie in the polytope generated. Hence Assumption (ii) amounts to a simple random selection of a polytope from the population of realized polytopes.

3. Efficient Generation of Random Polytopes

An operational definition of the class of random polytopes considered in this paper was given in the last section. We turn now to designing an efficient Monte Carlo procedure for their generation for a fixed F_o.

A direct implementation of the definition of Section 2 would be Algorithm A.

Algorithm A [Direct Generation of a Random Polytope]

A1. [Get hyperplanes]. Generate p^i, i = 1, 2, ..., m according to the multivariate distribution induced by F.

A2. [Select half-spaces]. Repeat this step for j = 1, 2, ..., m. Generate a uniform [0,1] random number u_j. If $0 \leq u_j < 1/2$, set $i_j = 0$; otherwise set $i_j = 1$.

A3. [Is polytope acceptable?]. If $S = \{(p^1, i_1), (p^2, i_2), \ldots, (p^m, i_m)\}$ is non-empty, stop. Otherwise, return to A2.

Unfortunately, Algorithm A is not practical. First, Step A3 requires an effort equivalent to solving a linear program with m constraints in R^n. Second, the expected number of passes through Steps A2 and A3 required to attain a non-empty region grows explosively in the number of constraints m. For example, for 100 lines in the plane, the expected number of iterations is on the order of 10^{26}.*

We could, of course, avoid the computational difficulties of generating a feasible polytope by simply forcing feasibility through a selection of $\leq$ for all constraints. This, however, does not represent a simple random selection from the regions generated, and violates our definition of a randomly generated polytope. This is not just a minor technical point, because a choice of the region covering the origin biases the choice toward larger regions, in an effect similar to the inspection paradox of renewal theory [2]. A more sophisticated approach would be to attempt to prune the tree of inequality choices. This can also introduce subtle biases in region selection. For example, in a depth first search where a node is fathomed if it corresponds to an infeasible partial selection of constraint directions, the ultimate selection is biased toward those regions bounded by constraints forming early branches of the tree.

The algorithm we propose for region selection is based on the idea of establishing a one-to-one correspondence between the N regions and a set of "tags" labeled 1 through N. A simple random selection of a tag then constitutes a simple random selection of a region. The tags to be employed will be certain distinguished vertices of the regions. A geometric description of the procedure is first presented below, followed by an algebraic description.

Algorithm B [Monte Carlo Procedure for the Generation of Random Polytopes (Geometric Description)]

B1. [Get hyperplanes]. Randomly generate a set of m hyperplanes H_1, $H_2, \ldots, H_m$.

---

*The expected number of iterations is, in general, given by $((\sum_{i=0}^{n} \binom{m}{i})/2^m)^{-1}$. The result follows from the fact that $\sum_{i=0}^{n} \binom{m}{i}$ represents the number of regions created (see [7]).

B2. [Divide the space in half]. Pass a randomly generated hyperplane H_o through vertex v simply randomly selected from the set of all vertices formed by the m hyperplanes.

B3. [Choose one of the two uncut polytopes]. Select the unique polytope P lying in a fixed half-space of H_o.

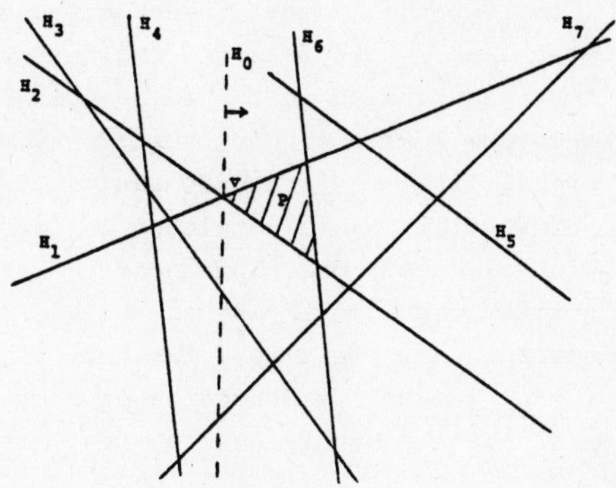

Figure 2

From the remarks above, to show that this procedure constitutes a simple random selection of a region, it suffices to show that a one-to-one correspondence has been established between the set of all vertices and the set of all regions. v is formed by exactly n hyperplanes, and these hyperplanes are in general intersection* with probability one. There are then 2^n regions sharing vertex v. The addition of hyperplane H_o passing through v creates new regions by passing through old regions. The total number of regions, new and old, is given by $2^{n+1} - 2$ [7]. Hence H_o passes through all but two of the old regions, and there is a unique region in each of its two half-spaces. Thus, to every vertex there is a unique associated region. Conversely, we must show that corresponding to every region there is a unique vertex. This is true for the bounded regions, but not all of the unbounded regions. If our interest were in generating only bounded polytopes through a rejection filter, there would be no problem. However, the general case requires enclosing the intersecting hyperplanes in a

---

*m hyperplanes in R^n are said to be in general intersection if any k of them intersect in a (n-k)-dimensional region, where the intersection is void if $n-k < 0$.

hypercube (<u>the problem domain</u>) large enough so that no vertices lie outside.* Then Step B2 should be altered to include in the class of vertices from which v is selected all those vertices formed by the m hyperplanes together with 2^n enclosing hyperplanes of the problem domain (those regions selected lying wholly outside the enclosing cube would be rejected). So we assume now that all regions are bounded. Let P be any region. Consider the linear program corresponding to an objective function hyperplane H_o and feasible region P with improvement lying in the direction opposite to the fixed half-space of H_o. Since P is bounded, an optimal extreme point solution v exists. Moreover, it is unique with probability one, since H_o was generated with an absolutely continuous distribution. This completes the argument.

The following algebraic description of the procedure includes a constructive algorithm for identifying the unique region that lies in the fixed half-space of H_o. This is a non-trivial problem, since the determination of each inequality direction is dependent on the joint configuration of all of the hyperplanes passing through v.

<u>Algorithm C [Monte Carlo Procedure for the Generation of Random Polytopes (Algebraic Description)]</u>

This algorithm makes a simple random selection from the polytopes formed by the m hyperplanes $(p^j)^T x = ||p^j||^2$, $j = 1, 2, \ldots, m$, where $m > n$.

C1. [Choose binding constraints]. Simply randomly select, without replacement, any n indices from the set $\{1, 2, \ldots, m\}$. Denote them by $\ell_1, \ell_2, \ldots, \ell_n$.

C2. [Find critical vertex]. Compute the critical vertex v by $v=B^{-1}b$, where $B = [p^{\ell_1}, p^{\ell_2}, \ldots, p^{\ell_n}]^T$ and $b = [||p^{\ell_1}||^2, \ldots, ||p^{\ell_n}||^2]^T$. Note that, with probability one, B is full rank, so that B^{-1} exists.

C3. [Determine cutting hyperplane]. Randomly generate $c \in R^n$. The cutting plane H_o through v is then given by $H_o = \{x | c^T x = c^T v\}$.

C4. [Set inequality relationships]. Let $y^* = c^T B^{-1}$. The unique polytope P_c lying in the half-space $c^T x \geq c^T v$ is given by setting the ℓj^{th} inequality to "$\geq$" if $y_j^* > 0$ and to "$\leq$" if $y_j^* < 0$. (The event $y_j^* = 0$ has probability zero.) The direction of the inequalities

---

*An alternative is to randomly choose a half-space of H_o in Step B3. If the region contained in that half-space is unbounded, accept it. If the region is bounded, then reject it with probability 1/2. The mean number of iterations required to generate a region is then at most 2.

corresponding to the hyperplanes in the index set $\{1, 2, \ldots, m\} - \{\ell_1, \ell_2, \ldots, \ell_n\}$ are chosen so as to make the critical vertex v feasible.

Step C4 is justified by the following argument. By construction, v is optimal for the linear program

$$\text{minimize } c^T x$$

$$\text{subject to } (p^{\ell_j})^T x \pm s_j = ||p^{\ell_j}||^2 \qquad j = 1, 2, \ldots, n$$

$$s_j \geq 0$$

where the $\pm$ sign in front of s_j is understood to be chosen so that the resulting feasible region is the polytope P_c. By the Strong Duality Theorem, there exists an optimal solution y^* to the dual program

$$\text{maximize } \sum_{j=1}^{n} ||p^{\ell_j}||^2 y_j$$

$$\text{subject to } B^T y = c \qquad j = 1, 2, \ldots, n$$

$$\pm y_j \leq 0.$$

In particular, $y^* = B^{-T}c$ must be dual feasible. This condition determines the signs of the slacks s_j in the primal problem and hence the inequality direction in Step C4.

4. The Class of Geometrically Random Polytopes

We have, until now, purposely left the class of random polytopes considered as general as possible. This maximizes the range of real world problems and randomly generated test problems that fall within our classification. Turning from the descriptive point of view to the normative, although the procedure for deciding constraint inequality directions is specific, the distribution F governing the positions of the hyperplanes has been assumed continuous but otherwise arbitrary.

From the standpoint of generating test problems, F should clearly be chosen to agree closely with the distribution exhibited by the class of real world problems for which the algorithm tested will ultimately be applied. The difficulty is that very little is known about the distribution of real world problems. The approach often used may be described as the generation of hyperplane positions in as "unbiased" a way as possible, followed by a filter that rejects all sampled feasible regions that fail to satisfy various structural properties that the real world problem class is known to possess. Our focus in this section will be on how to choose F so that the candidate population of feasible regions generated is "unbiased."

In the absence of any structural information, we interpret bias as geometric bias, and our search is for a distribution F that is geometrically random. For example, the likelihood of obtaining a plane nearly orthogonal to the x_1-axis should be the same as that of obtaining a plane nearly orthogonal to the x_2-axis (rotational symmetry). Also, the likelihood of the plane being near the origin should be the same as the likelihood of it being one unit away (translational symmetry). In short, all hyperplane positions and orientations should be equally likely. This property of space homogeneity may be more generally expressed in the following way: *The probability of all geometrically equivalent* events should be the same.* We can extend this definition of geometric randomness in the solution space to the parameter space by the following definition.

Definition: A random hyperplane parameterized by p will be said to be geometrically random if the probability measure M generating p is invariant under transformations of the parameter space induced by Euclidean transformations of the solution space.

Various authors have already noted the importance of requiring the distribution over p to be rotational symmetric (see, e.g., Liebling [6], Schmidt and Mattheiss [11], Van Dam and Telgen [12]). In addition, Van Dam and Telgen noted the significant geometric regularity induced by an uncritical introduction of randomness such as choosing the constraint coefficients independently and uniformly over the interval [0,1], as many mathematical programming test generators do (see, e.g., [1], and [8]). However, to our knowledge, no one has noted, in this context, the equally important requirement of translational symmetry. For example, selecting p uniformly within a hypersphere has rotational symmetry but fails to have translational symmetry.

The following theorem explicitly characterizes a probability density function that is invariant under translations and rotations. The theorem is a generalization to n dimensions of a result proven for n = 2 and 3 by Kendall and Moran [4]. Incidentally, Poincare [9] and Polya [10] have shown that the Euclidean invariant measure is unique for the cases n = 2 and 3. The proofs for Theorems 1 and 2 appear in [7].

---

*An event, in our context, is a (measurable) collection of hyperplanes in solution space R^n. Two events are geometrically equivalent if there exists a rotation and/or translation of R^n that will make one set of hyperplanes coincident with the other.

Theorem 1: The density function $f(p) = \frac{1}{||p||^{n-1}}$ defined over D = $\{p: \ ||p|| \leq 1\}$ induces a probability measure $M(E) = \int_E f(p)dp$ for $E \subseteq D$ that is invariant under transformations induced by Euclidean transformations of the solution space.

Theorem 1 characterizes the probability distribution that describes geometrically random hyperplanes. The next theorem spells out a Monte Carlo procedure for generating hyperplanes that satisfy such a distribution.

Theorem 2: A geometrically random hyperplane parameterized by p meeting the unit hypersphere D can be obtained by selecting the direction d of p as a point uniformly distributed on D and independently and uniformly selecting its length r along the line segment in D in the direction d.

The Monte Carlo procedure outlined in Theorem 2 requires a random deviate uniformly distributed on the unit sphere (i.e., a random direction). Any number of efficient techniques for its generation exist, the simplest of which is to generate n independent normal random variables and divide by their norm (see Knuth [5]).

References

[1] Charnes, A., Raike, W.M., Stutz, J.D., and Walters, A.S., "On Generation of Test Problems for Linear Programming Codes," Communications of the ACM, Vol. 17, No. 10, Oct. 1974, pp.583-586.

[2] Feller, W., An Introduction to Probability Theory and Its Applications, Volume 2, NY: John Wiley & Sons, 1971.

[3] Grunbaum, B., Convex Polytopes, NY: John Wiley & Sons, 1967.

[4] Kendall, M.G., and Moran, P.A.P., Geometrical Probability, Charles Griffin and Company, London, 1963.

[5] Knuth, D., The Art of Computer Programming Volume 2: Semi-numerical Algorithms, Reading, MA: Addison-Wesley, 1971.

[6] Liebling, T.M., "On the Number of Iterations of the Simplex Method", in: R. Henn, H.P. Kunzi and H. Schubert, eds., Operations Research, Verfahren XVII, 1972, pp. 248-264.

[7] May, J.H., and Smith, R.L., "Random Polytopes: Their Definition, Generation, and Aggregate Properties," Technical Report, Graduate School of Business, University of Pittsburgh, and Dept. of Industrial and Operations Engineering, University of Michigan, October 1980.

[8] Michaels, W.M., and O'Neill, R.P., "A Mathematical Program Generator MPGENR," ACM Transactions on Mathematical Software, Vol. 6, No. 1, March 1980, pp. 31-44.

[9] Poincare, H., Calcul des Probabilities, 2nd ed., Paris, 1912.

[10] Polya, G., "Uber Geometrische Wahrscheinlichkeiten," S-B. Akad. Wiss. Wien., 126, 1917, pp. 319-328.

[11] Schmidt, B.K., and Mattheiss, T.H., "The Probability That A Random Polytope is Bounded," Mathematics of Operations Research, Vol. 2, 1977, pp. 292-296.

[12] Van Dam, W.B., and Telgen, J., "Randomly Generated Polytopes for Testing Mathematical Programming Algorithms," Report 7929/0, Erasmus University, Rotterdam, 1979.

CONSTRUCTION OF NONLINEAR PROGRAMMING TEST PROBLEMS WITH KNOWN SOLUTION CHARACTERISTICS

Gideon Lidor*

Dept. of Computer Sciences, The City College
New York, N.Y. 10030

Abstract

The ability to supply realistic-looking test problems with known characteristics is essential for testing the efficiency and robustness of nonlinear programming algorithms and codes. This paper deals with methods of constructing (random) nonlinear programming test problems with known solutions and specified characteristics which usually affect the success of algorithms and their related software. In particular, it is shown how test problems with given multiple stationary points can be constructed, allowing for programs with several local optima thus providing for more rigorous testing.

1. Introduction

Testing and evaluation of mathematical programming software has in the past relied mostly on "hand picked" test problems selected from the literature and from known applications. Typical examples of non-linear programming (NLP) test problems are Rosenbrock's "valley" functions, Colville's well known collection of test problems and Dembo's geometric programming problems [4].

In recent years with the increased emphasis on sound statistical methods in performance evaluation, more attention has been diverted to machine-generated test problems. An early attraction of generating problems by machine was recognized in testing linear programming and network algorithms. There a machine generator can provide an unlimited source of large test problems with known solutions, bypassing the tedious task of collecting and coding such problems manually. The work of Charnes et al. [2] in linear programming, Klingman et al. [6] and Mulvey [10] in network problems and Rardin and Lin [11] in integer programming are good examples of such applications.

Due to the inherently more complex nature of nonlinear programming problems, the effort in machine generation of NLP problems has been much more limited. Michaels and O'Neill [9] have developed a portable package for generating convex NLP test problems with quadratic objec-

---

*This work was supported by CUNY PSC/BHE grant No. 13160 (1979-1980).

tive function and quadratic constraints. Schittkowski [13] has implemented a generator based on generalized polynomial (signomial) functions in the objective function and the constraints. Both approaches utilize a simple idea suggested by Rosen and Suzuki [12] in 1965 for generating nonlinear problems with known solutions.

This paper surveys the desired properties and the difficulties associated with machine generation of NLP test problems and suggests a generalization of the techniques mentioned above. This makes it possible to construct problems with multiple known critical points, thus providing for more interesting problems and more comprehensive testing.

The next section reviews the desired properties of machine generated test problems and outlines the special difficulties encountered in generating NLP problems. The proposed approach and an example are presented in section 3.

2. Design Criteria for NLP Test Problem Generators

The ultimate test of any algorithm or software is in the real life applications for which it was designed. As a result, there are some valid objections to testing methods which employ "random" or "artificial" test problems. Nevertheless, there are undeniable advantages to machine-generated problems in terms of controlling desired problem characteristics, availability, portability and the number, size and variety of test problems which can be constructed. In fact, given sufficient flexibility within a generator, one should be able to construct problems which will adequately represent a given population of real life problems.

A list of commonly accepted features of test problem generators (see for example [9] and [13]) will include:

- known optimal solutions and optimal values of the multipliers when they exist (these can either be specified by the user or generated by machine)
- user control of the range and distribution of problem parameters and coefficients. This would include, for example, the ability to have only integer valued parameters and to specify densities of matrices related to the problem.
- user-specified dimensions, including the number of variables, terms, constraints (equality and inequality) as well as the number of active constraints at the optimal point.

Additional features useful for certain types of problems include the ability to specify degeneracy, create problems with unique solutions or multiple solutions, local and global solutions, specify various starting points, scale factors, bounds on variables and many other special features. From a practical standpoint, a generator designed for wide general use should be portable and easy to use in different machine installations. Thus, the programs and random number generators should be machine independent, well documented and should require minimum effort by the user. Allowing default values for many of the aforementioned options can substantially reduce the user's effort.

Most of the features discussed so far can be implemented in a straightforward manner in generators designed for specific types of problems like linear programs or quadratic programs. When dealing with general nonlinear programs, however, serious difficulties may arise. Many of the algorithms for NLP are not general methods to begin with, their application being limited to specific types of problems or restrictive conditions (e.g. convex problems, differentiable functions). It would be necessary, therefore, to have special generators for the various classes of problems unless a completely general scheme could be devised. The whole concept of arbitrary test problems has been challenged by Lyness [8] who proposed a special family of test problems with certain invariant properties. Hillstrom [5] proposes studying families of problems with representative, known topographies. The question of what constitutes a representative sample of "real life" problems for a given algorithm remains unanswered.

Some of the questions arising in the design of test problem generators originate from basic differences in the definition of valid criteria for evaluating software. Although many real life problems do not exhibit pathological behavior, an algorithm for solving them should still be tested for the ability to handle pathologies. Lootsma [7] cites robustness and reliability as the primary considerations in evaluating software. A recent survey by Crowder and Saunders [3], which ranked CPU time as the major consideration, did not explicitly address reliability and robustness.

Despite the arguments against general purpose NLP test problem generators, they clearly provide some significant advantages, especially in extensive comparative studies. Still, the complex nature of the problems poses a number of difficulties in implementing NLP generators. Unlike linear programs, whose optimal points can be easily characterized, it is not always easy to identify an optimum in the nonlinear case. When all the functions involved are differentiable, one can construct the problem so that the desired solution point x^* will satisfy the Kuhn-Tucker necessary conditions for optimality. However, mild sufficient conditions are not readily verified (see for example Avriel [1] for discussion of second order sufficient conditions). The generator may thus be forced to use more restrictive conditions to ensure that x* is indeed a minimum. Even so, there is no guarantee that the problem does not possess other (unknown) local minima and possibly an unknown global minimum. One cannot hope, in general, to identify all the optimal solution points, although one certainly wishes to test convergence to the global optimum. It is usually impossible even to ascertain whether a given problem is convex over the feasible region, whether there are one or more solution points and whether the problem has singularities which may hinder the tested method. It may be relatively easy to deliberately introduce pathologies into the problem in order to test robustness, yet it may be very difficult to identify non-deliberate pathologies which arise in the course of constructing a problem artifically. From the designer's point of view, these difficulties mean that a NLP test problem generator will have to sacrifice both in generality and in the ability to control and know all the problem characteristics.

3. Construction of Nonlinear Programs with Multiple Solutions

Typical existing methods for generating NLP test problems can only ensure one known solution. However, Schittkowski [13] notes that his approach, based on generalized polynomials, may, and in fact does, occasionally generate problems with additional, unknown, solutions, local as well as global. As pointed out earlier, this situation is to be expected in a general NLP generator. Still, it would be useful to test the behavior of algorithms when a problem is known to have two or more given solution points. In what follows we consider one approach to constructing such problems without undue restrictions on the types of problems to be generated. The description refers only to a rather general scheme and not to any specific implementation.

We consider the nonlinear program

$$\text{minimize } f(x) = g_0(x) \tag{1}$$

$$\text{subject to } g_k(x) \geq 0 \qquad k = 1,2,\ldots,K \tag{2}$$

where x is a real n-vector and all the functions $g_k(x)$ are twice differentiable real functions.

Each of the functions $g_k(x)$ $k = 0,1,\ldots,K$ can be rewritten in the form

$$g_k(x) = \sum_{j \in J_k} c_j t_j(x) \tag{3}$$

where the c_j are constant coefficients, the $t_j(x)$ are terms, which may be constants or some functions of the variables. The index sets J_k

represent a partition of the set [1,2,...,M]. We assume that all the terms in the problem are indexed consecutively starting with the first term in the objective function and ending with the last term in the K-th constraint. J_k is the set of indices in $g_k(x)$.

The approach, which generalizes Rosen and Suzuki's method [12], is to prespecify (or to generate randomly according to given rules) the terms $t_j(x)$ and then compute the coefficients c_j so that desired conditions are satisfied. We shall assume for simplicity that the program is to have two (local) minima at the points x* and y*, with the corresponding Lagrange multipliers u* and v* (u* and v* are K-vectors). The results apply to any number of specified solution points.

Under our assumptions, the Lagrangian function

$$L(x,u) = f(x) - \sum_{k=1}^{k} u_k g_k(x) \tag{4}$$

has to satisfy

$$\left.\begin{aligned} D_x L(x^*,u^*) &= 0 \\ D_y L(y^*,v^*) &= 0 \end{aligned}\right\} \tag{5}$$

$$\left.\begin{aligned} u_k^* g_k(x^*) &= 0 \quad u_k^* \geq 0 \\ v_k^* g_k(y^*) &= 0 \quad v_k^* \geq 0 \end{aligned}\right\} \qquad k = 1,2,\ldots,K \tag{6}$$

where the notation D_x stands for the gradient vector w.r.t.x.

In addition to the necessary conditions (5) and (6), the points x* and y* must also be feasible, namely, satisfy the constraints (2). We may assume that u* and v* were chosen so that conditions (6) are satisfied. Rewriting (5) in terms of the terms and coefficients we obtain a system of linear equations in the coefficients c_j

$$\left.\begin{aligned} \sum_{j\varepsilon J_0} c_j D_x t_j(x^*) - \sum_{k=1}^{k} u_k^* \sum_{j\varepsilon J_k} c_j D_x t(x^*) &= 0 \\ \sum_{j\varepsilon J_0} c_j D_x t_j(y^*) - \sum_{k=1}^{k} v_k^* \sum_{j\varepsilon J_k} c_j D_x t(y^*) &= 0 \end{aligned}\right\} \tag{7}$$

The original constraints may be converted to equalities by adding unknown constants b_k to yield:

$$\left.\begin{aligned} \sum_{j\varepsilon J_k} c_j t_j(x^*) + b_k &= 0 \\ \sum_{j\varepsilon J_k} c_j t_j(y^*) + b_k &= 0 \end{aligned}\right\} \qquad k = 1,2,\ldots,K \tag{8}$$

If we specify p solution points x1*, x2*, ..., xp* for a problem with n variables, M terms and K constraints, then equations (7) and (8) amount to a linear system of (n + K)p equations in (M + K) unknowns (the cj's and the bk's). The success of this method hinges therefore on the existence of a sufficient number of "independent" terms in (7) and (8), so that a nontrivial solution exists to the homogeneous system of equations. The 'independence' of terms depends in turn on the form of the terms and on the solution points and cannot be guaranteed in advance. However, in highly nonlinear problems with a sufficiently large number of terms, the method is expected to succeed. Clearly, one cannot expect to have a large number of solution points in relatively simple looking problems (with some notable exceptions, e.g. trigonometric functions). It should be pointed out that one can easily increase the number of terms substantially by including a linear and quadratic form in the objective function and each of the constraints.

So far, the two points x* and y* were guaranteed to be Kuhn-Tucker points but not necessarily minima. To ensure that any selected solution is indeed a minimum, additional conditions may have to be added to (6), (7) and (8). In hindsight the additional conditions may prove unnecessary if the solution points prove to be minima without these conditions. However, one cannot usually forecast that such will be the case. A sufficient condition for a minimum is that in addition to satisfying (6), (7) and (8) the Hessian matrix of the Lagrangian be positive definite at each solution point, in other words,

$$\left.\begin{aligned} D^2_{xx}L(x^*,u^*) &= P \\ D^2_{yy}L(y^*,v^*) &= Q \end{aligned}\right\} \qquad (9)$$

where P and Q are some positive definite matrices.

The matrices P and Q can be generated (randomly) as products of a transposed upper triangular matrix (randomly generated) with the matrix itself. Conditions (9) which were employed by Schittkowski [13] add $(n + 1)n/2$ equations for each solution point (note that P and Q are symmetric), and are quite restrictive, however, milder sufficient conditions seem more difficult to implement. Manipulation of P and Q may offer some control of the behavior of the problem near the solution, for example, it may be possible to create "flat" behavior near x* by proper construction of P.

<u>Example</u>

Consider a problem of the form

minimize $f(x) = c_1x_1^2x_2 + c_2x_1^{-1}x_2^2 + c_3x_1x_2^{-1} + c_4x_1 + x_5x_2$

subject to $g_1(x) = c_6x_1 + c_7x_2 \geq 0$

$g_2(x) = c_8x_1^2x_2 + c_9x_1x_2^{-1} \geq 0$

with minimizing points at x* = (1,1) and y* = (2,1) and the corresponding multiplier vectors u* = (1,1) and v* = (1,1).

Both constraints are to be active at the optimum.

Conditions (7) and (8) yield the following linear homogeneous equations in the variables c_j and b_k (we ignored here conditions (9) to simplify the solution).

	c_1	c_2	c_3	c_4	c_5	c_6	c_7	c_8	c_9	b_1	b_2	
dL/dx_1^*	2	-1	1	1	0	-1	0	-2	-1	0	0	= 0
dL/dx_2^*	1	2	-1	0	1	0	-1	-1	1	0	0	= 0
dL/dy_1^*	4	-1/4	1	1	0	-1	0	-4	-1	0	0	= 0
dL/dy_2^*	4	1	-2	0	1	0	-1	-4	2	0	0	= 0
$g_1(x^*)$	0	0	0	0	0	1	1	0	0	1	0	= 0
$g_2(x^*)$	0	0	0	0	0	0	0	1	1	0	1	= 0
$g_1(y^*)$	0	0	0	0	0	2	1	0	0	1	0	= 0
$g_2(y^*)$	0	0	0	0	0	0	0	4	2	0	1	= 0

One can set arbitrarily $c_5 = c_7 = c_8 = 1$ and solve for the remaining unknowns, whereby the following solution is obtained: $c_1 = 1$, $c_2 = 0$, $c_3 = -3$, $c_4 = 0$, $c_5 = 1$, $c_6 = 0$, $c_7 = 1$, $c_8 = 1$, $c_9 = -3$, $b_1 = -1$, $b_2 = 2$. The original problem reduces to:

$$\text{minimize } f(x) = x_1^2 x_2 - 3x_1 x_2^{-1} + x_2$$

$$\text{subject to } g_1(x) = x_1 - 1 \geq 0$$

$$g_2(x) = x_1^2 x_2 - 3x_1 x_2^{-1} + 2 \geq 0$$

Although conditions (9) were not included, one can verify that both $x^* = (1,1)$ and $y^* = (2,1)$ are local minima with $f(x^*) = f(y^*) = -1$.

Comments

1. The idea of using generalized polynomials is especially attractive with this approach since generating a problem amounts only to generating the exponents of the x_i's in each term, after deciding on on the number of terms in the objective function and each constraint. The process can be completely preprogrammed. In addition, computation of gradients and Hessians can be easily done without any user-supplied routines. Signomial problems include as special cases linear programs and convex and nonconvex quadratic programs. They can provide very close approximations to many other types of functions.
2. One can view Schittkowski's approach [13] as a special case of the proposed approach, where the free coefficients c_j are limited to the coefficients of the quadratic and linear terms in f(x) and only a single solution is specified. His excellent work inspired

much of this current development.

3. There is no loss of generality in assuming that the original inequality constraints include the constants b_k. If a constraint is to be inactive at the optimum (with a dual variable at level 0), one should simply add a random positive constant to the right hand side of the appropriate equation.
4. The accuracy of the "known" solution is not absolute, but depends on the accuracy of computations of the coefficients. Still, the accuracy in solving a system of linear equations is usually far greater than that of the tested NLP algorithms. This is another advantage of machine generators relative to real life problems, whose "known" solutions are often limited in accuracy by the NLP methods used to compute them initially.

4. Summary and Conclusions

We have demonstrated that despite their limitations, machine-generated NLP test problems can play a major role in testing and evaluation of NLP algorithms and software. We have shown how to generate more general and more useful problems by extending well known existing techniques. Important further contributions in this area would be the development of simpler and milder conditions to replace the restrictive sufficient conditions and, hopefully, a method for insuring known global, as well as local solutions.

References

[1] Avriel, M., Nonlinear Programming, Prentice Hall, 1976.

[2] Charnes, A., Stutz, J.L., Raike, W.M. and Walters, A.S., "On Generation of Test Problems for L.P. Codes," Communications of the ACM, 17, 1974, pp. 583-586.

[3] Crowder, H.P. and Saunders, P.B., "Results of a Survey on MP Performance Indicators," COAL Newsletter, Math. Programming Society, January 1980.

[4] Dembo, R.S., "A Set of Geometric Programming Test Problems and Their Solutions," Mathematical Programming, 10, 1976, pp. 192-213.

[5] Hillstrom, K.E., "A Simulation Test Approach to the Evaluation of Nonlinear Optimization Algorithms," ACM TOMS, 3, 1977.

[6] Klingman, D., Napier, A. and Stutz, J.D., "NETGEN: A Program for Generating Large Scale Assignment, Transportation and Minimum

Cost Flow Network Problems," Management Sciences, 20, 1974.

[7] Lootsma, F.A., "Performance Evaluation of Non-Linear Programming Codes via Multi-Criteria Decision Analysis," COAL Newsletter, Math. Programming Society, Committee on Algorithms, January 1980.

[8] Lyness, J.N., "A Bench Mark Experiment for Minimization Algorithms," Argonne National Laboratory, Applied Math. Div., 1977.

[9] Michaels, W.M. and O'Neill, R.P., "NLPGNR User's Manual," Dept. of Computer Science, Louisiana State University, Baton Rouge, Louisiana, 1978.

[10] Mulvey, J.M., "Testing of a Large Scale Network Optimization Program," Harvard Business School, HBS 75-38, 1975.

[11] Rardin, R.L. and Lin, B.W.Y., "Developing of Parametric Generating Procedure for Integer Programming Test Problems," ORSA/TIMS meeting, Las Vegas, Nevada, 1975.

[12] Rosen, J.B. and Suzuki, S., "Construction of Nonlinear Programming Test Problems," Communications of the ACM, 8, 1965, p. 113.

[13] Schittkowski, K., "A Numerical Comparison of Optimization Software Using Randomly Generated Test Problems," Rechenzentrum Preprint no. 43, University of Wurzburg, West Germany, 1978.

A COMPARISON OF REAL-WORLD LINEAR PROGRAMS AND THEIR RANDOMLY-GENERATED ANALOGS

Richard P. O'Neill*

Abstract

The intent of the study is to determine the ability of randomly-generated problems to simulate real-world problems for the purposes of testing, benchmarking and comparing software implementations of solution algorithms, and to determine if the "degree of randomness" is related to the difficulty in obtaining a solution. Randomly-generated analogs are defined to be problems created with some of the characteristics of a real-world (actual application) problem, but containing data with random elements. These analogs fall into classes that can be characterized by "nearness" to the real-world problem. The first class is obtained by randomly perturbing the problem data. The second class is obtained by randomizing the ones in the Boolean image of the problem data. The third class consists of problems obtained from software generator that accepts the problem characteristics as input. The random elements are generated from uniform and normal distributions.

The measures of difficulty include central processor time, iterations and bumps in the optimal basis. Several optimizers are used in the study. In general, the results show that the difficulty increases with the "degree of randomness" and difficulty difference increases as a function of size.

Introduction

In today's computing milieu, software testing plays an important role. Its importance is, perhaps, best illustrated by a few examples. Anecdotal folklore carries several interesting stories about programs that were not sufficiently tested. One story tells of a linear programming code that was developed, but was unable to solve any problem save the one it was tested on. Another story describes an integer programming code that was tested on only a handful of problems. In its first field test, it failed. Subsequently, it had to be almost completely rewritten. In another case, software for linear programming parametrics was accepted on the basis of only one five-by-seven test problem.

---

*Director, Oil and Gas Analysis Division, EI-522, Room 4520, MS 4530, 12th & Pennsylvania Avenue, N.W., Washington, D.C. 20461

Error detection in code development is important. It becomes even more important as the economics of computing change; that is, as the relative cost of software development versus computing goes up. This relative cost difference and the importance of "error-free" software create a need for more testing. Additionally, testing should be done for comparison and benchmarking when choosing between two software packages that perform essentially the same function. This paper focuses on choosing test data. More specifically, it addresses the question of what inference can be made when using random or structured random programs in the testing of mathematical programs.

Random Testing

In code comparison randomly generated problems have been used extensively. NETGEN [6], a random network problem generator, has been used for code comparison and demonstration of tactical variations within codes. In nonlinear programming, most test problems are small (see Coleville [2]) and often designed to be pathological (e.g., the Rosenbrock function). Rosen and Suzuki [12] proposed a method of generating convex programs which was generalized and implemented by Michaels and O'Neill [9].

In mathematical programming, random test problems have many attractive features. They are easy to obtain, have a known optimal solution, have controllable characteristics that allow for scientific testing, and are virtually unlimited. But, software is usually developed to solve real problems. Real problems are problems arising outside mathematical programming whose solutions can be applied to answer questions in other disciplines (e.g., physics, engineering or economics). For this reason the question has persisted: What inference can be made from the solution of randomly generated problems to real problems?

Universe of Linear Programs

A linear program can be defined as:

$$\begin{aligned} &\min cx \\ &\quad Ax\ r\ b \quad \text{where } r(i)=(=,\ \le,\ \ge) \qquad (1) \\ &\quad\quad \text{for } i = 1,\ldots,m \end{aligned}$$

A, called the body of the program, has m rows and n columns, with a and s defined as the mean and standard deviation of the coefficients in A, excluding the zeros and ones (plus and minus). Most advanced codes that solve linear programs reorder the rows and columns of basis matrix at reinversion to create a sparse, stable basis representation for subsequent calculations. A measure of the complexity of the basis is the number of spikes after reordering. A spike is a basic column that has entries above the diagonal. Spikes are grouped into bumps. A bump consists of one or more spikes and is the minimal square submatrix containing a set of spike columns. In general, the complexity decreases as the number of bumps increases.

The simplex method solves linear programs by moving from vertex to vertex in polyhedron defined by the constraints. The universe of feasible linear programs contains problems that require the simplex method to visit every vertex before terminating with an optimal solution (see Klee and Minty [5]). This universe also contains problems that only need to examine the starting vertex before terminating with an optimal solution.

Real problems undergo a formulation or filtering process before they are seen by the mathematical programming community. This cycle usually has the following steps.

1. Formulate a linear program.
2. Attempt to solve the linear program.
3. If the program is solved satisfactorily, stop. The problem is now an application.
4. If the problem was not solved satisfactorily, (e.g., numerical instability or too much time is required), rescale and/or reformulate and go to 2, or abandon and stop.

Therefore, real problems are a subset of attempted applications that were deemed easy and economical enough to solve by the formulator.

Hypothesis

The basic hypothesis of this study is that real linear programs are easier to solve than their randomly generated analogs. This is, given linear programs that cannot be distinguished on the basis of a set of descriptive parameters, the programs that are more "random" are more difficult to solve. Examples of descriptive parameters are the number and type of rows, the number and type of columns, the average number of nonzeros and ones per column, and the distribution of the structural coefficients in A. To make this hypothesis more precise, some definitions must be introduced.

Randomly generated programs are subdivided into three categories: (1) perturbed, (2) randomized Boolean image, and (3) k-parameter randomly generated. A perturbed linear program is defined as:

min cx

P(A) r f(P(A)x*)

A and c are the same as in (1) and x* is an optimal solution to (1). The mapping, P, operates on a matrix element-by-element as follows:

$$P(a) = \begin{cases} a, \text{ if } a \text{ is } 0, +1, \text{ or } 1 \\ a*R(w1,w2), \text{ otherwise} \end{cases}$$

where R is a random variable with parameters w1 and w2. The function, f, is used to maintain feasibility and is defined element-by-element as:

$$f(b(i)) = \begin{cases} b(i), \text{ if } r(i) = \text{'='} \\ b(i) + R(w1, w2), \text{ if } r(i) = \text{'<='} \\ b(i) - R(w1, w2), r(i) = \text{'>'}. \end{cases}$$

The randomized Boolean image linear program is defined as:

min cx

B(A)x r f(B(A)x*)

The mapping, B, operates on a matrix element-by-element as follows:

$$B(a) = \begin{cases} a, \text{ if } a = 0, +1 \text{ or } -1 \\ R(w1, w2), \text{ otherwise.} \end{cases}$$

The k-parameter randomly generated linear program is one with at least k descriptive parameters specified and the remaining elements generated randomly. These problems were generated using LPGENR [8].

The Experimental Design

The experiment was conducted using five real linear programming problems obtained from IBM through Harlan Crowder. These five problems, named as they appear on the matrix file, and eight descriptive parameters are displayed in Table 1. These problems are small by commercial standards, but do represent applications and do not consume massive amounts of computing resources to solve.

	WEYERSHR	ADLITTLE	SHARPB18	BEACONFD	ISRAEL
Rows	41	56	96	173	174
E(=)	40	15	13	140	0
G(>=)	0	0	0	0	0
L(<=)	1	41	83	33	174
Columns	107	97	79	264	317
Matrix					
E(nonzeros)	7.0	3.8	9.0	13.0	16.0
E(ones)	2.7	1.0	2.3	1.0	1.3
Mean value*	.5	.5	1.0	.75	.5
standard deviation*	1.0	1.0	5.0	.25	2.0

Notes: E() = average per column
*nonzeros not including ones

Table 1 - Problem Statistics

For each of the real problems, a job was run which performed the following:

(1) Generated a perturbed, randomized Boolean image, and k-parameter randomly generated linear program.

(2) Solved each problem (including the real problem) using MIÑOS [11].

(3) Solved each problem (including the real problem) using WHIZARD [4].

The above run was then replicated five times using different seeds for the random number generated. For the perturbed problems, a uniform random variable was chosen on the interval (0, 0.10). For the randomized Boolean image problems, a normal distribution was choosen with the mean and standard deviation in Table 1. For the k-parameter (k=8) randomly generated linear program, the parameters were chosen from Table 1 using a normal distribution.

The data collected for each problem consisted of the number of iterations, central processor time, and the number of bumps and spikes at optimality. Further, the runs were made during non-peak processing to minimize the central processor time variations.

Analysis of the Results

Table 2 contains the mean and standard deviation of each of the five replications for the number of iterations, the central processor time, and the number of bumps at optimality. (All results were calculated using SAS [13].) In three cells, all the problems generated were infeasible. This indicates the possibility of a very "tight" formulation (i.e., small feasible region) so that slight perturbations "trick" the solver into thinking the problem was infeasible. In one cell all the problems were unstable numerically (i.e., the optimizer could not solve them due to numerical problems). In previous work, Layman and O'Neill have shown that the solver declared a problem with an optimal solution to be infeasible [7]. Those cells that contained infeasible or numerically unstable problems were eliminated from the subsequent analysis.

Since the purpose of the experiment was to assess the difference between real problems and their random analogs, the difference between the solution statistic (i.e., number of iterations or processor time) for the real problem and its paired analog was calculated and then divided by the statistic for the real problem. This statistic is a measure of the relative deviation of the random analog and the real problem.

	WEYERSHR			ADLITTLE			SHARPE18		
	NO. ITER	CP SECS	BUMPS	NO. ITER	CP SECS	BUMPS	NO. ITER	CP SECS	BUMPS
REAL	78.6	.52	12.4	123.2	1.15	27.4	96	.97	27
	(.5)	(.10)	(1.5)	(1.8)	(.06)	(2.2)	(0)	(1.02)	(0)
PERT	57.6	.58	12.8	122.4	1.12	21.8	*	*	*
	(7.8)	(.06)	(2.9)	(8.3)	(.06)	(1.9)			
BOOL	54.0	.54	10.0	78.4	.76	21.4	*	*	*
	(6.8)	(.08)	(3.4)	(9.4)	(.06)	(1.9)			
GENR	108.6	1.61	40.2	129.6	1.66	43.0	206	4.10	53.8
	(12.2)	(.38)	(.9)	(7.1)	(.25)	(9.1)	(19.7)	(.45)	(6.8)

	BEACONFD			ISRAEL		
	NO. ITER	CP SECS	BUMPS	NO. ITER	CP SECS	BUMPS
REAL	125	2.14	0	198.8	5.03	49.0
	(0)	(.07)	(0)	(1.6)	(.13)	(4.0)
PERT	139.6	2.51	4.0	330.4	9.62	29.2
	(15.5)	(.18)	(1.3)	(26.3)	(.85)	(2.8)
BOOL	313	7.08	26.2	*	*	*
	(28)	(.63)	(1.2)			
GENR	**	**	**	362.2	22.9	92
				(29.7)	(2.35)	(6.1)

Machine: IBM 370/168 MP
Op. Sys.: OS/MVS/SE
Notes: Sample Size = 5
( ) Indicates Std. Dev.
* Indicates Infeasible
** Numerically Unstable

Table 2 - Solution Statistics

First, means and t-statistics were calculated for the relative differences. The results are presented in Table 3. All are significantly different from zero at the .0001 level.

	Perturbed	Boolean	Generated
Iterations	.32	.72	1.00
	(3.7)	(3.3)	(10.8)
CP Time	.36	1.12	2.78
	(3.8)	(3.8)	(11.0)

Note: t-statistic in ( )

Table 3 - Mean of Relative Differences

The next step was to examine general linear regression models of the results.

The regression model chosen was the solution statistic as a function of a dummy variable representing the solver and the three random analogs--perturbed, Boolean and k-parameter--and the problem parameters. The problem parameters were number of rows, number of columns, average number of nonzeros, and average number of ones per column. The problem parameters are used to explain the variation due to the problem. Since the discussions for both central processor (cp) time and number of iterations are similar, the former is discussed below and results for the latter presented in the Appendix. Their associated coefficients all have high t-statistics as shown in Table 4. The coefficients for the dummy variables for the Boolean and generated analogs are both positive (the perturbed value is included in the intercept) and have large t-statistics.

From the plots of the residuals in Figure 1 (scatter plot), Figure 2, (stem-leaf and box plot), and Figure 3 (normal probability plot), the normal distribution assumptions of the residuals appear to be satisfied.

Next, variables representing the difference between the real and random analogy pair of number of bumps and spikes at optimality were added to attempt to explain additional variation. The results are presented in Table 5. The r-square value increased slightly and the coefficient of the spike term was significant and had the expected sign. The coefficient for the generated dummy changed signs and remained significant. The residual plots presented in Figures 4, 5, and 6 show that the model assumptions are satisfied.

The results of the number of iterations are very similar and are presented in the Appendix.

Conclusions

Since this study was limited to five real problems that were small by today's standards, broad sweeping conclusions are not justified. Several conclusions and conjectures will be made. First, the "more random" the problem the more difficult it is to solve. This is especially true in the generated problems. Second, problem structure not easily measurable before a problem is solved may be critical in explaining solution statistics of random test problems.

Acknowledgements: The author would like to thank Harvey Greenberg and Richard Jackson for many helpful suggestions and discussions.

Table 4. Regression Statistics
DEPENDENT VARIABLE: RELATIVE CPTIME

PARAMETER	ESTIMATE	T FOR H0: PARAMETER=0	PR > |T|
INTERCEPT	-3.854	-5.97	0.0001
BOCL	1.079	4.68	0.0001
GENR	2.266	9.84	0.0001
SCIVER	0.652	3.72	0.0003
M	0.073	4.97	0.0001
N	-0.006	-2.66	0.0092
ENZ	-0.626	-3.71	0.0003
E1	1.916	4.34	0.0001

F VALUE	PR > F	R-SQUARE	C.V.
37.74	0.0001	0.72	63.5

STD DEV	CPTIME MEAN
0.918	1.45

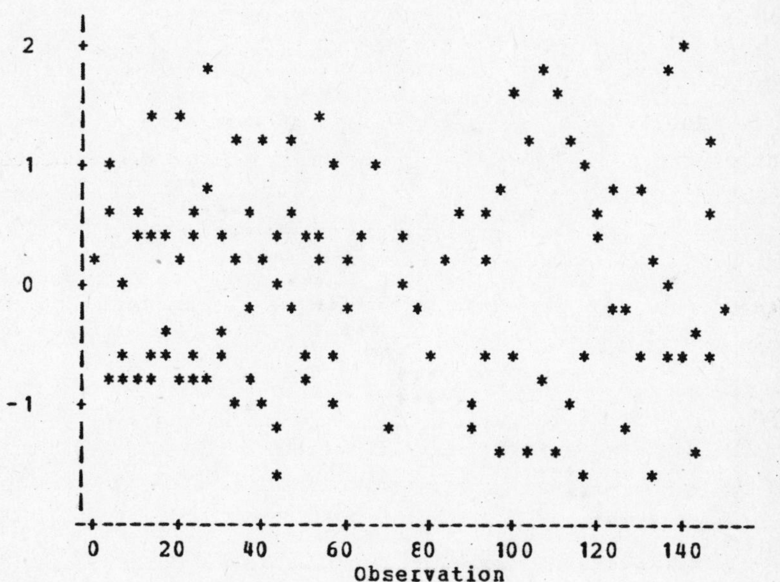

Figure 1. Scatter Plot of the Residuals

```
STEM LEAF                       #      BOXPLOT
  20 2                          1         |
  18 5                          1         |
  16 133                        3         |
  14 24                         2         |
  12 088078                     6         |
  10 6789                       4         |
   8 2669                       4         |
   6 1333917                    7      +-----+
   4 00113580289               11      |     |
   2 0000469003368             13      |     |
   0 282                        3      |  +  |
  -0 9753097                    7      *-----*
  -2 308                        3      |     |
  -4 887655333371              12      |     |
  -6 997664888420              12      +-----+
  -8 83320652                   8         |
 -10 9483                       4         |
 -12 84                         2         |
 -14 739854                     6         |
 -16 1                          1         |
     ----+----+----+----+
     Multiply Stem.Leaf By 10**-01
```

Figure 2. Stem Leaf and Box-Plot of the Residuals

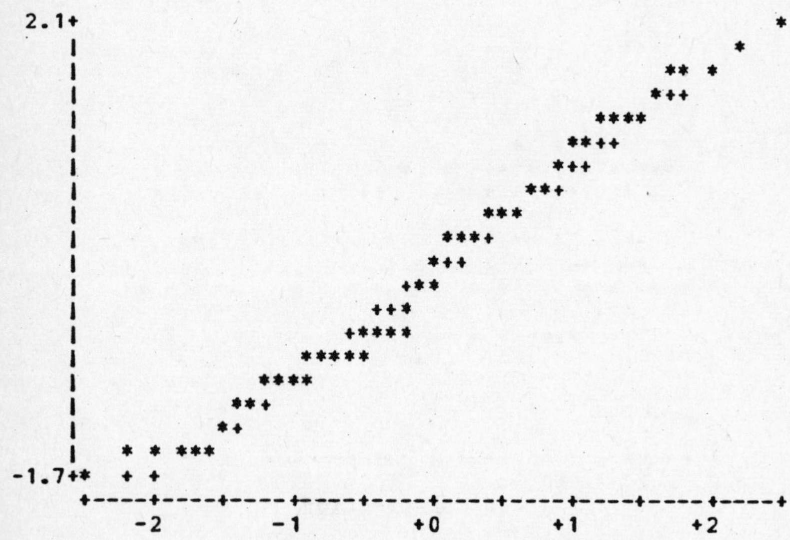

Figure 3. Normal Probability Plot of the Residuals

Table 5. Regression Statistics
DEPENDENT VARIABLE: RELATIVE CP TIME

PARAMETER	ESTIMATE	T FOR H0: PARAMETER=0	PR > |T|
INTERCEPT	-6.185	-11.18	0.0001
BCOL	0.348	1.80	0.0743
GENR	0.958	3.03	0.0031
SOLVER	0.500	3.56	0.0006
M	0.167	11.06	0.0001
N	-0.014	-6.21	0.0001
ENZ	-1.801	-9.86	0.0001
E1	4.912	10.75	0.0001
BUMPS	0.010	0.14	0.8905
SPK	0.122	7.62	0.0001

F VALUE	PR > F	R-SQUARE	C.V.
65.27	0.0001	0.85	46.3918

STD DEV	CPTIME MEAN
0.670	1.45

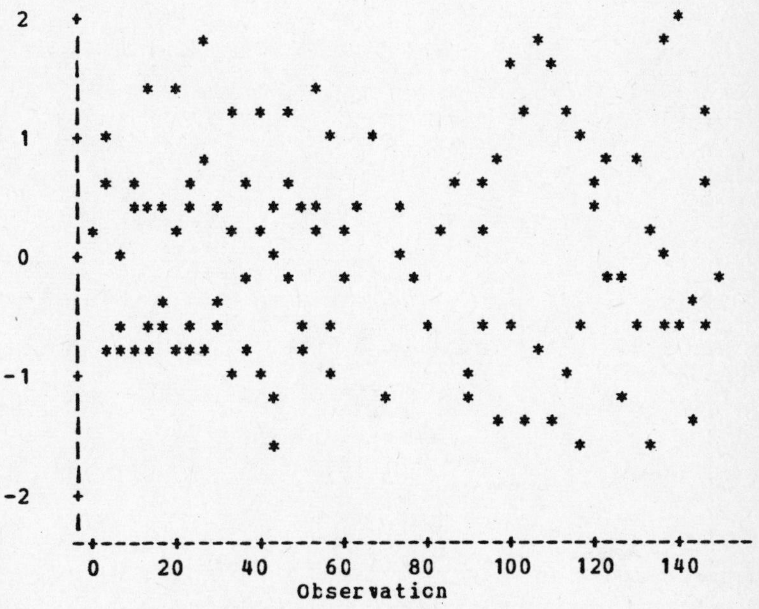

Figure 4. Scatter Plot of the Residuals

```
STEM LEAF                              #       BOXPLOT
  20 2                                 1          |
  18 5                                 1          |
  16 133                               3          |
  14 24                                2          |
  12 088078                            6          |
  10 6789                              4          |
   8 2669                              4          |
   6 1333917                           7       +-----+
   4 00113580289                      11       |     |
   2 0000469003368                    13       |     |
   0 282                               3       |  +  |
  -0 9753097                           7       *-----*
  -2 308                               3       |     |
  -4 887655333371                     12       |     |
  -6 997664888420                     12       +-----+
  -8 83320652                          8          |
 -10 9483                              4          |
 -12 84                                2          |
 -14 739854                            6          |
 -16 1                                 1          |
     ----+----+----+----+
     Multiply Stem. Leaf By 10**-01
```

Figure 5. Stem Leaf Box-Plot of the Residuals

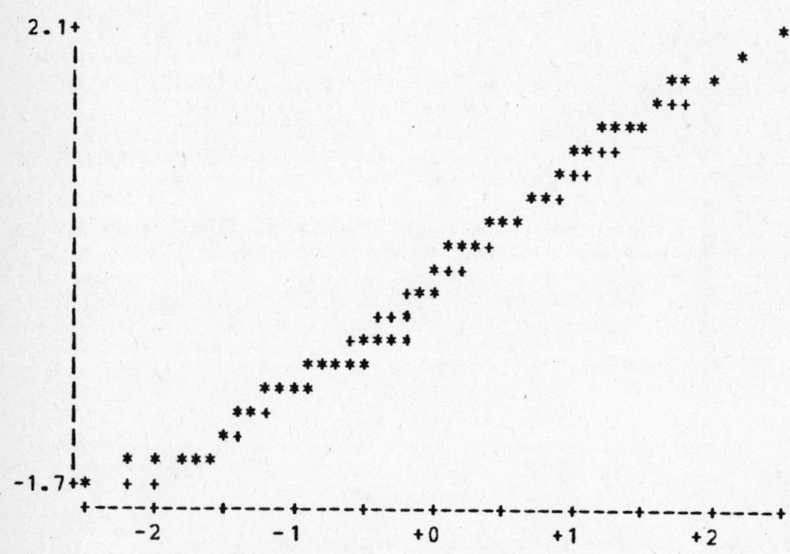

Figure 6. Normal Probability Plot

References

[1] Charnes, A., Raike, W.M., Stutz, J.D., and Walters, A.S. "On Generation of Test Problems for Linear Programming Codes," Comm ACM, 17, 10 (October 1974), pp. 583-586.

[2] Coleville, A.R. "A Comparative Study in Nonlinear Codes," Rep. No. 320-2949, IBM New York Scientific Center, June 1968.

[3] Goldman, A.J. and Kleinman, D., "Example Relating to the Simplex Method," Operations Research, Volume 12, pp. 159-160.

[4] Ketron Inc., MPS III Manual, Arlington, Virginia, 1980.

[5] Klee, V. and Minty, G.J., "How Good Is Simplex Algorithm?" Inequalies III, Editor O. Shisha, Academic Press, 1972.

[6] Klingman, D., Napier, A., and Stutz, J. "NETGEN, A Program for Generating Large Scale Capacitated Assignment Transportation, and Minimim Cost Flow Problems," Manage. Sci., 5 (January 1974), pp. 814-821.

[7] Layman, C.H. and O'Neill, R.P., "A Study of the Effects of LP Parameters on Algorithm Performance," Computers and Mathematical Programming, Editor, W.W. White, National Bureau of Standards, February 1978.

[8] Michaels, W.M., and O'Neill, R.P. "LPGENR User's Guide," Department of Comptr. Sci., Louisiana State University, Baton Rouge, La., March 1978.

[9] Michaels, W.M., and O'Neill, R.P. "A Mathematical Program Generator, MPGENR, ACM Transactions on Mathematical Software, Volume 6, No. 1, March 1980.

[10] Murtagh, B.A. and Saunders, M.A., "Minos Systems Guide," Technical Report SOL 77-31, December 1977.

[11] Murtagh, B.A., and Saunders, M.A., "Minos User's Guide," Technical Report SOL 77-9, February 1977.

[12] Rosen, J.B., and Suzuki, S., "Construction of Nonlinear Programming Test Problems," Comm. ACM, 8, 2 (February 1965), p. 113.

[13] SAS Institute Inc., SAS User's Guide, 1979 Edition, SAS Institute Raleigh, North Carolina.

Appendix

Table 6. Regression Statistics
DEPENDENT VARIABLE: RELATIVE NUMBER OF ITERATIONS

PARAMETER	ESTIMATE	T FOR H0: PARAMETER=0	PR > |T|
INTERCEPT	-1.58	-3.75	0.0003
BOCL	0.53	3.49	0.0007
GENR	0.74	4.89	0.0001
SCIVER	0.51	4.43	0.0001
N	-0.00	-0.33	0.7386
ENZ	-0.21	-1.86	0.0651
E1	0.62	2.13	0.0355

F VALUE	PR > F	R-SQUARE	C.V.
14.67	0.0001	0.50	89.2

STD DEV	ITER MEAN
0.60	0.67

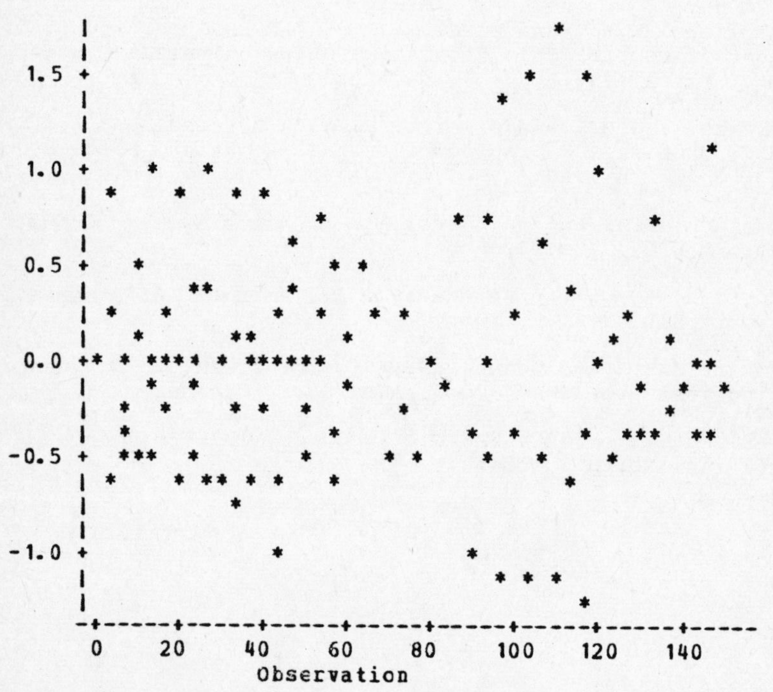

Figure 7. Scatter Plot of the Residuals

```
STEM LEAF                         #      BOXPLOT
  16 2                            1         0
  14 50                           2         0
  12 6                            1         0
  10 023                          3         |
   8 0156115                      7         |
   6 67                           2         |
   4 90678                        5         |
   2 023456902558                12      +-----+
   0 12245563457889              14      |     |
  -0 75410997665433221110        20      *--+--*
  -2 99773318774421              14      |     |
  -4 98753209765443210           17      +-----+
  -6 185552                       6         |
  -8 74                           2         |
 -10 663                          3         |
 -12 0                            1         |
     ----+----+----+----+
     Multiply Stem. Leaf By 10***-01
```

Figure 8. Stem Leaf Box-Plot of the Residuals

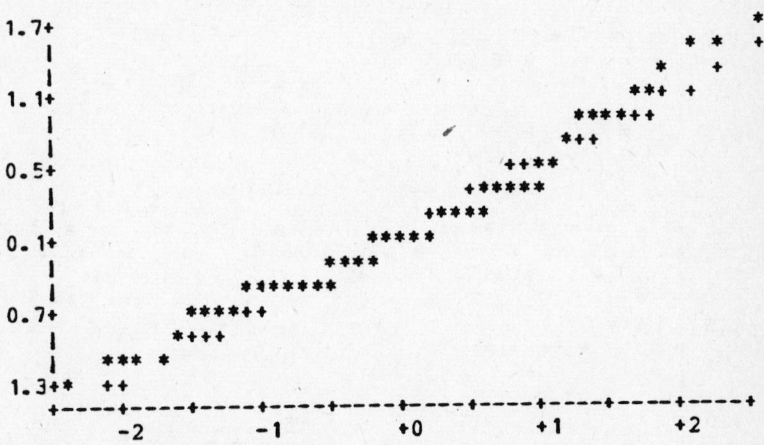

Figure 9. Normal Probability Plot of the Residuals

Table 7. Regression Statistics
DEPENDENT VARIABLE: RELATIVE NUMBER OF ITERATIONS

PARAMETER	ESTIMATE	T FOR H0: PARAMETER=0	PR > |T|
INTERCEPT	-1.845	-3.84	0.0002
BOOL	0.275	1.65	0.1029
GENR	0.820	2.99	0.0035
SCIVER	0.565	4.63	0.0001
M	0.052	4.00	0.0001
N	-0.004	-2.02	0.0462
ENZ	-0.553	-3.49	0.0007
E1	1.406	3.55	0.0006
BUMPS	0.122	1.89	0.0621
SPK	0.011	0.81	0.4195

F VALUE	PR > F	R-SQUARE	C.V.
13.26	0.0001	0.54	86.2135

STD DEV	ITER MEAN
0.581	0.67

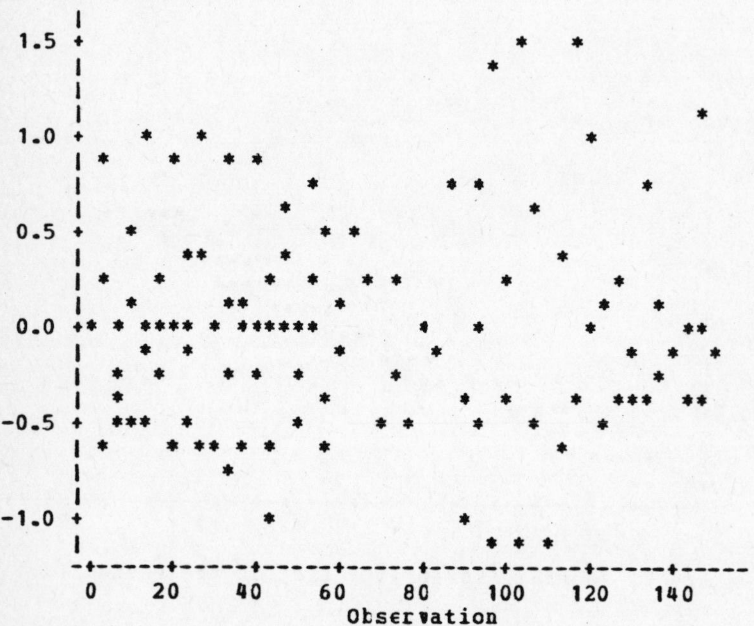

Figure 10. Scatter Plot of the Residuals

```
STEM LEAF                          #      BOXPLOT
  16 2                             1         0
  14 50                            2         0
  12 6                             1         0
  10 023                           3         |
   8 0156115                       7         |
   6 67                            2         |
   4 90678                         5         |
   2 023456902558                 12      +-----+
   0 12245563457889               14      |     |
  -0 75410997665433221110         20      *--+--*
  -2 99773318774421               14      |     |
  -4 98753209765443210            17      +-----+
  -6 185552                        6         |
  -8 74                            2         |
 -10 663                           3         |
 -12 0                             1         |
     ----+----+----+----+
   Multiply Stem. Leaf By 10**-01
```

Figure 11. Stem Leaf Box-Plot of the Residuals

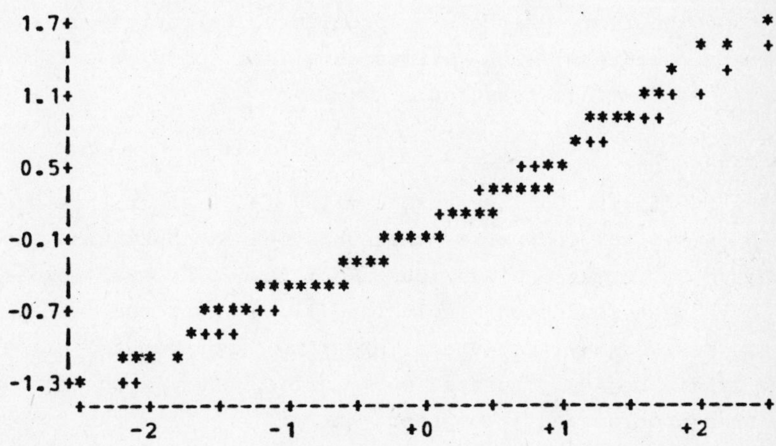

Figure 12. Normal Probability Plot of the Residuals

EVIDENCE OF FUNDAMENTAL DIFFICULTIES IN NONLINEAR OPTIMIZATION CODE COMPARISONS

E.D. Eason
Failure Analysis Associates
2225 E. Bayshore Rd.
Palo Alto, California 94303

ABSTRACT

Several nonlinear optimization code comparisons have been published, providing data for picking a code for a given application or developing new codes. The common performance measures (number of function evaluations, standardized computer time, number of problems solved) are intuitively machine independent, which encourages such use. Unfortunately, the relative performance of optimization codes does depend on the computer and compiler used for testing, and this dependence is evident regardless of the performance measure. In addition, relative performance measured on a single machine may depend significantly on the desired degree of accuracy, the choice of test problem(s), the chosen performance measure, and even the time of day (machine workload) when the tests are run. Numerical evidence of these difficulties is presented, based on tests of the same problem and algorithm decks on several different computers, with various compilers, problem sets, accuracy levels, and performance measures.

1.0 Introduction

Since the first nonlinear programming (NLP) algorithms were implemented on digital computers, there has been a continuing interest in the relative performance of various algorithms. Several comparison studies have been conducted, including [1]-[6], but the testing and comparison methodology, computers, programs, and problems differ in all of these studies. Any potential optimization user is faced in most cases with a problem and a computer that differs from the comparison study conditions, yet the user must choose a code to solve the problem. A reasonable question is how much should an optimization user rely on the reported performance results when choosing an optimization code?

This paper is the result of an accumulation of evidence of fundamental difficulties affecting nonlinear optimization code comparisons. Two of the computer codes and the 13 problems from the author's earlier comparison study [2] were run on several computers with various compilers as the author moved from one institution to another over a period of several years. The codes and problems were implemented in ANSI Standard FORTRAN so that the same program deck could be used on all computers with only trivial modifications to ensure comparable word length. In addition, several practical problems were solved by at least two algorithms each because of the author's lack of confidence in the results from any one optimization code.

It was clear at an early stage that the Colville time standardization procedure (used in [1]-[4]) does not adequately eliminate machine-to-machine variability [7]. The results were somewhat better with an alternate standardization benchmark, but machine-dependent differences in the relative performance of codes appeared that could not be explained by differences in average execution speed.

The purpose of this paper is to present the collected evidence for machine-dependent comparison difficulties and summarize the known difficulties that affect the validity of comparisons that involve only one machine. A theory is advanced to explain why optimization code performance should be so sensitive to seemingly minor factors. The basic conclusion is that the early published results are of limited value to the potential optimization user, particularly if his computer differs from the one on which test results were generated.

2.0 Comparison Difficulties on a Single Computer

If there were only one type of computer in the world, there would still be difficulties that would affect the validity of code comparison studies. The first difficulty is that the relative performance of NLP codes depends on the required accuracy. This fact is obvious in Fig. 1, where the convergence curves cross for several codes on a particular problem, and some codes appear to be unable to achieve high accuracy levels. For example, PATSH appears in Fig. 1 to be about an order of magnitude faster than DAVID, MEMGRD, SIMPLEX, and ADRANS at 1.0E-7 relative error, but the reverse is true at 0.1 relative error. This difficulty can be minimized by plotting convergence curves as in Fig. 1 and conducting the comparison at one or more accuracy levels that are of practical interest, as in [2], [5].

A second, well-known difficulty is the fact that the relative performance of NLP codes depends on the problem. As an example of the magnitude of the effect, consider the results of two codes, NMSERS and PATSH on two problems in [2]. The results, in CPU seconds interpolated from convergence curves at 0.0001 accuracy, using a CDC 6600 with FTN compiler (OPT=0), are as shown in Table 1.

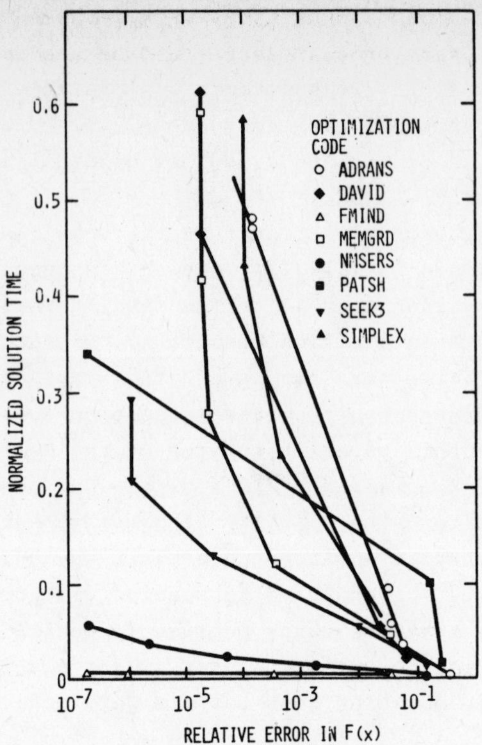

Figure 1 - Convergence Characteristics on Problem 2[2]

	NMSERS	PATSH
Problem 2	0.46	4.2
Problem 4	0.84	0.30

Table 1
Code Performance is Problem Dependent

Depending on the problem to be solved, these results suggest that NMSERS varies from 9 times faster to 3 times slower than PATSH. Similar results have been obtained in other comparison studies such as [3], where code NLP is 7 times faster than SUMT on problem 9, but 16 times slower on problem 7.

The usual method of avoiding this difficulty is to test the codes on a set of problems and compare the codes on the basis of average efficiency, robustness (number of problems solved), or both criteria as in [2], [5], [6]. There are two basic approaches to selecting problem sets - collections of real problems and computer generated sets [8]. Either approach may also include randomized starting points and parametric variations within a given problem class, [9], [10]. Collections of test problems are given in the previous comparison studies and in [11], [12].

The problem set approach is well established, and to the extent that the potential user's problem is represented in the tested sets, the results give a fair indication of which codes to try first. However, the choice of problems is generally over-emphasized, relative to equally important factors such as comparing at equal accuracy levels, tuning the algorithm constants, etc., so that the validity of the averaged performance results may still be questionable. Furthermore, since the practical problem has usually never been solved before, it may <u>not</u> be well represented in the solved test problems. This is especially true in engineering applications, where objective functions and constraints often incorporate simulation algorithms. Unlike the mathematical test problems with exact solutions, such problems are usually non-differentiable (because of actual discontinuities, inability to evaluate the functions accurately, and lack of an analytic form to manipulate), and they typically require significant computation per function and constraint evaluation.

The fourth difficulty is that many codes exhibit widely varying performance depending on the values of control parameters and heuristic constants. These parameters and constants are not always stated in reports of code tests, so the average performance by most users of the code may be quite different from the tested performance. If the measured performance and robustness was obtained by adjusting the parameters for optimum performance on each problem, it is important to meticulously follow this procedure for all tested codes. Any study that does not document the control parameters and adjust them fairly (or consistently not adjust them) for all codes is of limited practical value to future users and may be severely biased. This and other aspects of experimental design and reporting are discussed in [13].

Finally, the most common performance measure, computer execution time, is itself a variable on most machines that simultaneously run several programs. In such a multiprogramming environment, the measured time for the test program will vary over the day, week, and year depending on the load on the machine. In [2], a 6% variation in time to execute Colville's timer was observed for repeated passes through the same DO loop within a single program execution on an IBM machine. The author has, in one case, observed a 50% time variation for two identical runs on a CDC computer, and 15-100% variability is reported by others [14] for compute-bound programs. The cause of the variability is the somewhat arbitrary assignment of time to the several executing jobs by the computer accounting program. Timing variability can, of course, be handled by standard experimental design techniques such as blocking or randomization.

A common alternate performance measure is the number of function evaluations. However, the number of function evaluations does not fully reflect the relative time and cost of the codes, even in cases where function evaluation times are long. This is because the time spent in constraint evaluation, derivative evaluation, and algorithm overhead is often significant compared to function evaluation time. For example, the results in Table 2 for three codes on a mechanical design problem involving stress analysis and dynamic simulation show significant variation in the effective time per function evaluation, T/n.

	MECHANICAL DESIGN PROBLEM			PROBLEM 13 [15]	
	NMSERS	PATSH	PCON	SEEK1	ADRANS
Execution Time, T (sec)	168	210	493	46	73
Function Evaluations, n	294	382	335	76	76
Ratio, T/n	0.57	0.55	1.5	0.60	0.96

TABLE 2
Number of Function Evaluations
as a Predictor of Execution Time

The actual time per function and constraint evaluation for the design problem is about 0.5 sec on a CDC 6600, corresponding to about 1.5 million machine instructions. This is a much more time-consuming problem than the usual mathematical test function, yet it is clear that comparison of function evaluations would give a misleading ranking of execution time. The reason in this case is that the constraints require significant execution time, and they are evaluated more often than the objective function by PCON. A similar discrepancy between number of evaluations and execution time was noted on Problem 13 in [15], which involved a simulation of vehicle acceleration. The source of the discrepancy is unknown in this case, which was run on an IBM 370/165. An equivalent evaluation approach could potentially reduce the discrepancies, but the function, constraint, and derivative counts must be somehow weighted relative to each other. An empirical study of an equivalent evaluation approach for function, gradient, and Hessian evaluations concluded that such an approach is inferior to direct time measurement [16].

3.0 Comparison Difficulties with Results Obtained on a Different Computer

It is often necessary to choose between optimization codes that have been tested only on machines that are unavailable to the potential user. In addition, there is great interest in comparing the results of several comparison studies, since they inevitably include different codes and problems. For either of these purposes, a machine-independent measure of performance is essential. The number of function evaluations, though not a particularly good estimator of computing time, is often considered such a machine-independent measure.

Unfortunately, the number of function evaluations does vary when the same code and problem decks are run on different machines. As an example, the NMSERS code solved Problem 5 in [2] to nearly the same accuracy on three computer and compiler combinations. In all cases, the standard non-optimizing compiler was used, and the card decks were identical except for the declaration cards required to implement Implicit Double Precision on the IBM machine. The results show a factor of ten difference in the number of function evaluations, as shown in Table 3. A smaller discrepancy was found at low accuracy on Problem 10 in [15] and at intermediate accuracy on Problem 4 in [2].

	Problem 5 [2]		Problem 10 [15]	
Computer, Compiler	Function Evaluations	Accuracy ERRX	Function Evaluations	Accuracy ERRX
CDC 6400, RUN	2982	3.2E-14	40	0.11
CDC 6600, FTN, OPT = 0	4961	3.2E-14	31	0.11
IBM 370/165, G	535	7.9E-15	31	0.11

TABLE 3

Variation of Measured Function Evaluations for NMSERS

Word length may be the controlling difference between the IBM and CDC machines. Operations are carried to about 14 digits in CDC single precision and 16 digits in IBM double precision. More importantly, Implicit Double Precision was used to avoid modifying the card decks, and constants are only accurate to about 7 digits on an IBM machine when this feature is used. Thus, the IBM machine solved a problem which differed in the seventh digit from the problem solved on the CDC machine. If this is the explanation for the difference, the performance of at least one optimization code is extremely sensitive to practically insignificant differences in the problem statement, algorithm constants, or round-off level.

The word length is identical on the two CDC machines, and they use the same set of machine instructions. Thus, a logical explanation for the observed difference in number of function evaluations is that the two compilers produced slightly different sequences of machine instructions from the same FORTRAN deck. The 6400 and 6600 runs are equally accurate, but the path taken on the 6400 required 40% fewer function evaluations in one case, 30% more in the other.

The results in Table 3 show the largest of several discrepancies in only twenty-six code/problem combinations. Thus, an order of magnitude variability in function evaluations due to computer and compiler differences is probably not an extremely rare event. If codes are compared on different machines, two codes with an actual factor of ten difference in performance could appear to be equally fast or two orders of magnitude different, depending on which code is tested on which machine.

If execution time is used instead of function evaluations for comparisons on different machines, the relative speed of the machines must be normalized. The most common procedure is to divide solution times by the time to execute Colville's timing routine [1]. An improved approach [7] is to divide the solution times by the time to execute the Curnow and Wichmann Synthetic Benchmark [17], which was designed to simulate a typical FORTRAN Program. Results for both procedures are shown in Table 4, where the times are the average time to solve a set of 5 problems (2, 4-6, 8 in [2]) on the computers and compilers listed. The accuracy is controlled at 0.0001 relative parameter error in all cases, and the effect of workload is somewhat controlled by basing all results on at least two runs.

The first observation is that on any one machine and compiler, PATSH would appear to be between 1.5 and 2.6 times slower than NMERS on these problems. The ratio of performance does depend on the machine and compiler, but at least one code is consistently faster. However, if the codes are run on different machines, say a 6400 and 7600, and the times are not standardized, PATSH would appear to be 9 times faster or 26 times slower than NMSERS, depending on which code is run on which machine.

Computer, Compiler	Execution Time* (Sec) NMSERS	PATSH	Colville Standardization NMSERS	PATSH	Synthetic Benchmark Standardization NMSERS	PATSH
CDC 6400, RUN	0.71	1.4	0.013	0.025	0.039	0.076
CDC 6600, FTN, OPT = 0	0.43	1.1	0.024	0.061	0.063	0.16
OPT = 2	0.28	0.55	0.061	0.12	0.057	0.11
CDC 7600, FTN, OPT = 0	0.080	0.16	0.022	0.043	0.057	0.11
OPT = 2	0.053	0.079	0.072	0.11	0.055	0.082
IBM 370/165, G (Implicit Double Prec.)	0.19	0.42	0.028	0.061	0.029	0.065

*Mean time on problems 2, 4-6, 8 [2], where time on each problem is linearly interpolated on plot of execution time vs. log (error) at relative parameter error ERRX = 1.0E-4.

Table 4

Effect of Time Standardization by Colville's Method and Synthetic Benchmark

When Colville Standardization is used, the uncertainty is not significantly reduced. If PATSH is tested on the 6400 and NMSERS is tested on the 7600 with optimizing compiler, PATSH appears to be 3 times faster. If NMSERS is tested on the 6400 and PATSH on the 6600 with optimizing compiler, PATSH appears to be 9 times slower. The standardized times for either PATSH or NMSERS vary on these machines by about a factor of 5, where the values would be identical if the standardization worked perfectly. Others have found a factor of 8:1 variation in standardized times for the same code on different machines [4]. These findings cast doubt on the validity of Colville's study where the factors discussed in Section 2.0 were not controlled and each code was run on a different machine.

The same standardization procedure using the Synthetic Benchmark works somewhat better. Comparisons made on different machines indicate that the two codes are comparable, or that PATSH is up to 5.5 times slower, depending on which machines are involved. The range of variability in the standardized time for a single code on various machines is also reduced, to 2.5:1 for PATSH, for example.

The results with Synthetic Benchmark Standardization may be the best that can be expected for any standardization technique, because the relative execution time for two codes does vary with the computer. Table 5 shows an example where PATSH is three times faster than NMSERS when both are tested on the CDC 7600 machine, but the reverse is true when the identical comparison is made on the IBM 370/165. The accuracy is 0.0001 in all cases in this table.

	Execution Time (Sec)	
Computer, Compiler	NMSERS	PATSH
CDC 7600, FTN, OPT = 0	0.12	0.040
IBM 370/165, G	0.045	0.16

Table 5
Example Where Relative Execution Time
Depends on the Computer (Problem 5 [2])

This result is for a single problem, and mean results for five problems (including this one) show less of a range. However, even for the set of problems, the ratio of times for the two codes varied by a factor of almost two (2.6:1.5, see Table 4), depending on the computer. It is also important to note that the effect of optimizing compilers is not included in the results in Table 5, and simply switching optimizing options can have an additional effect on some codes. For instance, the Colville timer runs 260% slower than the Synthetic Benchmark on a CDC 7600 with the non-optimizing compiler OPT = 0, but 23% faster than the Benchmark when both use the optimizing compiler OPT = 2. It is clear that the relative performance as measured by execution time actually does depend on the computer and compiler, and no standardization procedure can eliminate this dependence.

If number of function evaluations and execution time are rejected as machine-independent performance measures, the only remaining common measure is robustness (ability to solve a wide variety of problems). Unfortunately, robustness also varies with the computer and compiler. PATSH solved nine of the test problems in [2] on an IBM 370/165, G compiler, but it solved only six of the same problems to the same accuracy (0.0001) on CDC 6600 or 7600 machines. The codes and problems were identical except for minor word length differences and the precision of constants as discussed in Section 2. As a second example, NMSERS did not solve Problem 2 [2] on the CDC 6400, but it did on the CDC 6600 and 7600 and the IBM 370/165. Here word length and precision of constants cannot be the cause; instead the difference in the RUN and FTN (non-optimizing) compilers is the suspected cause. Whatever the cause, the fact that the ability to solve a problem to a modest level of accuracy is affected by the computer and/or compiler in four problem/code pairs out of 26 tested indicates that robustness is not a reliable machine-independent performance measure.

4.0 Discussion

The results presented above suggest that any minor perturbation in the problem and/or algorithm, as translated into machine instructions, can cause an optimization code to take a shorter or longer path to the optimum or not even converge. Seemingly minor differences in the word length, the precision of constants, or the sequence of instructions derived from the FORTRAN algorithm are apparently sufficient to obtain order-of-magnitude changes in measured performance, and these differences are always present in different machines and compilers. In addition, there are significant difficulties in conducting comparisons that are valid for even one machine.

These rather limited results argue for statistical experimentation as practiced in any empirical science. The factors that affect the measured performance must be identified and treated by standard experimental design techniques, including blocking, factorial analysis, replication, and randomization. For example, a replicated set of runs for PATSH and NMSERS, on a wider variety of problems and computers, blocked by accuracy level and randomized over time of day and starting point, might establish that the codes can be expected to solve about the same fraction of problems posed, with a coefficient of variation of 50%, and that PATSH execution time will average 200% that of NMSERS, with 95% confidence bounds at 30% and 1000%. Though less satisfying than a clear-cut "A is twice as fast as B," such statistical results appear to be the best one can do since the algorithms seem to be inherently sensitive to uncontrollable factors. Such a statistical approach is being applied to varying degrees in recent studies [6], [18], [19].

The presence of large uncertainties in measured performance of optimization codes is not a severe drawback, because the important features of any new problem to be solved may not be well represented in the test problems, in which case measured performance is not relevant. Furthermore, differences of a factor of ten in execution speed and cost are often irrelevant compared to the large labor cost for problem preparation and interpretation of results in practical applications. Preparation time can be reduced by skillful programming to nearly the same level for any algorithm within a given class (non-derivative, first derivative, second derivative), so the variation in total cost to solve practical problems is much less than the variation in computer cost.

5.0 Conclusions

A limited sample of code comparison results obtained on several computers and compilers indicates that relative optimization code performance is not machine-independent. This fact is evident whether performance is measured by number of function evaluations, execution time, or fraction of problems solved. Furthermore, optimization code performance is sensitive to many factors which have not been well controlled in the early code comparison studies. Thus, results from these previous studies should be used with extreme caution, particularly if the studies were conducted on a different machine or compiler than the one available to the user. A possible reason for some of these difficulties is the inherently path-dependent nature of most optimization algorithms, but this theory requires further support. The practical solution is to adopt a statistical testing approach, as in several recent studies, and accept broad confidence bands on measured relative performance.

Acknowledgement

Financial support from University of Toronto, University of California, Berkeley, National Science Foundation and Department of Energy, Contract AT(29-1)-789, is gratefully acknowledged.

References

[1] Colville, A.R., "A Comparative Study on Nonlinear Programming Codes," IBM New York Scientific Center Report No. 320-2949, June 1968, summarized in Proceedings of the Princeton Symposium on Mathematical Programming, Kuhn, H.W., ed., Princeton, NJ, 1970, pp. 487-501.

[2] Eason, E.D. and Fenton, R.G., "A Comparison of Numerical Optimization Methods for Engineering Design," Journal of Engineering for Industry, Trans. ASME Series B, Vol. 96, No. 1, February 1974, pp. 191-196.

[3] Stocker, D.C., A Comparative Study of Nonlinear Programming Codes, MS Thesis, University of Texas at Austin, 1969.

[4] Himmelblau, D.M., Applied Nonlinear Programming, McGraw-Hill, New York, 1972.

[5] Sandgren, E., The Utility of Nonlinear Programming Algorithms, PhD Thesis, Purdue University, 1977.

[6] Schittkowski, K., "A Numerical Comparison of 13 Nonlinear Programming Codes with Randomly Generated Test Problems," Preprint, March 1979.

[7] Eason, E.D., "Validity of Colville's Time Standardization for Comparing Optimization Codes," ASME Paper No. 77-DET-116, September 1977.

[8] Dembo, R.S. and Mulvey, J.M., "On the Analysis and Comparison of Mathematical Programming Algorithms and Software," Proceedings of the Bicentennial Conference on Mathematical Programming, Gaithersberg, Maryland, 1976.

[9] Hillstrom, K.E., "A Simulation Test Approach to the Evaluation and Comparison of Unconstrained Nonlinear Optimization Algorithms," Argonne National Laboratories Report No. ANL-76-20, 1976.

[10] Nazareth, L. and Schlick, F., "The Evaluation of Unconstrained Optimization Routines," Proceedings of the Bicentennial Conference on Mathematical Programming, Gaithersberg, Maryland, December 1976.

[11] Dembo, R.S., "A Set of Geometric Programming Test Problems and Their Solutions," Working Paper #87, Department of Management Sciences, University of Waterloo, Ontario, June 1974.

[12] Hock, W. and Schittkowski, K., "Test Examples for the Solution of Nonlinear Programming Problems," Parts 1 and 2, Preprints 44, 45, Institut fur Angewandte Mathematik und Statistik, Universitat Wurzburg, 1979.

[13] Crowder, H.P., Dembo, R.S., and Mulvey, J.M., "Reporting Computational Experiments in Mathematical Programming," Mathematical Programming, Vol. 15, #3, November 1978, pp. 316-329.

[14] Bell, T.E., "Computer Performance Variability," Computer Performance Evaluation, National Bureau of Standards Special Publication 406, August 1975.

[15] Eason, E.D. and Fenton, R.G., "Testing and Evaluation of Numerical Methods for Design Optimization," University of Toronto Report UTME-TP 7204, September 1972.

[16] Miele, A. and Gonzalez, S., "On the Comparative Evaluation of Algorithms for Mathematical Programming Problems," presented at Nonlinear Programming Symposium - 3, Madison, Wisconsin, July 1977.

[17] Curnow, H.J. and Wichmann, B.A., "A Synthetic Benchmark," The Computer Journal, Vol. 19, No. 1, February 1976, pp. 43-49.

[18] Schittkowski, K., Nonlinear Programming Codes - Information, Tests, Performance, Lecture Notes in Economics and Mathematical Systems, No. 183, Springer Verlag, New York, 1980.

[19] Sandgren, E., "A Statistical Review of the Sandgren-Ragsdell Comparative Study," Proceedings, U.S. Bureau of Standards/Mathematical Programming Society Conference on Testing and Validating Algorithms and Software, Boulder, Colorado, January 1981.

A STATISTICAL REVIEW OF THE SANDGREN-RAGSDELL COMPARATIVE STUDY

Eric Sandgren
IBM Corporation
Information Systems Division
Lexington, Kentucky

ABSTRACT

A statistical analysis of the solution times of the algorithms in the Sandgren-Ragsdell study is conducted. An analysis of variance is performed to demonstrate that there is statistical evidence that selected codes are superior to others on the basis of their relative solution times. A logarithmic transformation is used to produce a semi-normal distribution of the solution times with a variance assumed to be equal for all of the algorithms. A paired comparison is then conducted on the differences in the mean logarithmic solution times for each of the algorithms over the entire test problem set. The selected confidence level for all comparisons was fixed at 95%. The factors contributing to the success of this analysis are discussed as well as the additional data which would be required to conduct this type of analysis in general.

1. Introduction

On the surface the comparison of computational algorithms does not appear to be a Herculean task. Statistically the problem reduces to nothing more than comparing the treatment means of the codes on an appropriate subset of the problem population. Classes of algorithms which are structured to solve a well defined set of problems are easily adapted to this approach. This is mainly due to the fact that the problem populations are so well defined. Take for example the case of linear programming algorithms. The general linear programming problem may be expressed as:

$$\text{Minimize} \quad y = \sum_{i=1}^{N} C_i X_i \tag{1}$$

$$\text{subject to} \quad \sum_{i=1}^{N} a_{ij} x_i - b_j \geq 0; \; j = 1,\ldots,M \tag{2}$$

$$\text{with} \quad X_i \geq 0; \; i = 1,\ldots,N \tag{3}$$

For a specified number of variables and constraints a statistically valid subset of problems may be generated by selecting the coefficients a_{ij}, c_{ij} and b_j with a random number generator. A random starting point or a set of random starting points may be selected for each problem and all codes to be tested could then be applied to each of the test problems. The results, whether they are the solution times, the iteration count or some other measureable quantity, could be evaluated through an analysis of variance. This type of approach has been applied to not only linear programming algorithms but to quadratic, integer, network and shortest path problems as well [1,2,et al.].

With this background let us now turn to the comparison of nonlinear programming algorithms. The general nonlinear programming problem may be stated as:

$$\text{Minimize} \quad f(\bar{x}); \qquad \bar{x} = \{x_1,\ldots,x_N\} \tag{4}$$

$$\text{subject to} \quad g_j(\bar{x}) \geq 0; \qquad j = 1,\ldots,J \tag{5}$$

$$\text{and} \quad h_k(\bar{x}) = 0; \qquad k = 1,\ldots,K \tag{6}$$

$$\text{with} \quad a_i \leq x_i \leq b_i; \qquad i = 1,\ldots,N \tag{7}$$

This formulation introduces substantial complexity to the comparison procedure. The basic problem which must be dealt with is that it is no longer an easy task to select a problem subset which adequately represents the problem population. The problem formulation is simply too general in nature since the functions f, g and h may be any mathematical functions. Some of the difficulties introduced due to the problem generality are the limited ability of current nonlinear programming codes to solve problems, the possibility of local minima, extreme solution time discrepancies for problems of similar size and the necessity of selecting program input parameters. Many of these problems may be eliminated by restricting the comparison to a specific class of nonlinear programming problems. This approach was implemented in a comparison of seven algorithms on a randomly generated set of quadratic, cubic and quartic objective functions and constratins [3]. Unfortunately, the results of such a study are limited to the type of problem considered and the extrapolation of the results to the general problem is simply not possible.

Another approach is to select a test problem set which is to represent the total problem population. This is the approach which appears most frequently in the literature [3,4,5,6, et al.]. Most of these studies consider the number of problems solved to be the major comparative criterion. Invariably, however, an additional comparison is made based upon the CPU times of the codes. It is this aspect of comparison which will be investigated in some detail. In particular, the data from the Sandgren-Ragsdell study [3,7] will be used as the data base for an analysis of variance on the solution times of the nonlinear programming codes tested in the study. This study was selected for two basic reasons. First of all the data was readily available and in a convenient form. Secondly, the comparative results show such a clear distinction between the various algorithm classes that it should be possible to make statistical statements about the mean solution times of the various algorithms.

2. The Sandgren-Ragsdell Study

The Sandgren-Ragsdell study was conducted over the period from June 1973 to the end of 1977 at Purdue University with the support of a grant from the National Science Foundation. The goal of the study was to discern the utility of the world's leading nonlinear programming methods in an engineering design environment. The study initially involved thirty-five algorithms and thirty test problems. The algorithms and problems contained in the study are listed in Tables 1 and 2 respectively. Complete descriptions of the codes and problems are given by Sandgren [3]. During the initial testing, eleven codes were eliminated due to poor performance leaving a total of twenty-four codes to be compared. The test problem set consisted of problems taken from previous comparative studies as well as additional engineering applications: Due to the fact that fewer than five of the codes were able to solve seven of the test problems, these problems were not included in the final results. Again, the question of how well these twenty-three test problems represent the entire population of engineering design problems is a valid one. Therefore, the discussion will be limited to the results on this test problem set and no attempt will be made toward generalization. One can expect, however, the same difficulties in analyzing any comparative data from nonlinear programming algorithms as are found in the analysis of this test data. What this test problem set does have is a wide variety in size and structure. The number of variables range from two to forty-eight, the number of constraints vary from zero to nineteen and the number of variable bounds range from three to seventy-two. The computational time required to solve each problem ranged from several tenths of a second of

CODE NUMBER	NAME AND/ OR SOURCE	CLASS	UNCONSTRAINED SEARCH METHOD
1	BIAS	exterior penalty	variable metric (DFP)
2	SEEK1	interior penalty	random pattern
3	SEEK3	interior penalty	Hooke-Jeeves
4	APPROX	linear approximation	none
5	SIMPLX	interior penalty	simplex
6	DAVID	interior penalty	variable metric (DFP)
7	MEMGRD	interior penalty	memory gradient
8	GRGDFP	reduced gradient	variable metric (DFP)
9	RALP	linear approximation	none
10	GRG	reduced gradient	variable metric (BFS)
11	OPT	reduced gradient	conjugate gradient (FR)
12	GREG	reduced gradient	conjugate gradient (FR)
13	COMPUTE II (0)	exterior penalty	Hooke-Jeeves
14	COMPUTE II (1)	exterior penalty	conjugate gradient (FR)
15	COMPUTE II (2)	exterior penalty	variable metric (DFP)
16	COMPUTE II (3)	exterior penalty	simplex/Hooke-Jeeves
17	Mayne (1)	exterior penalty	pattern
18	Mayne (2)	exterior penalty	steepest descent
19	Mayne (3)	exterior penalty	conjugate direction
20	Mayne (4)	exterior penalty	conjugate gradient (FR)
21	Mayne (5)	exterior penalty	variable metric (DFP)
22	Mayne (6)	exterior penalty	Hooke-Jeeves
23	Mayne (7)	interior penalty	pattern
24	Mayne (8)	interior penalty	steepest descent
25	Mayne (9)	interior penalty	conjugate direction
26	Mayne (10)	interior penalty	conjugate gradient
27	Mayne (11)	interior penalty	variable metric (DFP)
28	SUMT IV (1)	interior penalty	Newton
29	SUMT IV (2)	interior penalty	Newton
30	SUMT IV (3)	interior penalty	steepest descent
31	SUMT IV (4)	interior penalty	variable metric (DFP)
32	MINIFUN (0)	mixed penalty	conjugate directions
33	MINIFUN (1)	mixed penalty	variable metric (BFS)
34	MINIFUN (2)	mixed penalty	Newton
35	COMET	exterior penalty	variable metric (BFS)

TABLE 1. Codes in Study

PROBLEM NUMBER	NAME AND/ OR SOURCE	N	J	K	NUMBER OF VARIABLE BOUNDS
1	EASON #1	5	10	0	5
2	EASON #2	3	2	0	6
3	EASON #3	5	6	0	10
4	EASON #4	4	0	0	8
5	EASON #5	2	0	0	4
6	EASON #6	7	0	4	12
7	EASON #7	2	1	0	4
8	EASON #8	3	2	0	6
9	EASON #9	3	9	0	4
10	EASON #10	2	0	0	4
11	EASON #11	2	2	0	4
12	EASON #12	4	0	0	8
13	EASON #13	5	4	0	3
14	COLVILLE #2	15	5	0	15
15	COLVILLE #7	16	0	8	32
16	COLVILLE #8	3	14	0	6
17	DEMBO #1	12	3	0	24
18	DEMBO #3	7	14	0	14
19	DEMBO #4	8	4	0	16
20	DEMBO #5	8	6	0	16
21	DEMBO #6	13	13	0	26
22	DEMBO #7	16	19	0	32
23	DEMBO #8	7	4	0	14
24	WELDED BEAM	4	5	0	3
25	COUPLER CURVE	6	4	0	6
26	WHIRLPOOL	3	0	1	6
27	SNG	48	1	2	72
28	FLYWHEEL	5	3	0	10
29	AUTOMATIC LATHE	10	14	1	20
30	WASTE WATER	19	1	11	38

TABLE 2. Problems in Study

CPU time on a CDC 6500 to several hundred seconds of CPU time. The computational time required to run all of the codes on each of the test problems was well in excess of 200 hours of CPU time. The financial burden of a larger test problem set becomes prohibitive rather quickly.

The test codes were modified to run on Purdue University's CDC 6500 computer system. All codes were converted to single precision arithmetic. Any code which required analytical gradient information was altered to accept a numerical gradient approximation via forward differences. All print instructions were removed from the basic iteration loop so as not to influence the measured solution time. All program input parameters such as initial penalty parameters and line search convergence criteria were set at the recommended values from the user's manual for the initial run. If a normal, satisfactory termination was not obtained after this run, an ordered step-by-step change was made to the input parameters using the information contained in the user's manual as a guide to the adjustment. No attempt was made to decrease the solution time once a satisfactory solution was found. The individual runs were made in a random order to remove any bias in the results due to inherent sources of variation such as the system load or an operating system change during the course of the study. In order to compare the solution times on problems which required such different amounts of computational time the solution time for each problem was normalized by the average solution time for all of the algorithms on that problem. This means that time comparisons will be based on the fraction of the average solution time and a number less than one represents a problem solved by a code in less than the average time.

The major conclusion from the study was that the generalized reduced gradient codes were generally more effective than any of the penalty function algorithms. These codes solved a greater percentage of the test problems and were found to be significantly faster than any other type of method in the study. Solving a greater number of problems is an easy result to quantify but the comparison of the normalized solution times is a little more difficult. If indeed the generalized reduced gradient codes are faster then it should be possible to generate this result from a statistical analysis of the data.

3. Preliminary Analysis

The goal of this analysis is to compare the treatment means of the twenty-four algorithms to determine if indeed there is evidence to support the statement that the generalized reduced gradient codes are computationally faster than the other codes tested. The treatment means will initially be based on the normalized solution times of the algorithms on the twenty-three problem test set. The normalized solution time data is presented in tabular form in Table 3. Notice that only one code has twenty-three data points. This is because only one code solved all of the test problems. A blank space in Table 3 indicates that a code failed to solve that particular problem. No account will be taken in the analysis for the failure of a code to solve a problem. The comparison of means will be done only on the set of problems solved by each individual code. Since the problems themselves could not be randomly generated from the set of all nonlinear programming problems related to engineering design, the problems were used as a block. That is to say that all of the codes solved the same set of test problems. As mentioned previously, this may limit the ability to make concrete statistical statements regarding the performance of each algorithm on the general design problem but it adds greatly to the precision within the experiment.

In order to perform an analysis of variance the data will be treated as samples taken in a random order from twenty-four normal populations having the same variance σ^2 and differing if at all only in their means. Right away we have a potential problem. The assumption that all of the nonlinear programming codes tested have the same variance must be checked This may be accomplished by plotting the residuals in the data versus the calculated mean value for each of the algorithms. The residuals are the quantities remaining after the treatment averages have been removed from the data. Mathematically, the residuals may be defined as $y_{ti} - \hat{y}_t$ where y_{ti} is the ith data point from the tth treatment and $\hat{y}_t$ is the calculated mean value of treatment t. A plot of the residuals versus the calculated treatment means is given in Figure 1. From the funnel-shaped appearance of the data in Figure 1, one can see that as the mean solution time increases so do the residuals. This indicates that the codes which require a longer average time to solve a problem tend to have a larger variance also. The easiest way to rectify this problem is to find a transformation which may be applied to the data which will tend to equalize the variances. This may be done in the following manner: first assume that the standard deviation of each code is proportional to some power of the mean normalized solution time of that code. That is to say:

$$\sigma_t \propto N_t^B \tag{8}$$

The power B may then be determined by finding the best straight line fit through a log-log plot of the calculated variance versus the

PROBLEM	CODE 1	12	11	10	9	8	21	15	13	27	33	35	31	32	3	14	16	19	20	22	26	28	29	34
1	.376	.097	.091	.052	.110	.169	.252	.799	1.099	.699	1.312	1.439	.953	1.825	----	1.198	1.307	2.271	----	2.542	1.230	1.346	1.422	1.408
2	.661	.043	.031	.019	.082	.192	.476	.508	.944	.997	1.270	1.561	2.881	2.122	1.690	.433	1.168	1.420	.686	----	1.178	1.764	2.057	.803
3	.366	.043	.018	.008	.020	.070	.091	.163	.170	.353	1.959	1.672	.505	2.462	.840	.139	.199	.877	.254	8.140	.453	----	2.681	1.522
4	.387	.431	.151	.272	9.150	1.580	.137	.591	----	.209	1.063	.256	----	.606	----	----	----	.359	.229	.278	.297	----	----	----
5	.518	.568	.440	.387	----	2.561	.525	.397	.213	----	1.330	.465	2.519	1.774	1.164	.692	.376	.635	.575	.986	----	1.838	2.040	----
6	----	.247	.240	.124	----	.350	----	----	1.373	----	----	3.666	----	----	----	----	----	----	----	----	----	----	----	----
7	.329	.205	.156	.092	.485	.458	.251	.761	.913	.707	.419	5.511	1.438	1.291	1.906	1.352	1.638	1.508	----	.281	.927	1.071	.971	.331
8	.577	.091	.062	.041	.163	.076	.262	.575	.806	1.131	1.847	3.580	1.342	2.976	1.040	----	1.012	.906	.259	.322	.932	1.604	2.418	----
9	.595	.351	.090	.262	1.802	.568	.342	.234	.459	1.514	1.126	.703	2.928	2.477	1.472	.181	.541	1.042	.322	.517	1.347	1.705	2.324	1.128
10	.537	.325	.115	.190	----	.144	1.013	.608	.692	.716	----	.832	2.109	1.277	2.769	----	.755	1.269	1.285	.667	.824	2.010	1.862	----
11	.299	.233	----	.124	4.121	.268	.189	.049	.401	1.688	2.759	.329	----	2.306	----	.338	.146	1.112	.576	----	2.059	----	----	----
12	.283	.103	.068	.043	.116	.138	----	----	----	.330	.884	3.536	.507	2.652	----	----	----	----	----	----	----	1.944	2.157	1.243
13	.909	.092	.061	.030	.133	.160	.843	----	1.087	----	3.058	1.658	2.093	----	----	1.881	----	----	----	----	----	----	----	----
14	1.370	.154	.088	.066	1.528	.251	.288	----	.466	.308	1.521	2.010	4.694	1.211	.502	----	.260	2.809	.308	.223	.428	----	----	1.518
15	----	.380	----	----	.148	1.348	----	----	----	.190	----	1.599	----	2.334	----	----	----	----	----	----	----	----	----	----
16	1.457	.154	.197	----	.089	.548	----	.749	.435	.288	.483	3.888	1.079	.778	.947	----	----	2.684	----	1.643	----	----	----	.582
17	.322	.080	.100	.065	.134	----	.184	.594	----	----	.558	3.626	1.135	1.042	1.453	----	----	----	----	----	----	2.002	2.901	.802
18	1.613	.081	.059	----	.103	.119	.330	----	.216	.198	----	1.279	.879	----	----	----	----	----	----	----	6.115	----	----	----
19	.329	.366	.336	----	----	----	----	.297	----	----	.634	.793	.439	1.342	----	----	----	2.436	.607	----	----	3.321	----	1.097
20	.392	.423	.106	.068	.081	.163	.223	.469	----	.562	.965	.750	1.687	1.022	----	----	----	----	----	----	----	3.087	2.907	3.093
21	.911	.832	.491	.093	----	.280	.173	----	1.651	.536	.755	1.837	2.099	1.586	2.429	----	----	----	----	.407	----	----	----	----
22	----	.138	.095	.042	.051	.107	----	.215	----	----	2.421	3.708	----	----	----	.540	.410	----	----	----	----	----	----	3.272
23	1.542	.837	.330	.441	.955	----	----	----	----	1.895	----	----	----	----	----	----	----	----	----	----	----	----	----	----

TABLE 3. Normalized Solution Time Data

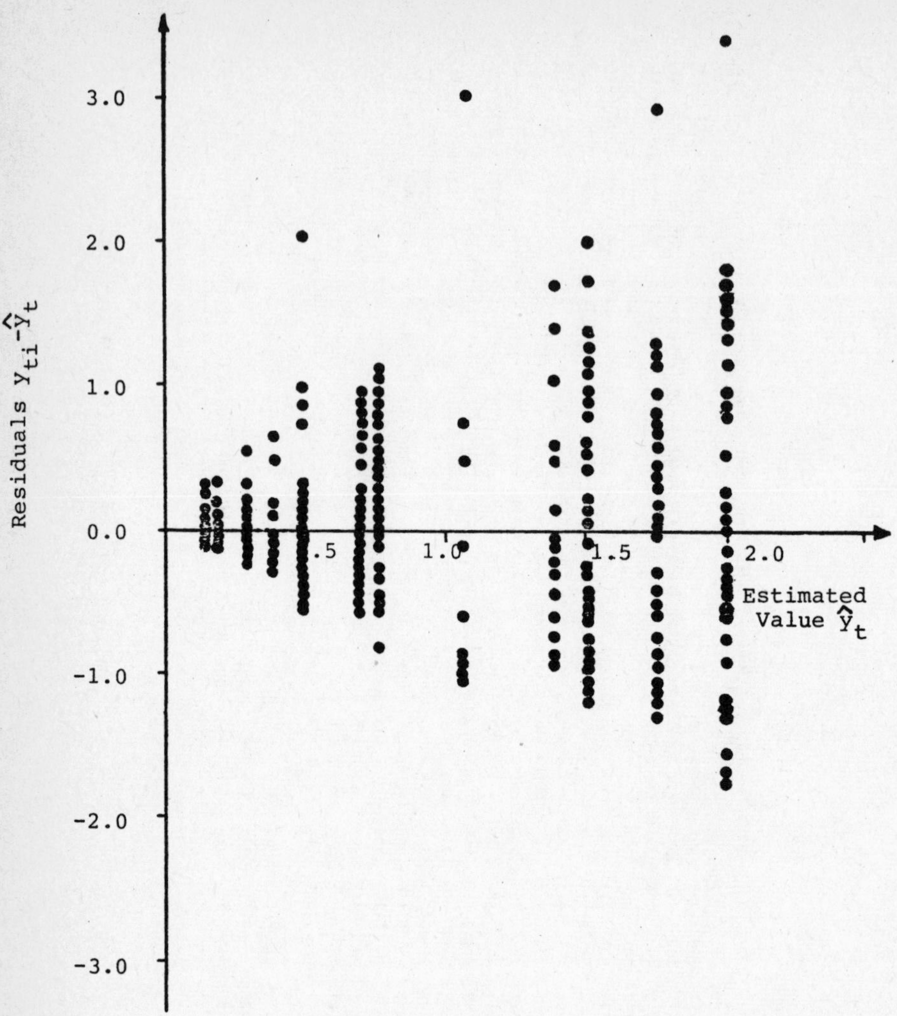

FIGURE 1. Plot of Residuals vs. Estimated Values for Normalized Solution Time Data

calculated mean value for each code. Such a plot is given in Figure 2. From this plot, the best fit line is one with a slope of approximately 1. This suggests that a logarithmic transformation will stabilize the variance [8]. This means that each data point is replaced by the logarithm to the base 10 of its value. This will not add any additional complication to the interpretation since any conclusions drawn on the logarithmic data may be just as easily applied to the original data. A plot of the residuals of the log transformed data versus the calculated treatment means is shown in Figure 3. No obvious trend is evident in this data. It can still be shown from Bartlett's test that all of the variances of the treatments are most likely not equal even using the transformed data. Experiments have shown, however, that a small to moderate difference in the variance of the treatments has a small effect on the analysis [9]. The log transformed data is shown in tabular form in Table 4. The dot diagram of the transformed data demonstrating the normal nature of the data is presented in Figure 4.

4. Analysis of Variance

An analysis of variance will be used to test the null hypothesis that all of the treatments have the same mean value. The alternate hypothesis is that there is indeed a difference in the treatment means. The acceptance or rejection of the null hypothesis will be based on whether the variation between treatments is significantly greater than the variation within treatments. This is accomplished by constructing an analysis of variance table. First consider the variation within treatments. The within treatment variation is an estimate of the internal consistency of the data. It is the collection of the individual variances calculated for each of the individual treatments. The within treatment sum of squares is given by:

$$S_R = \sum_{t=1}^{k} \sum_{i=1}^{n_t} (y_{ti} - \hat{y}_t)^2 \tag{9}$$

Here again y_{ti} is the ith observation within the tth treatment and $\hat{y}_t$ is the calculated average of the data points for the tth treatment. In addition, k represents the number of treatments and n_t represents the number of data points which are available for treatment t. The number of within treatment degrees of freedom is given by:

$$v_R = \sum_{t=1}^{k} (n_t - 1) \tag{10}$$

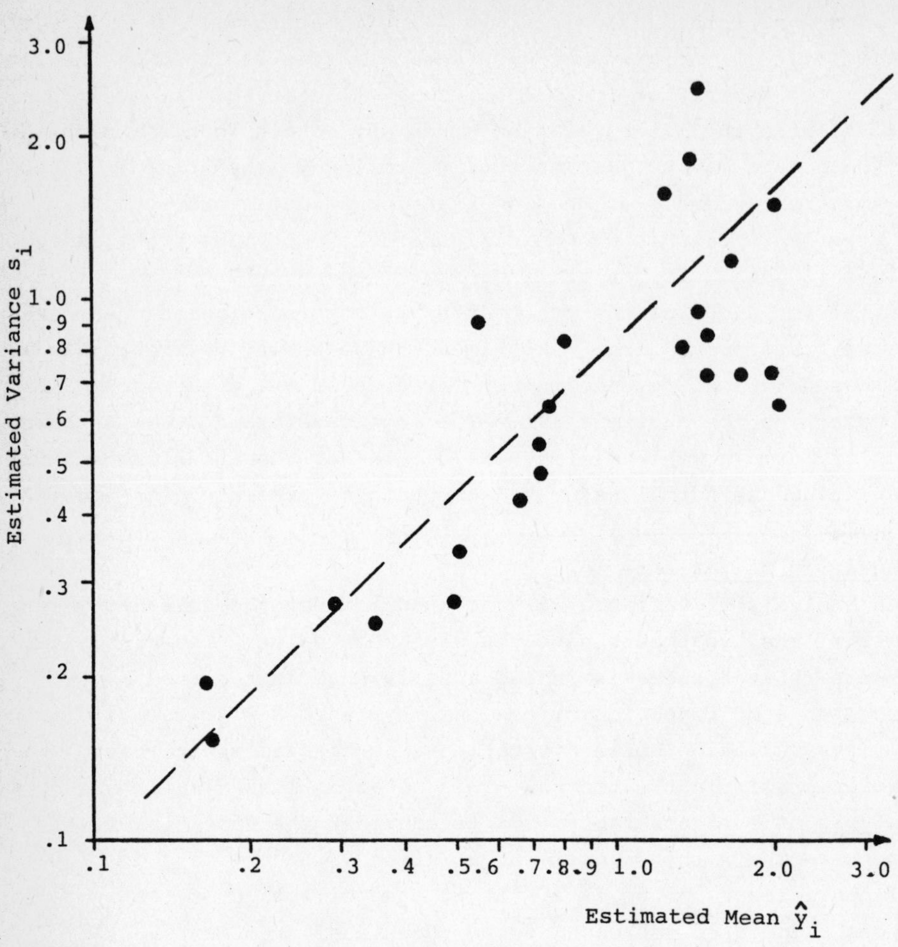

FIGURE 2. Plot of the Estimated Variance Versus the Estimated Mean Value for the Normalized Solution Time Data

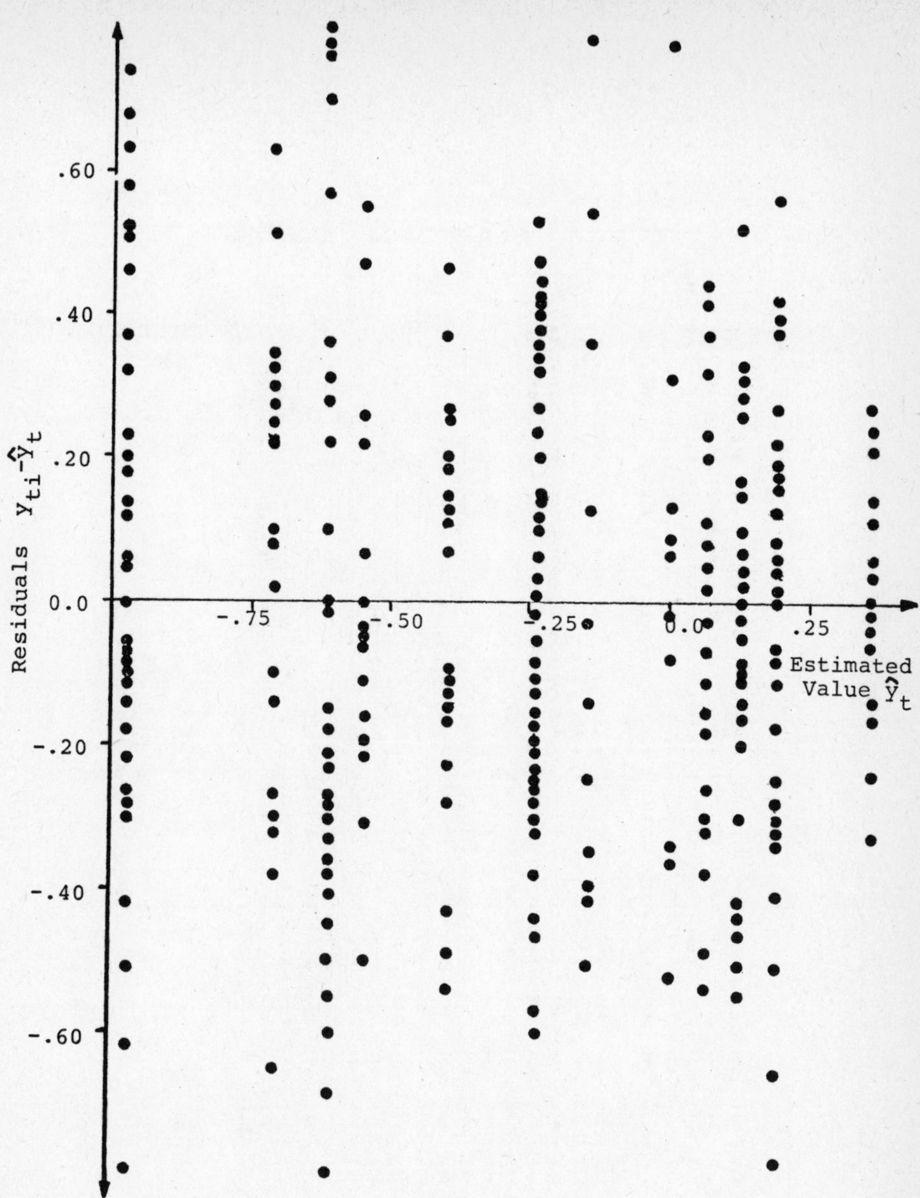

FIGURE 3. Plot of Residuals Versus Estimated Values for Log Transformed Solution Time Data

PROBLEM	CODE 1	12	11	10	9	8	21	15	13	27	33	35	31	32	3	14	16	19	20	22	26	28	29	34
1	-.425	-1.01	-1.04	-1.28	-.959	-.772	-.599	-.098	+.041	-.156	+.118	+.158	-.021	+.261	----	+.079	+.116	+.356	----	+.405	+.090	+.129	+.153	+.149
2	-.180	-1.37	-1.51	-1.72	-1.09	-.717	-.322	-.294	-.025	-.001	+.104	+.193	+.460	+.327	+.228	-.364	+.067	+.152	-.164	----	+.071	+.246	+.313	-.095
3	-.437	-1.37	-1.74	-2.10	-1.70	-1.15	-1.04	-.788	-.770	----	+.292	+.223	-.297	+.391	-.076	-.857	-.701	-.057	-.595	+.911	-.344	----	+.428	+.182
4	-.412	-.366	-.821	-.565	+.961	+.199	-.863	-.228	----	----	+.027	-.592	----	-.218	----	----	----	-.445	-.640	-.556	-.527	----	----	----
5	-.286	-.246	-.357	-.412	----	+.408	-.280	-.401	-.672	-.452	+.124	-.333	+.401	+.249	+.066	-.160	-.425	-.197	-.240	-.006	----	+.264	+.310	----
6	----	-.607	-.620	-.907	----	-.456	----	----	+.138	-.680	----	+.564	----	----	----	----	----	----	----	----	----	----	----	----
7	-.483	-.688	-.807	-1.04	-.314	-.339	-.600	-.119	-.040	-.151	-.378	+.741	+.158	+.111	+.280	+.131	+.214	+.178	----	+.551	-.033	+.030	-.013	-.480
8	-.239	-1.04	-1.21	-1.39	-.788	-1.12	-.582	-.240	-.094	+.054	+.266	+.554	+.128	+.474	+.017	----	+.005	-.043	-.587	-.492	-.031	+.205	+.383	----
9	-.225	-.455	-1.05	-.582	+.256	-.246	-.466	-.631	-.338	+.180	+.052	-.153	+.467	+.394	+.168	-.742	-.267	+.018	-.492	-.287	+.129	+.232	+.366	+.052
10	-.270	-.488	-.939	-.721	----	-.842	+.006	-.216	-.160	-.145	----	-.080	+.324	+.106	+.442	----	-.122	+.103	+.109	-.176	-.084	+.303	+.270	----
11	-.524	-.633	----	-.907	+.615	-.572	-.703	-1.31	-.397	+.227	+.441	-.483	----	+.363	----	-.471	-.836	+.046	-.240	----	+.314	----	----	----
12	-.548	-.987	-1.17	-1.37	-.936	-.860	----	----	----	-.481	-.054	+.549	-.295	+.424	----	----	----	----	----	----	----	+.289	+.334	+.095
13	-.041	-1.04	-1.21	-1.52	-.876	-.796	-.074	----	+.036	----	+.485	+.220	+.321	----	----	+.274	----	----	----	----	----	----	----	----
14	+.137	-.912	-1.08	-1.18	+.184	-.600	-.541	----	-.332	-.511	+.182	+.303	+.672	+.083	-.299	----	-.585	+.449	-.511	-.652	-.369	----	----	+.181
15	----	-.420	----	----	-.830	+.130	----	----	----	-.721	----	+.204	----	+.368	----	----	----	----	----	----	----	----	----	----
16	+.163	-.812	-.706	----	-1.05	-.261	----	-.126	-.362	-.541	-.316	+.590	+.033	-.109	-.024	----	----	+.429	----	+.216	----	----	----	-.235
17	-.492	-1.10	-1.00	-1.19	-.873	----	-.735	-.226	----	----	-.253	+.559	+.055	+.018	+.162	----	----	----	----	----	----	+.301	+.463	-.096
18	+.208	-1.09	-1.23	----	-.987	-.924	-.481	----	-.666	-.703	----	+.107	-.056	----	----	----	----	----	----	----	+.786	----	----	----
19	-.483	-.437	-.474	----	----	----	----	-.527	----	----	-.198	-.101	-.358	+.128	----	----	----	+.387	-.217	----	----	+.521	----	+.040
20	-.407	-.374	-.975	-1.17	-1.09	-.788	-.652	-.329	----	-.250	-.016	-.125	+.227	+.009	----	----	----	----	----	----	----	+.490	+.463	+.490
21	-.091	-.080	-.309	-1.03	----	-.553	-.762	----	+.218	-.271	-.122	+.264	+.322	+.200	+.385	----	----	----	----	-.390	----	----	----	----
22	----	-.860	-1.02	-1.38	-1.29	-.971	----	-.668	----	----	+.384	+.569	----	----	----	-.268	-.387	----	----	----	----	----	----	+.515
23	+.188	-.077	-.481	-.356	-.020	----	----	----	----	+.278	----	----	----	----	----	----	----	----	----	----	----	----	----	----

TABLE 4. Log Transformed Solution Time Data

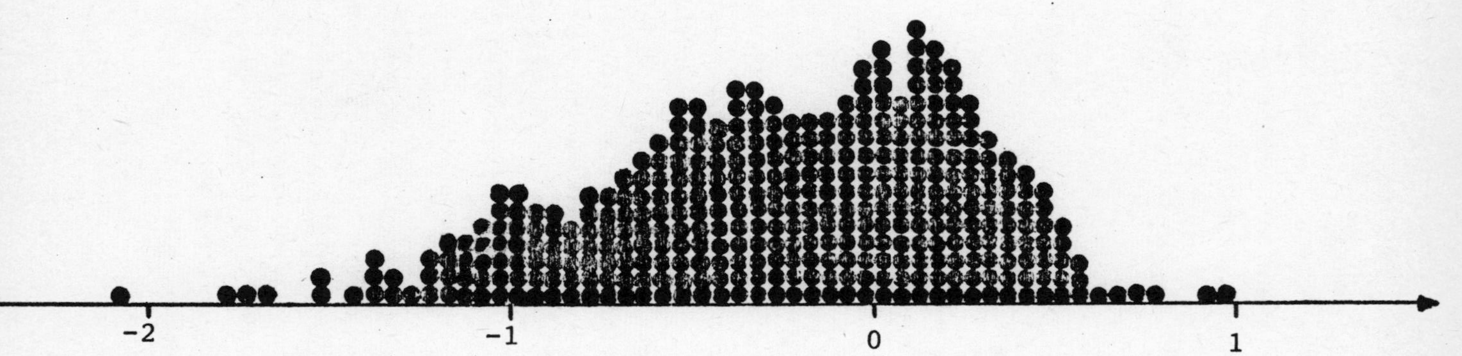

FIGURE 4. Dot Diagram of the Log Transformed Solution Time Data

and the estimate of the variance is given by

$$S_R^2 = \frac{S_R}{V_R} \tag{11}$$

Now if there were actually no real differences in the treatment means and the null hypothesis is correct another estimate of the variance could be calculated using the grand average of all the observations, $\hat{y}$. The sum of squares could be expressed as:

$$S_T = \sum_{t=1}^{k} n_t(\hat{y}_t - \hat{y})^2 \tag{12}$$

The associated number of degrees of freedom is given by:

$$v_T = k - 1 \tag{13}$$

Finally the estimate of variance between treatments is given by:

$$S_t^2 = \frac{S_T}{V_T} \tag{14}$$

The results of these calculations for the comparative study data are listed in Table 5. The null hypothesis may now be verified or rejected by comparing the ratio of the between and within treatment variance estimates with the F distribution with 23 and 345 degrees of freedom. The ratio of the estimates of variance for the data is found to be on the order of 20. This means that the null hypothesis may be rejected with 99.99+% certainty. This is good news indeed for it means that at least some of the codes have differing means which one would simply have to expect.

The analysis may now be taken one step farther to a multiple comparison of the treatment means. Through the application of a multiple comparison the individual means of any two codes in the study can be compared and it can be determined within a certain level of confidence whether the mean of the log normalized solution times for one code can be said to be greater or less than that of another code. Tukey [10] developed such a paired comparison procedure. The confidence limits for the comparison of two treatment means is given as:

$$\pm \frac{q_{k,v,\alpha/2}}{2} \left[\left(\frac{1}{n_i} + \frac{1}{n_j}\right) S^2\right]^{1/2} \tag{15}$$

SOURCE OF VARIATION	SUM OF SQUARES	DEGREES OF FREEDOM	MEAN SQUARE
BETWEEN TREATMENTS	$S_T=58.351$	23	$s_T^2=2.537$
WITHIN TREATMENTS	$S_R=43.815$	345	$s_R^2=0.127$
TOTAL ABOUT THE GRAND AVERAGE	$S_D=102.166$	368	$s_D^2=0.278$

TABLE 5. Analysis of Variance Table

where $q_{k,v,\alpha/2}$ is the upper significance level of the studentized range for comparing k means with v degrees of freedom in the estimate of variance with a confidence level of $1-\alpha$. The differences in the treatment means for the comparative data is given in Table 6. The differences which exceed the 95% confidence limit as expressed by the equation 15 are circled.

5. Results

From the data presented in Table 6, one can see that there are quite a few significant differences in the treatment means among the algorithms based on the log normalized solution time data. All the algorithms in the study may be said to have a higher or lower mean solution time than several other algorithms. The question is can the statement be made that the Generalized Reduced Gradient algorithms will on the average produce lower computational solution times than the penalty function codes? The answer is yes but with some qualification. The Generalized Reduced Gradient algorithms are codes 8, 10, 11 and 12. For code number 10 the results are straightforward. This code has a mean solution time which is lower than any other code in the study with the exception of the other GRG codes at a 95% level of confidence. The same can be said for code number 11 with two exceptions. Code 9, a repetitive linear programming algorithm and code 21, an exterior penalty function algorithm, do not fall within the 95% confidence limit. For codes 8 and 12 additional codes fall outside the confidence interval. So only for two of the GRG codes can the statement be made that they are computational faster than the other methods. The failure of the other two GRG codes to fall within this category also simply points out the difficulty of making statements concerning the relative solution times of the various algorithms. The fact remains, however, that for codes 10 and 11, some strong statements can be made which reinforces a major conclusion from the Sandgren-Ragsdell study.

6. Discussion

The analysis performed was to enable a comparison of the mean solution times of algorithms on a set of test problems. It can easily be argued that the percentage of problems solved is a far more important indication of performance. Of the codes which solve a large percentage of the problems, however, the relative solution times become an important factor in selecting the best codes. The results from this analysis indicate that the comparison of the mean solution times of dif-

	CODE	1	12	11	10	9	8	21	15	13	27	33	35	31	32	3	14	16	19	20	22	26	28	29	34
Y_t		-.242	-.711	-.940	-1.10	-.599	-.562	-.543	-.413	-.228	-.254	+.063	+.179	+.149	+.199	+.123	-.264	-.265	+.106	-.358	-.144	+.000	+.274	+.316	+.067
$Y_i - Y_j$	1	*	+.469	**+.698**	**+.858**	+.357	+.320	+.301	+.171	-.014	+.012	-.305	-.421	-.391	-.441	-.365	+.022	+.023	-.348	+.116	-.098	-.242	**-.516**	**-.558**	-.309
	12		*	+.229	+.389	-.112	-.149	-.168	-.298	**-.483**	-.457	**-.774**	**-.890**	**-.860**	**-.910**	**-.834**	-.447	-.446	**-.817**	-.353	**-.567**	**-.712**	**-.985**	**-1.03**	**-.778**
	11			*	+.160	-.341	-.378	-.397	**-.527**	**-.712**	**-.686**	**-1.00**	**-1.12**	**-1.09**	**-1.14**	**-1.06**	**-.676**	**-.675**	**-1.05**	**-.582**	**-.796**	**-.940**	**-1.21**	**-1.26**	**-1.01**
	10				*	**-.501**	**-.538**	**-.557**	**-.687**	**-.872**	**-.846**	**-1.16**	**-1.28**	**-1.25**	**-1.30**	**-1.22**	**-.836**	**-.835**	**-1.21**	**-.742**	**-.956**	**-1.10**	**-1.37**	**-1.42**	**-1.17**
	9					*	-.037	-.056	-.186	-.371	-.345	**-.662**	**-.778**	**-.748**	**-.798**	**-.722**	-.335	-.334	**-.705**	-.241	-.455	**-.599**	**-.873**	**-.915**	**-.666**
	8						*	-.019	-.149	-.334	-.308	**-.625**	**-.741**	**-.711**	**-.761**	**-.685**	-.298	-.297	**-.668**	-.204	-.418	**-.562**	**-.836**	**-.878**	**-.629**
	21							*	-.130	-.315	-.289	**-.606**	**-.722**	**-.692**	**-.742**	**-.666**	-.279	-.278	**-.649**	-.185	-.399	**-.543**	**-.817**	**-.859**	**-.609**
	15								*	-.185	-.159	-.476	**-.592**	**-.562**	**-.612**	**-.536**	-.149	-.148	**-.519**	-.055	-.269	-.413	**-.687**	**-.729**	**-.479**
	13									*	+.026	-.291	-.407	-.377	-.427	-.351	+.036	+.037	-.334	+.130	-.084	-.228	**-.502**	**-.544**	-.294
	27										*	-.317	-.433	-.403	-.453	-.377	+.010	+.011	-.360	+.104	-.110	-.254	**-.528**	**-.570**	-.321
	33											*	-.116	-.086	-.136	-.060	+.327	+.328	-.043	+.421	+.207	+.063	-.211	-.253	-.003
	35												*	+.030	-.020	+.056	+.443	+.444	+.073	**+.537**	+.323	+.179	-.095	-.137	+.113
	31													*	+.099	+.026	+.413	+.414	+.043	**+.507**	+.293	+.149	-.125	-.167	+.083
	32														*	+.076	+.463	+.464	+.093	**+.557**	+.343	+.199	-.075	-.117	+.133
	3															*	+.387	+.388	+.017	**+.481**	+.267	+.123	-.151	-.193	+.057
	14																*	+.001	-.370	+.094	-.120	-.264	**-.538**	-.580	-.331
	16																	*	-.371	+.093	-.121	-.265	**-.537**	-.579	-.330
	19																		*	+.464	+.250	+.106	-.168	-.210	+.040
	20																			*	-.214	-.358	**-.632**	-.674	-.425
	22																				*	-.144	-.418	-.460	-.211
	26																					*	-.274	-.316	-.066
	28																						*	-.042	+.208
	29																							*	+.250
	34																								*

A circled value indicates a 95% confidence level

TABLE 6. Results from Paired Comparison Proceedure

ferent algorithms is possible but that unless there is a significant difference in the means a distinction between the means will not be possible. This is due in part to the large variance in the solution times of all of the algorithms. The only way to increase the sensitivity of the analysis is to increase the number of test problems. Whether it is feasible to increase the test problem set enough to decrease the confidence limit to the point where a complete comparison can be made is a matter of economics. Doubling the number of problems would only reduce the confidence limits by approximately 30%. Therefore, in order to differentiate between two mean solution times which are even fairly close, several hundred test problems would be required.

8. References

1. Lee, H. K., and Ravindran, A., "A Comparison of Five Algorithms for Solving Quadratic Programming Problems," Working Paper, School of Industrial Engineering, Purdue University, August 1975.

2. Kuhn, H. W., and Quandt, R. E., "An Experimental Study of the Simplex Method," Proc. Sympos. Appl. Math. 15, 1963, pp. 107-124.

3. Sandgren, E., "The Utility of Nonlinear Programming Algorithms," Ph.D. Dissertation, Purdue University, December 1977.

4. Colville, A. R., "A Comparative Study of Nonlinear Programming Codes," Technical Report No. 320-2949, IBM New York Scientific Center, June 1968.

5. Eason, E. D., and Fenton, R. G., "Testing and Evaluation of Numerical Methods for Design Optimization," UTME-TP 7204, University of Toronto, September 1972.

6. Fattler, J. E., Sin, Y. T., Root, R. R., Ragsdell, K. M., and Reklaitis, G. V., "On the Computational Utility of Polynomial Geometric Programming Solution Methods," Presented at the Tenth International Symposium on Mathematical Programming, Montreal, August 1979.

7. Sandgren, E., and Ragsdell, K. M., "The Utility of Nonlinear Programming Algorithms: A Comparative Study- Parts I and II," ASME Journal of Mechanical Design, Vol. 102, No. 3, July 1980, pp. 540-551.

8. Box, E. P., Hunter, W. G., and Hunter, J. S., Statistics for Experiments, Wiley, New York, 1978.

9. Gibra, I. N., Probability and Statistical Inference for Scientists and Engineers, Prentice-Hall, N.J., 1973.

10. Tukey, J.W., "Comparing Individual Means in the Analysis of Variance," Biometrics, 5. 99, 1949.

A METHODOLOGICAL APPROACH TO TESTING OF NLP-SOFTWARE

by

Jacques C.P. Bus

Mathematical Centre, Amsterdam

Abstract

Prior to detailed testing of software there should be: (1) knowledge of the software and the underlying algorithms; (2) knowledge of the class of problems to be solved by this software; and (3) goals of testing.

We present a methodological setup of testing nonlinear programming software which is based on prior knowledge. This leads to a discussion of the following topics: representativity of test problems, use of test problem generators, performance measures, presentation of test results, and presentation of final conclusions.

The concepts will be illustrated with a project which has been performed to evaluate a set of Newton-like methods for solving systems of nonlinear equations.

1. Introduction

The name of this conference "Testing and Validating Algorithms and Software" clearly expresses the distinguishable aspects of this field. First, both testing and validating are listed; these notions should be distinguished carefully (see also LYNESS (1979)). We can describe testing as collecting data or producing evidence about the performance of algorithms or software by applying them to test problems. It yields numbers which should be of a permanent and objective nature. On the other hand, validating is the process of selecting "good" programs or algorithms, based on evidence collected during testing and on other evidence about the algorithms or software programs (e.g. theoretical convergence, worst-case performance, documentation, well-structuredness, portability and length of program). The definition of "good" programs is subjective and based on the user's priorities. This subjectivity will not be removed, either by tests producing concrete numbers with thorough statistical analysis, or by stating precisely (in mathematical terms) other evidence. Moreover, even a mathematical definition of the decision process will not remove subjectivity. Such decision strategies help the user to make proper decisions based on his own priorities (see LOOTSMA (1979, 1980)). For instance, there will be circumstances that lead to a high priority for robustness and a commensurate low priority for efficiency, as well as circumstances that lead to the opposite choice.

The conference title also suggests a distinction between algorithms and programs. "Algorithm" often refers to the set of mathematical rules defining the arithmetical process that is implemented in a program. In fact, different programs can be implementations of the same algorithm. Clearly, the quality of the implementation may greatly influence test results. Today we can provide quality standards for programs which are relatively independent of the underlying algorithm (MORE (1979)). Furthermore, performance results generated by programs which use identical subroutines for equivalent arithmetical actions are more useful than results that don't. Unfortunately, it is difficult to separate performance due to programming skills, particularly for complex iterative mathematical programming methods.

In this paper we use the terms "algorithm" and "program" as follows.

An *algorithm* is an arithmetical process, described fully in some formal unambiguous way, which takes non-exact arithmetic into account and which terminates in finite time.

Upon termination, the possibility exists that the algorithm is successful or has failed. Iterative algorithms terminate due to satisfaction of convergence or failure criteria. These criteria are assumed to be part of the algorithm.

A *program* is an implementation of an algorithm in a certain programming language and for a certain computer system (with known machine constants and behavior).

In the following sections we shall illustrate the concepts with an example, i.e. testing a set of Newton-like algorithms for solving systems of nonlinear equations (BUS (1980)). For simplicity, we shall give here a definition of the class of problems considered in these examples.

Class *PI*

Let the following be given:

1. A function $F : D \to R^n$ for some n and some open domain $D \subset R^n$, with F differentiable on D and F' Lipschitz continuous on D;
2. $x_0 \in D$ (called the *initial guess*);
3. an upper bound on the error in the numerically computed value of F(x), dependent on the precision of arithmetic used;
4. tolerance values δ_F, δ_{rx} and δ_{ax}.

Find $\bar{x}$ such that

$$||F(\bar{x})|| \leqq \delta_F, \text{ and } ||\bar{x} - \bar{x}^*|| \leqq \delta_{rx}||\bar{x}|| + \delta_{ax},$$

with $\bar{x}^*$ such that $F(\bar{x}^*) = 0$.

We refer to this class of problems by *PI*. The algorithms considered for this class are iterative Newton-like algorithms.

In the next two sections we give a general discussion of testing and test sets. Following this, we consider efficiency and robustness, particularly for iterative algorithms in the field of nonlinear programming.

2. Testing

Typically, testing objectives are stated in the following form:

> *Give a prediction of some (relative) performance of certain algorithms if applied to a certain class of problems, together with information about the reliability of the prediction.*

Hence we consider a set of algorithms $A = \{A_1, A_2, \ldots, A_m\}$, and a class of problems P where P is a subset of the class of problems for which the given algorithms are designed (possibly the whole class). Finally, there is a description of the relevant (relative) performance and a distinction between good and bad performance. This last distinction is obviously subjective. Test results will not answer the question of which algorithm shows best performance. Instead, testing yields an ordered collection of evidence, based on numerical experiments about the performance of the algorithms. These test results are input to the user whereby he makes a decision regarding which algorithm to use in a certain situation.

The first step in the testing process is to formalize the description of relevant performance. A *performance measure* is a mapping $m : A \times P \to R$, where $m(A_k, p)$ expresses some performance of algorithm A_k $(k=1, \ldots, m)$ if applied to problem $p \varepsilon P$. Examples of performance measures for the Newton-like algorithms for solving problems from class *PI* are:

$$m_1(A,p) = \begin{cases} 1 & \text{if A solves p within required accuracy;} \\ 0 & \text{otherwise.} \end{cases}$$

$m_2(A,p) =$ "time" (in some time scale required by A to solve p successfully, if A fails then ∞.

$$m_3(A,p) = m_2(A,p) \;/\; (-{}^{10}\log ||\bar{x} - \bar{x}^*||),$$

where $\bar{x}^*$ is a solution and $\bar{x}$ its computed approximation. We refer to these as "example performance measures." Herein the performance measures are real valued mappings. Of course there are criteria for evaluating the performance of an algorithm or program which are difficult to quantify, e.g. complexity of algorithm, and well-documentedness of program. We shall not discuss these criteria in this paper, since evidence about these will not be obtained by numerical tests.

Since an algorithmic performance depends upon problem features, it would be desirable to obtain a characterization of P. We define the following notation:

A *problem indicator* is a mapping $i : P \to R$, from the class of problems into the real numbers.

Examples of problem indicators for class PI are:

$$i_1(p) = n \text{ (the number of variables);}$$

$$i_2(p) = \begin{cases} 1 & \text{if } F'(x) \text{ is available analytically for all } x, \\ 0 & \text{otherwise;} \end{cases}$$

$$i_3(p) = \max_{x \in S} ||F'(x)|| \; ||(F'(x))^{-1}||,$$

$$i_4(p) = |S|,$$

where $S = \{x \mid x \in D, F(x) = 0\}$ and $|S|$ the number of elements in S if S is finite, otherwise $|S| = \infty$;

$$i_5(p) = \max_{x \in D} \text{ (cost in some time scale to evaluate } F(x)),$$

$$i_6(p) = \max_{x \in D} (\; ||fl_\varepsilon(F(x)) - F(x)||),$$

where $fl_\varepsilon(F(x))$ is the numerically computed value of $F(x)$ using precision of arithmetic ε;

$$i_7(p) = \delta_F, \quad i_8(p) = \delta_{rx}, \quad i_9(p) = \delta_{ax};$$

$$i_{10}(p) = \max_{x \in D} \text{ (Lipschitz constant for } F' \text{ at } x).$$

In a given test we want to choose those problem indicators that influence performance. For instance, we expect $i_1, \ldots, i_{10}$ to influence the performance measures given earlier, m_1, m_2, m_3. Now we might ask to what extent the problem class is characterized by the selected problem indicators.

A finite set of problem indicators $\{i_k\}_{k=1}^r$ is a *complete characterization* of P if for all $p_1, p_2 \in P$:

(2.1) $P_1 = P_2 \iff i_k(p_1) = i_k(p_2) \qquad (k = 1, \ldots, r).$

For instance, for the subset P of PI of problems of fixed order n and F a linear function, we can construct a complete characterization by using the matrix and right hand side entries as problem indicators, together with indicators defining precisions and tolerances. However, a complete characterization gives a one-to-one correspondence of P with some subset of R^r, which is in fact more than is needed for practical purposes. Since PI is not finite-dimensional as the space of continuously differentiable functions is not, we can not find a finite complete characterization of PI. For testing, the following notion lies closer to our purposes.

Let A be a set of algorithms and $\{m_k\}_{k=1}^{s}$ be a set of performance measures. We say that a set of problem indicators $\{i_k\}_{k=1}^{r}$ is a *sufficient characterization* of problem class P with respect to A and $\{m_k\}_{k=1}^{s}$, if for all $A \varepsilon A$ and all $p_1, p_2 \varepsilon P$:

(2.2) $$i_k(p_1) = i_k(p_2) \quad (k = 1, \ldots, r) \quad \Longrightarrow \quad m_k(A,p_1) = m_k(A,p_2) \quad (k = 1, \ldots, s).$$

As algorithms are finite, there always exist sufficient characterizations of a given problem class relative to a finite set of algorithms and performance measures. Hence the notion of sufficiency is essentially weaker than completeness. Nevertheless, in many practical applications it remains difficult, if not impossible, to prove sufficiency of a characterization. The number of problem indicators in a sufficient characterization may be far too large for practical purposes.

<u>Example</u>: Consider the strict Newton algorithm, A, (see BUS (1980)) for PI. We consider only m_1 of the example performance measures. As problem indicators we use the example problem indicators given above with i_5 replaced by

$$i_5(p) = ||F'(x_0)|| \; ||(F'(x_0))^{-1}||$$

and we add to this set:

$$i_{11}(p) = \sup_{x \varepsilon D} ||F'(x)|| \; ||(F'(x))^{-1}||,$$

$$i_{11+k}(p) = \bar{x}_0^{(k)} \qquad (k = 1, \ldots, n),$$

$$i_{11+n+k}(p) = \bar{x}^{*(k)} \qquad (k = 1, \ldots, n),$$

for some uniquely defined solution $\bar{x}^* \varepsilon S$ and with the notation $\bar{x} = (x^{(1)}, \ldots, x^{(n)})^T \varepsilon R^n$.

Now consider problems p with $i_4(p) = 1$, $i_3(p)$ and $i_5(p)$ bounded and $i_{11}(p) = \infty$. Now choose p_1 and p_2 problems having only one singularity of the derivative in the domain and having the same values of the problem indicators. However, choose this singularity for p_1 such that it does not influence the Newton iteration and $m_1(A,p_1) = 1$, and for p_2 at one of the iteration points such that $m_1(A,p_2) = 0$. Hence, the given characterization is insufficient. In fact a sufficient characterization requires the values of the function and its derivative at all iteration points. This clearly puts the cart before the horse. So, how should we proceed in practice in order to get meaningful test results?

Given a set of algorithms and performance measures, together with a class of problems P, we can imagine two ways to proceed:

1. Find subclasses of P with a sufficient characterization (e.g. the linear functions of certain order in PI) and perform testing for these (sometimes very small) subclasses.
2. Find a characterization for P which is reasonably sufficient. For instance, by testing condition (2.2) and coupling that with the use of theoretical knowledge about these algorithms. Then perform testing P with this characterization.

The first choice gives reliable evidence about the dependence of the performance measures on the problem indicators for the subclasses of problems considered. It is attractive from the viewpoint of a numerical analyst or program designer. However, it might be insufficient from the users' viewpoint. The user's problem most likely does not belong to the small subclasses tested. Moreover, it might be impossible to determine whether the user's problem belongs to the subclasses tested. The second approach yields evidence about the performance of the algorithms for solving the test problems. However, any conclusions drawn from this evidence about the relationship between performance measures and problem indicators is based on the assumption that the characterization is sufficient, which is only approximately true.

Once we have chosen one of the above approaches, there remains the problem of finding the relationships among the performance measures and the problem indicators from the (approximately) sufficient characterization. Two designs are possible:

A. Construct functions, dependent on parameters, which model the relations among performance measures and problem indicators. Perform tests for suitable values of the problem indicators--for instance, the grid points of some grid in problem indicator space. Regression analysis can provide us with estimates for the model parameters, with confidence intervals. These estimates complete the modelling of the performance measures as functions of the problem indicators. Such a test is sometimes called *profiling*. This combined with approach 1 above, is performed in LYNESS & GREENWELL (1977).

B. Create representative test sets for P or subclasses of P. Test the algorithms for these sets and compute mean values and deviations of performance measures for suitable subsets of problems.

Usually one requires a combination of the two designs. For example the time required by a Newton-like algorithm to solve a problem $p \varepsilon PI$ can be modelled (in some given time scale) by

$$t(A,p) = \sum_{k=0}^{3} c_k(A,p)\, n^k,$$

where the factors $c_k(A,p)$ can be found, for each algorithm and certain classes of problems, using curve fitting techniques. As data for fitting such curves, one may use mean times, obtained by testing problems of a given size from a representative test set. For examples of testing design which follow these principles, see LYNESS & GREENWELL (1977), HILLSTROM (1977) and STETTER (1979).

It should be noted that some of the concepts here are related to those appearing in NELDER (1979) and STETTER (1979). Our notions of performance measures and problem indicators agree with Nelder's notions of "response variables" and "stimulus variables".

3. Test Problems

Let us take a class of problems P, a set of algorithms A for solving these problems, and a set of performance measures $\{m_k\}_{k=1}^{s}$. Assume that, based on the probability of occurrence of $p \varepsilon P$ in the user world (as subjectively measured) we have been given a probability distribution function on P. We can then provide the following description of representative test sets.

A test set $T = \{p_1, \ldots, p_N\} \subset P$ is a *representative test set* for P with respect to the given distribution on P and to the set of algorithms A and performance measures $\{m_k\}_{k=1}^{s}$, if there is a sufficient characterization $\{i_k\}_{k=1}^{r}$ for P with respect to A and $\{m_k\}_{k=1}^{s}$ such that, with the notation:

$$\mu(p) = (i_1(p), \ldots, i_r(p))^T, \quad p \varepsilon P,$$

and $\mu(P)$ the image of P in R^r, T is obtained by selecting a random sample $x_1, \ldots, x_N \varepsilon \mu(P)$, given the induced distribution on $\mu(P)$, and selecting p_k $(k = 1, \ldots, N)$ such that $\mu(p_k) = x_k$.

For the sake of description, we have used a distribution on P. This is a rather abstract approach. In practice, one is interested in the distribution of the chosen problem indicator values, i.e. the induced distribution of $\mu(P)$, about which certain assumptions can be made. For example, let P be sufficiently characterized by two problem indicators: i_1 indicating the number of variables with $2 \leq i_1(p) \leq 50$ and equal distribution within this interval, and i_2 indicating whether the problem has a solution or not, with the assumption that 80% of the user world problems have a solution. Then i_2 is a discrete random variable with binomial distribution. Hence the distribution on $\mu(P) = \{(i,j) \mid i=2, \ldots, 50;\ j=0, 1\}$ is given. Then we take N random variables from $\mu(P)$ for the given distribution and construct N problems with indicator values equal to these random variables. This process yields a representative test set.

In practice one finds two methods for creating test sets:

1. Collecting problems from the literature and user work (hand-picked).
2. Creating generators which depend on problem parameters that influence the values of one or more indicators.

It is often claimed that the first method automatically yields a test set which is more or less representative. However, there is little evidence to support this claim, whereas there are arguments suggesting the opposite. Most problems from the literature are very particular in structure and often constructed to make algorithms fail. Real-world problems may have little value as test problems because they can be very difficult to analyze such that the values of the problem indicators are unknown (e.g. the solution). We do not suggest rejecting all hand-picked problems. Many of these are well analyzed and can be used to design problem generators. Yet the use of problem generators is almost indispensable for creating representative test sets or for performance profiling. How one must construct such generators heavily depends on the problem class under consideration. Construction of an ideal problem generator, defined in the following sense, would be most desirable.

We say that a problem generator is *ideal* for a class of problems P, with respect to a set of problem indicators $\{i_k\}_{k=1}^{r}$ if, for each $v \in R^r$ belonging to the image of the mapping $(i_1(.), \ldots, i_r(.)) : P \to R^r$, it generates a problem p with $(i_1(p), \ldots, i_r(p)) = v$.

For small-problem indicator sets, ideal problem generators can sometimes be constructed easily.

4. Efficiency and Robustness

In this section we discuss aspects of performance measures which are related to efficiency and robustness. We restrict attention to the field of nonlinear programming, in particular, iterative algorithms which require user-supplied software defining problem functions.

Let $\mathcal{A}$ denote a finite set of algorithms for solving problems from class P. Let $t(A,p)$ $(A \in \mathcal{A},\ p \in P)$ denote the total time required by A to solve p, using some time scale related to the total CPU-time required to solve p by some implementation of A. Then, for $A_1, A_2 \in \mathcal{A}$, we say that A_1 is *more efficient* than A_2 for solving p if

$$t(A_1,p) \leq t(A_2,p).$$

Assume that we have chosen a time scale such that the CPU-time measurements for an implementation of A relates to algorithmic performance. We first show that choosing t as a performance measure might be naive and costly. Suppose $t(A,p)$ $(A \in \mathcal{A},\ p \in P)$ is measured by switching the clock on and off at the start and termination of the algorithmic process (and transformed the required time scale). Next, assume that the definition of p requires definition of ℓ problem functions defined by user-supplied software with mean evaluation times $i_1(p), \ldots, i_\ell(p)$, respectively. By the choice of t as a performance measure, we are forced to choose i_k $(k = 1, \ldots, \ell)$ in a characterization as these heavily affect t, meaning that test sets have to be representative with respect to these problem indicators i_k $(k = 1, \ldots, \ell)$. As very expensive problems are common in the real world, such testing will require an enormous amount of computer resources. However, this approach is superfluous. Let us write $t(A,p)$ as follows:

$$(4.1) \qquad t(A,p) = t^a(A,p) + \sum_{k=1}^{\ell} c_k(A,p)\, i_k(p),$$

where t^a denotes the total overhead time (total time minus the time required for user-supplied software) and $c_k(A,p)$ the number of calls of the k-th problem function if A is applied to p $(k = 1, \ldots, \ell)$. Note that $c_k(A,p)$ and $t^a(A,p)$ do not depend on $i_j(p)$ $(k, j = 1, \ldots, \ell)$. Use of (4.1) enables us to drop the performance measure t from our set and introduce performance measures t^a and c_k $(k = 1, \ldots, \ell)$. This increases the number of performance measures, but we can remove i_k $(k = 1, \ldots, \ell)$ from the set of problem indicators. Hence, considerably less expensive experiments are required to obtain sufficient information about performance. Afterwards, if t^a and c_k $(k = 1, \ldots, \ell)$ are predicted, then we can predict t by applying (4.1) for appropriate choices of $i_k(p)$. Further simplifications are possible. If algorithm A has constant overhead time per iteration, which is independent of the problem and the same holds for the overhead required in the initial and terminating phase of the algorithm, then we can rewrite (4.1) yielding:

$$(4.2) \qquad t(A,p) = t^i(A) + c^s(A,p)\ t^s(A) + \sum_{k=1}^{\ell} c_k(A,p)\ i_k(p),$$

where $t^i(A)$ and $t^s(A)$ denote the overhead per iteration in the initial and terminating phase, respectively. Use of (4.2) is often possible if we take problem classes with fixed dimensions (number of variables and constraints). Using (4.2), we avoid CPU-time measurements during testing. The measurement of $t^i(A)$ and $t^s(A)$ can be performed once for each algorithm. During testing we only have to count the number of iterative steps and function evaluations. Note that we have assumed that the evaluation time of a problem function is approximately constant with respect to the variables. Usually, the precision of the test results allow such assumptions.

We shall now return to the choice of the time scale. It is well known that the results of CPU-time measurements for two programs implementing the same algorithm may vary widely due to

- the choice of the computer, including the machine arithmetic;
- the operating system of the computer;
- the number of computer users during the tests, if it is a multi-user machine;
- the choice of the programming language;
- the choice of the compiler (-options);
- the choice of a software library;
- the structure of the program;
- the quality of the program and the use of various programming tricks.

A number of papers have noted these complications (e.g. PARLETT & WONG (1975), GILSINN et al. (1977), BUS (1980)). Although efficiency, of course, is related to software programs, we would like to know if the algorithmic choices can be made independent of the programming choices. In order to address this problem, the following conjecture is often assumed:

The relative efficiency of implementations of algorithms are representative for these algorithms if, throughout the experiment,

- the same machine and running system are used,
- the same language and compiler (-options) are used,
- the same software library is used,
- the implementations have the same structure, quality and programming

tricks,

- a dedicated machine (i.e. a machine without other users) is used.

(The last condition can be replaced by: use (4.2) and measure $t^i(A)$ and $t^s(A)$ on a dedicated machine).

Unfortunately, this conjecture is not strictly correct. For example, the ratio of times required for performing two arithmetic actions, e.g. matrix decompositions, may vary from one environment to another, due to arithmetic, language, and library. These difficulties can be solved to a certain extent for algorithms which have roughly the same structure and which allow use of (4.2) by defining standard time.

A *standard time unit* for a class of algorithms which allow use of (4.2) is defined to be equal to the CPU-time overhead in one iteration step of a high-quality implementation of a typical algorithm in the class.

Such a standard time unit depends on all environment parameters. Note that use of (4.2) is usually allowed only in experiments for problem classes with constant dimensions. So we obtain standard time units dependent on these dimensions.

The definition of standard time unit is rather vague. We shall illustrate it with the example of Newton-like algorithms (BUS (1980)). Given a complete environment (machine, language, compiler, and library), we implement one step of the strict Newton algorithm. Then the overhead times per iteration step for several orders can be measured on a dedicated machine. This yields the number of CPU-seconds in a standard time unit dependent on the number of variables of the problem for that particular environment.

If we express $t^i(A)$ and $t^s(A)$ in standard time and obtain predictions for $c^s(A,p)$ and $c_k(A,p)$ $(k = 1, \ldots, \ell)$ experimentally, we can then give predictions for $t(A,p)$ for values of the function evaluation times (in standard time). It is conjectured that this approach yields an environment-independent efficiency measure, hence an efficiency measure for algorithms. This conjecture is not really very important, as it can be checked for each environment. If conclusions are given in the above way, based on test results obtained experimentally, then in another environment we can measure the standard time unit and the values of $t^i(A)$ and $t^s(A)$ in standard time. If they agree with the times reported in the first environment, then the test conclusions can be used, otherwise new test conclusions can be obtained using the experimental results from the first environment and the times from the second.

The above reasoning assumes independence of $c^s(A,p)$ and $c_k(A,p)$

(k = 1, ..., ℓ) from the environment, which is also not completely accurate. Poor arithmetic in one environment may cause an algorithm to fail to solve a problem which is solved by an implementation of that algorithm in an environment with enhanced arithmetic (smaller machine precision). Hence, to insure transportability of test results, we also have to check the assumptions about the arithmetic in the various environments.

The last issue to be discussed is robustness. Intuitively, we call an algorithm *robust* if it solves relatively difficult problems adequately. This description contains two vague notions: 'difficult problem' and 'adequately'. To begin with the first, it can hardly be specified when a problem is difficult to solve. Moreover, it will generally depend on the algorithms under consideration. Therefore, robustness is usually quantified as the percentage of successfully solved problems from a representative test set or as a function expressing success dependent on some problem indicators. The notion 'adequately' is interpreted here as follows:

An algorithm *solves* a problem *adequately* with arithmetic precision ε, if an implementation of this algorithm (with this precision) terminates within finite (in practice: reasonable) time, yielding a solution satisfying the required tolerance criteria.

Note that the prior definition of algorithm takes into account the finite arithmetic and stopping criteria. The actual value of the arithmetic precision is defined by the implementation. We emphasize that we define an algorithm as successful only if it solves problems adequately, i.e. within required precision, since this is what the user wants. One might also consider performance measures which express the gain of accuracy per iteration step or the deflection of the result from the exact solution. Such results provide practical information about convergence behavior.

We conclude by stating that by no means have we tried to give a complete mathematically defined methodology for testing NLP-software. Too many practical and theoretical difficulties are left untouched. Instead, we have sketched the contours of testing methods in nonlinear programming.

References

BUS, J.C.P. (1980), Numerical solution of systems of nonlinear equations, Mathematical Centre Tracts, 122, Amsterdam.

FOSDICK, L.D. (ed.) (1979), Performance evaluation of numerical software, North-Holland.

GILSINN, J. et. al. (1977), Methodology and analysis for comparing discrete linear L_1 approximation codes, Commun. Statist. - Simula. Computa., B6, 399-413.

HILLSTROM, K.E. (1977), A simulation test approach to the evaluation of nonlinear optimization algorithms, ACM TOMS, 3, 305-315.

LOOTSMA, F.A., (1979), Performance profiles and software evaluation, in Performance evaluation of numerical software, ed. L.D. Fosdick, North-Holland, (1979)

LOOTSMA, F.A. (1980), Weights for the efficiency and the robustness of non-linear optimization codes, Paper pres. at Workshop on Linear and Nonlinear programming , Brussels.

LYNESS, J.N. (1979), Performance profiles and software evaluation, in Performance evaluation of numerical software, ed. L.D. Fosdick, North-Holland, (1979)

LYNESS, J.N. & C. GREENWELL (1977), A pilot scheme for minimization software evaluation, Argonne Natl. Lab. TM-323.

MORE, J.J. (1979), Implementation and testing of optimization software, in Performance evaluation of numerical software, ed. L.D. Fosdick, North-Holland, (1979).

NELDER, J.A. (1979), Experimental design and statistical evaluation, in Performance evaluation of numerical software, ed. L.D. Fosdick, North-Holland, (1979).

PARLETT, B.N. & Y. WANG (1975), The influence of the compiler on the cost of mathematical software, ACM TOMS, 1, 35-46.

STETTER, H.J. (1979), Performance evaluation of O.D.E. software through modelling, in Performance evaluation of numerical software, ed. L.D. Fosdick, North-Holland (1979).

A COMPUTATIONAL COMPARISON OF FIVE HEURISTIC ALGORITHMS FOR THE EUCLIDEAN TRAVELING SALESMAN PROBLEM

William R. Stewart, Jr.
School of Business
College of William and Mary
Williamsburg, VA 23185

ABSTRACT

This paper presents a computational comparison of five heuristic algorithms for the traveling salesman problem (TSP). Each of these algorithms is a composite algorithm consisting of a simple tour construction algorithm and a post-processor. The five tour construction procedures considered are all insertion algorithms which may be easily implemented. The post-processor used in the five composite algorithms is a modified 3-opt procedure that is shown to outperform both the 3-opt and 2-opt branch exchange procedures. The final result of the computational tests is the conclusion that there are simple and easily implemented heuristic procedures that will produce high quality solutions to the TSP in a moderate amount of computer time.

1. Introduction

This paper compares several algorithms for the Euclidean traveling salesman problem (TSP). A TSP is Euclidean when the nodes that must be visited all lie on the same plane and the cost of traveling between any pair of nodes is the Euclidean distance between them. The algorithms are compared on a series of test problems, and these tests indicate that simple and effective heuristic algorithms can be constructed to handle the Euclidean TSP.

In the following discussion, a feasible solution or route for the TSP will be referred to as a tour. Any permutation of the nodes in a TSP will constitute a tour when the nodes are visited in the order that they are listed, and the first node is revisited after the last node in the list. A very high percentage of these tours or feasible solutions will be undesirable since the total tour cost will be many times greater than the length of the optimal tour.

The algorithms discussed in this paper are similar to those discussed by Golden et al. [5]. The algorithms are composite algorithms that have two stages. The first stage involves the construction of a reasonable TSP solution by means of a simple and easy-to-use heuristic. The second stage takes that solution and attempts to improve it by means of a post-processor.

The algorithms used for the first stage are insertion algorithms similar to those described by Rosenkrantz, Stearns, and Lewis [13]. These algorithms get reasonable solutions for the TSP very quickly. The algorithms used in the second stage are adaptations of the branch exchange procedures of Croes [3] and Lin [8]. These branch exchange procedures will usually improve upon the solutions generated in the first stage. Insertion and branch exchange procedures are described at length below.

The next section of this paper describes the general format for the insertion algorithms and the composite algorithms to be tested. It also describes the branch exchange procedures that were considered in the construction of the composite algorithms. The third section presents computational comparisons for three branch exchange post-processors and five insertion algorithms. The best post-processor is combined with each insertion algorithm, and the resulting five composite algorithms are compared. The final section reviews the results of the computational tests and draws some conclusions from these tests.

2. Description of Composite Algorithms

The TSP will be referred to in terms of the following notation. Given a graph $G = \{N,A\}$ composed of a set of nodes N, a set of arcs A connecting those nodes, and a cost (distance) c_{ij} associated with each arc (i,j) in A, the TSP is the problem of finding the minimum cost tour of the nodes in N.

2.1 Insertion Algorithms.

Many straightforward heuristics have been suggested for the TSP. Among these are several that are classified as insertion algorithms. Due to their simplicity, these algorithms produce reasonable solutions to the TSP very quickly. The better ones often produce good solutions (i.e. within 5% of optimality). An insertion algorithm constructs a feasible tour by successively adding one node to an existing subtour. The general form of such algorithms is as follows:

Step 1. Obtain a TSP tour for a subset of the nodes $N' \subset N$ in G.

Step 2. Identify a node $k \in N-N'$ which is to be added to the existing subtour (Selection Criterion).

Step 3. Identify an arc (i,j) connecting nodes i and j in the subtour on N'. Insert node k between nodes i and j, and add k to N'. (Insertion Criterion).

Step 4. Stop if N'=N. Otherwise return to Step 2.

Steps 2 and 3 may be interchanged to read:

Step 2. For each node $k \in N-N'$, identify an arc (i,j) connecting nodes i and j in the subtour on N' where k may best be inserted (Insertion Criterion).

Step 3. Identify a node $k \in N-N'$ which is to be added to the existing subtour (Selection Criterion), and add it where it may best be inserted according to Step 2.

The insertion algorithms discussed in this paper will be presented in one or the other of these two basic formats.

The heuristic elements in the above format include the choice of an initial subtour in Step 1, and the choices of Selection and Insertion Criteria. To these three can be added a fourth heuristic element when considering composite algorithms. The fourth element is the choice of a post-processor. Only branch-exchange procedures like those described below are considered in this paper.

2.2 Branch-Exchange Procedures. The second stage of the composite algorithms requires a tour improvement procedure. The best known procedures of this type for the TSP are the brahch-exchange procedures introduced by Croes [3] and Lin [8] and later extended by Lin and Kernighan [9].

In the general case, r branches in a current feasible TSP solution are exchanged for r branches not in that solution as long as the result of that exchange is still a tour and the length of that tour is less than the length of the current tour. Exchange procedures are referred to as r-opt procedures where r is the number of branches exchanged at each iteration.

In an r-opt algorithm, all exchanges of r branches are tested until there is no feasible exchange that improves the solution. This solution is said to be r-optimal. In general the larger the value of r, the more likely it is that the final solution will be optimal. Unfortunately, the number of operations necessary to test all r exchanges is proportional to n^r, where n is the number of nodes in the TSP. Due to this complexity, values of $r = 2$ and $r = 3$ are most commonly used.

Several approaches have been suggested to get around the rapid increase in computer time as r is increased. Christofides and Eilon [2] suggest a procedure that implicitly eliminates from consideration exchanges that cannot improve the current tour. Recently, Or [12] has described a modified 3-opt that considers only a small percentage of 3-branch-exchanges. This modified 3-opt (OROPT) considers only those branch exchanges that would result in a string of one, two, or three adjacent nodes being inserted between two other nodes in the current tour. By thus limiting the number of exchanges that may be considered, OROPT requires many fewer calculations than a full 3-opt.

2.3 General Composite Algorithm. The general composite algorithm is a combination of the insertion algorithm and branch-exchange procedures just discussed. It is obtained by appending a branch-exchange procedure to the insertion algorithm as a post-processor.

The new Steps 4 and 5 are:

Step 4. If N' = N, go to Step 5. Otherwise return to Step 2.

Step 5. Apply a branch-exchange procedure to the solution produced by the insertion algorithm. Stop when no further improvements can be made.

2.4 Selection of Heuristic Elements.

2.4.1 Initial Subtour. Rosenkrantz, Stearns, and Lewis [14] discuss several insertion algorithms. These algorithms all use a single node selected at random as the initial subtour. To produce better solutions, these algorithms are repeated using different nodes, and the best solution over several trials is used as the final solution.

When the TSP is Euclidean, better alternatives exist for the initial subtour. Or [12], Stewart [15], and Norback and Love [10], [11] all present insertion algorithms that make use of the convex hull of the set of nodes N. The convex hull of a set of nodes is the smallest convex set than includes N. So, by definition, the boundary of the convex hull is made up of arcs that connect nodes of N and form a subtour. Hardgrave and Nemhauser [5] have shown that there exists an optimal tour for every Euclidean TSP in which the relative order of the nodes on the boundary of the convex hull is preserved. This means that the optimal tour visits nodes on the boundary of the convex hull in the same order as if the boundary itself were followed. In the discussions that follow, references to the convex hull will mean the nodes on the boundary of the convex hull, and a subtour on those nodes will mean the subtour consisting of the boundary itself.

2.4.2 Selection Criteria. Many criteria have been suggested for the selection of the node to be next inserted in the current subtour. The criteria proposed by Stewart [15], Or [12], and Norback and Love [10], [11] were chosen for the tests conducted in this paper.

These criteria have all performed well in previous tests. The criteria include:

1. Choosing the node k that may be inserted at minimal increased cost (cheapest insertion - Rosenkrantz, Stearns and Lewis [14]). Choose $k \in N-N'$ and $i,j \in N'$ such that $c_{ik} + c_{kj} - c_{ij}$ is a minimum.

2. Choosing the node k such that the proportional increase in cost is minimal (ratio insertion - Stewart [15]). Choose node $k \in N-N'$ and $i, j \in N'$ such that $(c_{ik} + c_{kj})/c_{ij}$ is a minimum

3. Choosing the node k such that the product of ratio and distance is minimized (Or [12]). Choose node $k \in N-N'$ and $i,j \in N'$ such that $((c_{ik} + c_{kj})/c_{ij}) \times (c_{ik} - c_{kj} - c_{ij})$ is minimized.

4. Choosing the node $k \in N-N'$ and $i, j \in N'$ such that the angle formed by the two arcs (i,k) and (k,j) is a maximum (greatest angle - Norback and Love [10]).

These are just a few of many alternatives that might be employed as Selection Criteria.

2.4.3 Insertion Criterion. The insertion criteria that have been used generally fall into two categories:

1. Cheapest insertion. Insert the node $k \in N-N'$ between those two connected nodes, $i,j \in N'$ that minimize the quantity $c_{ik} + c_{kj} - c_{ij}$.

2. Insertion and Selection Criteria identical. Insert the node k that has been selected between those two nodes $i,j \in N'$ that caused this selection.

Stewart uses the first criterion while Or and Norback and Love use the same criterion for insertion as for selection. The cheapest insertion criterion can be used for Selection as well as Insertion Criterion, as is the case with number 1 above.

2.4.4 Post-Processor. Three post-processors were considered in the construction of the composite algorithms, a 2-opt, a 3-opt and an OROPT. The OROPT was finally chosen as the post-processor. Justification for this selection is presented at the beginning of the next section.

3. Computational Results

All of the algorithms tested in this section were coded by the author and were run on an IBM 370-158 computer with 4.0 megabytes of storage under OS/VS2 release 1.7K. The algorithms were coded in FORTRAN and compiled using IBM's FORTRAN IV G compiler level 21. The computer times for the insertion and composite algorithms include the input of the data and the calculation of the Euclidean distance matrix for the TSP. The times reported varied less than 1% when a particular computer pass was rerun with all parameters the same. (This was done several times at various times of the day).

The data input consisted of the x-y coordinates of the test problems. The $[c_{ij}]$ matrix was calculated from these coordinates according to the following rule:

c_{ij} = SQRT $((x_i - x_j)^2 + (y_i - y_j)^2)$ * 10000., where the value stored in c_{ij} is the integer portion of the distance calculation on the right hand side of the equation. Integer arithmetic was used throughout for computational efficiency. The factor of 10000 was used to carry greater accuracy throughout the calculations. This procedure allows the comparisons between the algorithms presented here. To accurately compare the results of these algorithms to existing published results requires the use of a distance matrix that conforms to those published results.

Unfortunately, it has not always been clear how these distance matrices have been calculated. While different methods of calculating the distance matrices may lead to different solution values, these differences should only be slight for the test problems treated here.

3.1 Insertion Algorithms Tested. The insertion algorithms tested are listed below:

1. Convex hull, cheapest selection, cheapest insertion (CCC - Stewart).

2. Convex hull, minimum ratio times distance, for both selection and insertion (CRD - OR's first algorithm [12]).

3. Convex hull, minimum ratio times distance extended, for both selection and insertion (CRDX - Or's second algorithm [12], all nodes eligible for insertion).

4. Convex hull, largest angle, for both selection and insertion (CGA - Norback and Love [10], [11]).

5. Convex hull, cheapest insertion, ratio selection (CCR - Stewart [15]).

The second algorithm of Or differs from the first only in that it considers all nodes in the problem for insertion, even those nodes that are in the current subtour. Other than that the two algorithms of Or are identical. When a 2-opt, 3-opt, or OROPT post-processor is attached to these insertion algorithms, the resulting composite algorithms will have the same name with a 2, 3 or α appended (e.g. CCC would become CCO when an OROPT is added as a post-processor).

3.2 Selection of a Post-Processor

The OROPT procedure was chosen as the post-processor for the composite algorithms. Table 1 presents the results of the CCC, CCC2, CCCO, and CCC3 algorithms on eight test problems ranging in size from 50 up to 150 nodes. Larger problems were not considered due to the extensive amount of computer time required by the 3-opt post-processor.

Problem/ Source		No. of Nodes	Best Known Solution	CCC		CCC2		CCCO		CCC3	
				(%)	(sec)^a	(%)	(sec)^a	(%)	(sec)^a	(%)	(sec)^a
#8	[1]	50	430	5.80	.87	3.62	.14	2.79	.56	3.02	8.27
#9	[1]	75	553	5.96	1.82	3.06	.87	.94	2.06	0.00	31.88
#24	[7]	100	21282^b	8.29	2.77	.89	2.62	0.00	4.79	.47	73.44
#25	[7]	100	22141^b	5.20	2.74	3.10	1.53	2.60	3.14	2.60	74.83
#26	[7]	100	20749^b	4.26	2.57	2.98	.82	.50	2.64	.50	74.46
#27	[7]	100	21294^b	1.96	2.59	1.58	.89	.97	2.02	1.35	70.25
#28	[7]	100	22068^b	3.63	2.52	3.25	.67	2.08	2.15	1.52	72.78
#30	[7]	150	26736	7.78	5.34	5.67	4.36	.75	11.24	1.12	246.45

[a] Seconds of C.P.U. time on an IBM 370/158.
[b] Optimal

TABLE 1. Comparison of Post-Processors
(Per Cent Over Best Known and C.P.U. Times Added)

The C.P.U. times under each post-processor algorithm represent the amount of time that post-processor added to the CCC computation time. In other words, the CCC algorithm required 2.77 C.P.U. seconds for problem 24, but when a 2-opt was used as a post-processor, the CCC2 algorithm required 5.39 C.P.U. seconds, 2.77 for the insertion algorithm plus 2.62 for the 2-opt. The same relation holds for the CCCO and CCC3 algorithms.

Two conclusions are immediately clear from Table 1. First, the 3-opt post-processor requires substantially more time than either the 2-opt or the OROPT (15-20 times more time than OROPT). Second, the 2-opt post-processor is dominated by the OROPT and 3-opt in quality of solution. In all eight problems, OROPT produces better solutions than the 2-opt. The OROPT also produces slightly better solutions than the 3-opt (four better, two ties, two worse). Although the OROPT does not always outperform the 3-opt, tests of these two post-processors using the other insertion algorithms show no tendency for one to consistently outperform the other in quality of solution.

Given the excessive amount of time consumed by the 3-opt and the fact that it fails to produce solutions superior to those of the OROPT, the OROPT was chosen over the 3-opt. Given the inferior solutions produced by the 2-opt and the fact that an OROPT requires only three times as much computer time to produce much better solutions, the OROPT was also preferred to the 2-opt.

3.3 Comparison of Insertion and Composite Algorithms. Table 2 presents the results of applying the five insertion algorithms to fifteen test problems ranging in size from 50 up to 318 nodes. Table 3 presents the results of the five composite algorithms on the same fifteen problems. The times in Table 3 represent the total execution time for the composite algorithms (both insertion and post-processor).

The fifteen problems are all drawn from the literature, although not all of them have been previously treated as TSP's. Only the problems from Krolak, Felts and Marble [7] (problems 4-8 and 10-15) have been solved as TSP's. The 318-node problem has been solved as a Hamiltonian path problem. It is solved here as a TSP. The optimal solutions for problems 4-8 (#24-#28) are from Crowder and Padberg [4]. The best known solutions for the other problems are from Krolak et al. or were generated in the course of this study by using a 3-opt or composite algorithm. These solutions and the method of their generation are available from the author on request.

Several conclusions can be drawn from the data presented in Tables 2 and 3. Except for the angle insertion algorithm, which requires a lot of calculations at each iteration to find the largest angle, the insertion algorithms produce TSP solutions very quickly. These solutions are generally within 10% of the best known solutions with the exception of CRDX which performed very poorly.

The cheapest ratio insertion algorithm appears to have performed better than the other four algorithms. This is supported by the summary measures in Table 4. While these measures cannot be used for a statistical comparison of the algorithms (the data summarized is neither identically nor normally distributed), they do summarize the performance of the algorithms.

The addition of the OROPT post-processor to form the composite algorithms substantially improves the solutions. Based on the results in

Problem Source	Number of Nodes n	Best Known Solution	CCC Percent over (%)	CCC CPU (sec)	CRD Percent over (%)	CRD CPU (sec)	CRDX Percent over (%)	CRDX CPU (sec)	CCA Percent over (%)	CCA CPU (sec)	CCR Percent over (%)	CCR CPU (sec)
1. [1] # 8	50	429.73	5.80	.87	8.09	1.06	3.83	1.21	6.39	2.52	2.67[b]	1.01
2. [1] # 9	75	552.93	5.96	1.82	7.09	2.15	1.57[b]	2.53	9.40	6.60	4.29	1.93
3. [1] #10	100	640.9	5.90	2.55	5.48	3.33	14.60	3.83	7.91	10.38	3.23[b]	3.09
4. [7] #24	100	21282.0[c]	8.29	2.77	3.75	3.08	0.89[b]	4.49	5.81	10.17	3.62	2.76
5. [7] #25	100	22141.0[c]	5.00	2.74	1.57[b]	3.57	17.55	4.01	9.30	9.31	2.52	2.91
6. [7] #26	100	20749.0[c]	4.26	2.57	2.90	3.40	17.04	4.11	9.08	11.46	2.54[b]	2.91
7. [7] #27	100	21294.0[c]	1.96	2.59	1.78[b]	3.25	20.00	4.19	11.00	10.46	2.35	3.07
8. [7] #28	100	22068.0[c]	3.63	2.52	4.64	3.31	2.12[b]	4.02	6.59	9.41	3.45	2.79
9. [9]	105	14383.0	3.68[b]	2.83	4.42	3.74	15.86	4.33	6.29	10.52	5.73	4.10
10. [7] #30	150	26735.6	7.78	5.34	6.68[b]	7.11	18.15	8.21	7.86	24.71	6.84	5.94
11. [7] #31	150	26216.4	4.80	5.53	5.67	7.21	20.92	9.06	16.74	26.19	3.49[b]	6.06
12. [7] #32	200	29563.1	7.54	·9.67	8.77	12.96	11.98	15.22	12.50	47.76	4.67[b]	10.99
13. [7] #33	200	29678.2	8.24	9.90	7.79	11.69	21.21	16.07	8.13	41.40	6.09[b]	11.17
14. [16]	249	2363.6	10.36	13.74	8.04	24.64	46.83	27.32	9.15	88.00	6.39[b]	18.13
15. [9]	318	43864.7	6.93	22.71	6.89	34.44	22.77	37.40	7.50	158.20	6.51[b]	31.75

[a] IBM 370/158.

[b] Marks the best solution in each row.

[c] Optimal

TABLE 2. Tests for Insertion Algorithms

Problem Source	Number of Nodes n	Best Known Solution	CCCO Percent over (%)	CCCO CPU (sec)	CRDO Percent over (%)	CRDO CPU (sec)	CRDXO Percent over (%)	CRDXO CPU (sec)	CGAO Percent over (%)	CGAO CPU (sec)	CCRO Percent over (%)	CCRO CPU (sec)
1. [1] # 8	50	429.73	2.79	1.43	2.31	1.53	2.17	1.56	1.74[a]	3.39	2.29	1.44
2. [1] # 9	75	552.93	0.94	3.88	4.09	4.19	0.16[a]	3.41	1.85	9.11	1.74	2.89
3. [1] #10	100	640.9	3.09	6.10	1.33[a]	7.24	3.02	8.95	2.64	14.63	1.57	5.84
4. [7] #24	100	21282.0[b]	0.00[a]	7.56	0.06	6.18	0.00[a]	6.24	0.85	15.05	0.43	6.76
5. [7] #25	100	22141.0[b]	2.60	5.88	0.41[a]	6.13	1.70	10.23	1.45	13.78	0.97	4.92
6. [7] #26	100	20749.0[b]	0.50	5.21	0.00[a]	5.55	5.32	8.81	1.39	15.23	0.50	5.21
7. [7] #27	100	21294.0[b]	0.97[a]	4.61	0.97[a]	5.15	3.32	12.83	2.26	17.64	1.66	5.06
8. [7] #28	100	22068.0[b]	2.08	4.67	2.98	5.61	2.04	5.21	2.03[a]	12.47	2.58	4.55
9. [9]	105	14383.0	0.08[a]	6.21	0.78	8.75	7.98	8.71	0.62	15.96	1.40	8.75
10. [7] #30	150	26735.6	0.75	16.58	3.05	16.44	3.81	27.78	0.57[a]	38.53	1.60	14.92
11. [7] #31	150	26216.4	0.81	15.15	0.00[a]	23.16	3.84	30.90	3.21	48.58	0.79	16.92
12. [7] #32	200	29563.1	0.83[a]	38.57	3.76	38.05	7.02	35.79	4.49	93.69	2.27	35.05
13. [7] #33	200	29678.2	2.26	47.42	1.99	47.31	4.97	69.50	0.13[a]	73.64	0.69	32.95
14. [16]	249	2363.6	3.75	79.27	4.00	80.62	5.46	179.81	1.83	165.71	1.74[a]	76.11
15. [9]	318	43864.7	0.92	211.68	0.69	174.59	3.52	267.91	0.00[a]	331.40	2.01	115.11

[a]Marks best solution in row

[b]Optimal

TABLE 3. Tests for Composite Algorithms with OROPT

Insertion Algorithm	Stand Alone		With OROPT	
	$\overline{X}$ (%)	S (%)	$\overline{X}$ (%)	S (%)
CCC	6.01	2.21	1.49	1.16
CRD	5.57	2.33	1.76	1.51
CRDX	15.69	11.54	3.62	2.28
CGA	8.91	2.83	1.67	1.19
CCR	4.29	1.62	1.48	.68

TABLE 4. Mean and Standard Deviation of Performance (in percent)

Table 3, solutions within 3% of the best known solution are easily obtained. The CCRO algorithm is within 3% on all fifteen problems. It also requires the least computer time to produce a final solution on nine of the fifteen problems. CCRO and CCCO appear to have about the same average performance both in quality of solution and in computer time used, and both appear to perform better than the other algorithms on these measures.

In summary, CCR and CCRO appear to perform slightly better than their competitors on the problems tested. More extensive tests are necessary to show a statistical superiority. In addition, a general observation can be made based on the data in Tables 2 and 3. The OROPT appears to require less time the better the starting position it is given. Since it also appears that the total execution time for the composite algorithms will be dominated by the post-processor as the problem size grows, the better the solution in stage 1, the quicker the algorithm will run.

4. Conclusions

The composite heuristic algorithms presented and tested here are easily coded and implemented. They are capable of producing good solutions (usually within 2% of the best known solution) to very large Euclidean TSP's quickly. While not conclusive, the cheapest ratio insertion composite algorithm appeared to produce better solutions more consistently than the other algorithms tested. Much of the performance of these composite algorithms can be attributed to the modified 3-opt (OROPT) that was adopted as a post-processor. This procedure was shown to produce solutions of comparable quality to those produced by a 3-opt in time comparable to that required by a 2-opt.

There are several avenues for future research opened up by this paper. A more comprehensive set of computational tests on a large set of similarly structured problems is necessary if the composite algorithms are to be compared statistically. There are also opportunities for improving on the algorithms treated here by proposing and testing new Insertion and Selection Criteria or by streamlining OROPT still further.

5. References

[1] N. Christofides and S. Eilon, "An Algorithm for the Vehicle Dispatching Problem," Operational Research Quarterly, 20, 309-318 (1969).

[2] N. Christofides and S. Eilon, "Algorithms for Large Scale TSP's," Operational Research Quarterly, 23, 511-518 (1972).

[3] G. Croes, "A Method for Solving Traveling Salesman Problems," Operations Research, 6, 791-812 (1958).

[4] H. P. Crowder and M. W. Padberg, "Large-Scale Symmetric Traveling Salesman Problems," Management Science, 26, 495-409 (1980).

[5] B. Golden, L. Bodin, T. Doyle, and W. Stewart, Jr., "Approximate Traveling Salesman Problems," Operations Research, 28, 694-711 (1980).

[6] W. W. Hardgrave and G. L. Nemhauser, "On the Relation Between the Traveling Salesman Problem and the Longest Path Problem," Operations Research, 10, 647-657 (1962).

[7] P. Krolak, W. Felts, and G. Marble, "Efficient Heuristics for Solving Large Traveling Salesman Problems," paper presented at the 7th Mathematical Symposium, Hague, Netherlands (1970).

[8] S. Lin, "Computer Solutions of the TSP," Bell Systems Technical Journal, 44, 2245-2269 (1965).

[9] S. Lin and B. Kernighan, "An Effective Heuristic Algorithm for the Traveling Salesman Problem," Operations Research, 21, 498-516 (1973).

[10] J. P. Norback and R. F. Love, "Geometric Approaches to Solving the Traveling Salesman Problem," Management Science, 23, 1208-1223 (1977).

[11] J. P. Norback and R. F. Love, "Heuristic for the Hamiltonian Path Problem in Euclidean Two Space," Operational Research, 30, 363-368 (1979).

[12] I. Or, "Traveling Salesman-Type Combinatorial Problems and Their Relation to the Logistics of Blood Banking," Ph.D. Thesis, Dept. of Industrial Engineering and Management Sciences, Northwestern University (1976).

[13] F. P. Preparata and S. J. Hong, "Convex Hulls of Finite Sets of Points in Two and Three Dimensions," Communications of ACM, 20, 87-92 (1977).

[14] D. J. Rosenkrantz, R. E. Stearns, and P. M. Lewis, "Approximate Algorithms for the Traveling Salesperson Problem," Proceedings of the 5th Annual IEEE Symposium on Switching and Automata Theory, 33-42 (1974).

[15] W.R. Stewart, "A Computationally Efficient Heuristic for the Traveling Salesman Problem," Proceedings of the 13th Annual Meeting of Southeastern TIMS, Myrtle Beach, S.C., 75-83 (1977).

[16] B. E. Gillett and J. G. Johnson, "Mult-Terminal Vehicle-Dispatch Algorithm," Omega, 4, 711-718 (1976).

IMPLEMENTING AN ALGORITHM: PERFORMANCE CONSIDERATIONS AND A CASE STUDY*

Uwe Suhl
Freie Universität Berlin
Fachrichtung Wirtschaftsinformatik
Garystraße 21, D-1000 Berlin 33

ABSTRACT

After a general discussion of program performance and its influence factors we consider a case study. A specific version of TOYODA's heuristic algorithms for solving multidimensional knapsack problems was implemented in three different FORTRAN programs. Starting from a straightforward program, we introduce increasingly sophisticated data structures to improve the virtual execution time of the programs. All programs were compiled with the enhanced version of the FORTRAN H compiler available on IBM/370, 43xx, 303x machines, using its four optimization options. A computational study was performed with a series of randomly generated test problems with up to 3200 variables and constraints and about 70000 nonzeros. The purpose of this study was to measure virtual central processor times and working set sizes as a function of problem size and to study the influence of code optimization performed by the compiler. Since all three programs are representations of the same algorithm, conclusions can be drawn as to the relative importance of the influence factors on program performance. As is also demonstrated, the potential performance of a representation of an abstract algorithm may be difficult to project without a careful implementation.

1. Program Performance Evaluation

One of the most fundamental concepts of computer-science is the notion of an algorithm. Loosely, an (abstract) algorithm can be defined as a finite ordered sequence of operations, which precisely specifies how given elements of a set of inputs have to be transformed into elements of a set of outputs. For more details see [5]. Algorithms in this sense can be expressed in a number of ways. Mathematical operations can be used, as well as precise formulations in a natural language. Research in the development and complexity analysis

---

*Part of this work was done while the author was on a visitor's stay at the IBM Thomas J. W. Research Center.

of computer algorithms has had enormous growth in the last decade and has provided new insights, theoretical and practical, in what determines computational efficiency. Such studies now constitute one of the most important disciplines of computer science. For an overview see [1,5,8].

A (finite) computer program essentially consists of two parts, its underlying abstract algorithm expressed in any programming language and some body of data on which the algorithm operates. A program has the property that it can be executed by a processor either directly (interpretation) or after translation (assembler, compiler), linking, and loading. Our notation implies that a program is an algorithm; however, an algorithm may require substantial human work to be expressed in a program. One major problem area concerning the conversion of an algorithm into a program is the design and representation of appropriate data structures [4]. This is so, because abstract algorithms normally operate on unordered or ordered sets of data, i.e. on abstract data types for which various data representations exist which may have a strong influence on program performance.

Program performance evaluation studies the computer system resource consumption of a program during its execution on a processor. It is a special discipline of computer systems performance evaluation [3]. From a conceptual point of view a program transforms (during its execution on a computer system) input data into output data using up certain systems resources. Let D be the set of all admissible input data for the program. We assume that D is finite. The performance of a program P which has to be executed on a given computer system can be characterized by a mapping f(P):D ---> RPn, where RPn is the n-dimensional set of nonnegative real numbers. That is, for a given instance of input data x of D. f(P)(x) is an n-dimensional vector of nonnegative real numbers which describes the performance of the program executing on a specific computer system. This approach assumes that we are only interested in the intrinsic performance behavior of the program. We have also assumed that this behavior is deterministic in nature and therefore reproducible. Let f(P)(i) be the i-th projection of f(P) on RP, i.e. f(P)(i):D ---> RP. Each mapping f(P)(i), i=1, i=1,...,n is a performance index which evaluates a specific performance aspect. It is assumed, that performance indices relate to resource requirements of the program as a function of size and characteristics of its input data. Such indices may be space or time oriented.

Formal program performance evaluation concentrates on time or space analysis based on a model of the computation. It can be considered as a special branch of complexity analysis of algorithms in which one studies a particular implementation of an algorithm. In time analysis specific elementary operations are counted as a function of the input data. Asymptotic analysis determines only the growth rate of a performance index as a function of the 'size' of the input data. Complexity analysis can further be distinguished in worst-case and average case analysis. In worst-case analysis one wants to determine $\max\{f(P)(i)(x) \mid x \varepsilon D\}$ for a given $i \varepsilon I$. Expected analysis has as objective the study of the expected performance behavior of P for a given $i \varepsilon I$. This requires knowledge of the probability of the occurrence of each $x \varepsilon D$. Formal program analysis is useful if a program has not yet been implemented on a computer system. Frequently the effort involved in improving the performance of the wrong program can be saved by a complexity analysis. On the other hand too many important details of the computation are ignored in a formal analysis. An empirical evaluation of programs will almost always be necessary for performance comparisons.

Empirical program performance evaluation is based on executing a program on a computer system. A program executed by a processor is called a task. Performance is measured by execution of the task with a set $S \subset D$ of test data. Since the size of D is usually extremely large, only a comparatively small set S can be used for sampling. There is always the problem of selecting a set S, using appropriate measurement tools, and interpreting measurement results. Observed performance of a task on a computer system is determined by the input data and the workload of the system. To measure intrinsic performance behavior, then, it is, necessary to eliminate workload influence. The simplest way of doing this would be execution in a uniprogramming environment. Even this would only give an approximation, because time oriented performance indices on modern computer systems are stochastic in nature. In most cases measurement tools such as hard- or software monitors provide sufficient accuracy for progmatic program evaluation.

Computer systems with paged virtual memory are now widely used. Although program development is greatly reduced, programs which reference large arrays have to be carefully designed. From a macroscopic point of view we may have the following events during execution of a task on such systems:

```
I===I.......I===x-----I....I==*====,,,,,,....I===I....I===x-----I......I

 CPU         CPU  page    CPU I/O-op.     CPU      CPU  page
                  fetch                                 fetch
```

=== CPU executes considered task

... Task waits for subsequent CPU processing

x page fault occurred

--- missing page is fetched

* start I/O-operation

,,, nonoverlapped portion of an I/O-operation

Thus the task must be in exactly one of four (macroscopic) states:

-- executed by the central processor. Execution is interrupted whenever a page fault occurred, or the processor has to wait for the completion of an I/O operation issued by the task, or the time slice has been used up or the process has been finished.

-- page fetch state. The task is in this state after a page fault occurred. It is finished after the missing page has been fetched from its external slot and stored in a page frame.

-- I/O wait state. Since the execution of I/O operations can be overlapped with execution of the task by the central processor, this state corresponds only t the nonoverlapped portion of the I/O time.

-- waiting for the processor. After resolving the previous interrupt, the task has to wait in one of the system queues for subsequent execution.

The most important time oriented performance index is the execution time of a task, i.e., the elapsed time between execution of the first and the last instruction. However, the execution time in a multiprogrammed environment depends very much on the workload of the system during the considered time interval. It can, therefore, not be used as a reliable index. To restrict workload influence, a virtual clock is introduced which runs only if the task is in one of the first three states. The corresponding times are called virtual CP-time, virtual page fetch time and virtual nonoverlapped I/O time. The virtual execution time is the sum of these times and can be considered as one of the most important performance indices. An approximation of this index could be achieved by executing the task in uniprogramming mode. The following discussion gives a brief overview of the major factors influencing virtual execution time, and how it can be potentially reduced.

-- Reduction of the virtual I/O time. Let us first take a more detailed look at a single I/O operation of a moving head disk. Note, that I/O operations for a fixed head disk, drum or a tape drive are conceptually similar. For example, for a drum t1 would be equal to t2, for a tape drive the 'seek' time corresponds to the (constant) start/stop time and t2 is equal to t3.

```
start I/O        seek                    latency          transfer
I-------I---------------------I------------------I-----------I-----I
t0      t1                    t2                 t3          t4    t5
```

At time t0, the program executes a start I/O command. A channel program is initiated, and after some delay (which may be substantial if several I/O requests for the same device are queued) executed by the channel (I/O processor) and interpreted by the storage controller. At time t5 the processor is notified (interrupted) that the I/O operation is complete. Associated with any I/O operation is a relatively high time penalty regardless of how many words are transferred from or to main memory. This is so because for current disk/drum technology seek and latency times are a significant portion of the I/O time. It is, therefore, important as a general principle to reduce the number of I/O operations by taking advantage of the fast transfer rate, and increase the block size. Note, that a careful analysis is necessary, since larger block sizes may increase central processor time spent on de/blocking. Also, larger blocks require larger buffers in main memory. If buffers reside in a paged area of main memory page faults have to be considered too.

-- Overlapping of I/O operations and execution of the task by the central processor, also called asynchronous I/O. One technique to achieve this, is an interrupt driven processor which executes in problem mode the instructions of the program as long as the processor is not interrupted. After an I/O operation has been completed the processor is interrupted and control transferred to an I/O interrupt handler. After resolving the interrupt the processor resumes execution of the program. Asynchronous I/O is to a certain extent possible in some FORTRAN extensions. The BUFFER IN/OUT statement for sequential unformatted I/O is based on these ideas. However, in some systems this construction serves only as a vehicle for portability, i.e. there is no overlap between processor and channel. Multiple buffering is another technique to increase the overlap between central processor and I/O.

-- The virtual CP-time on a given processor is essentially a function of the number and type of instructions executed. Algorithms and data structures/representations have the most significant impact on CP-time.

For programs written in high level language compilation and interpretation have a strong influence on virtual CP-time. Experience shows that most programs spent more than 90% of their virtual CP-time in less than 5% of their code. As a consequence assembly language should only be considered for this part of the program.

-- The virtual page fetch time is a function of the number of page faults caused by the task. It depends in general on 1. the memory reference behavior of the task, 2. the page replacement algorithm, 3. the number of page frames allocated in real memory as a function of time, 4. the page size and 5. on technological factors. Most systems use a global replacement strategy, i.e. the number of page frames allocated to a task not only depends on its reference behavior but also on that of other active tasks. The number of page faults and, therefore, the virtual execution time is in such a case workload dependant. Scientific programs which reference large arrays have often poor data locality, i.e. the reference strings of operand fetches cover a large area of the virtual address space. Program performance can then be improved by replacing paging through I/O operations, taking advantage of larger block sizes and knowledge which data will be referenced next by the CPU, allowing a potential overlap between I/O and CPU processing.

The working set size of a task is also very important. In most computer systems with virtual memory a task is only activated if its more or less accurate estimated working set fits in the remaining real memory. Thus, the execution time of a task with large working set size depends critically on the amount of real memory available during its execution and provides an explanation why it is so much work load dependant.

The remaining part of the paper discussed the implementation of a mathematical programming algorithm and shows some of the influence factors on program performance.

2. TOYODA's Algorithm

We consider the 0-1 optimization problem (P'): max cx, s.t. to Bx $\leq$ b, where c is a positive n-dimensional row vector, B is a nonnegative real mxn matrix, b is a positive m-dimensional vector of reals and x is a n-dimensional (0-1)-vector. If B is a (0-1)-matrix and all coefficients of b are one, (P') is the so called set packing problem. It is well known that (P') is NP-complete. TOYODA [9] developed a heuristic algorithm for determining a 'good' integer solution for (P'). We used his algorithm with origin moving method III (AIII) as a subject for an implementation, to study various performance aspects. It is not intended to discuss algorithmic aspects of solving this type of problem. Mathematically AIII operates on an equivalent problem (P), which we obtain from (P') by dividing each row by b(i). Thus, in (P) the coefficients of the right hand side are all ones. We briefly outline the algorithm. For a more detailed discussion see [9].

J	Index set of the (0-1)-variables, i.e. J={1,2,...,n}.
I	Index set of the constraints, i.e. I={1,2,...,m}.
S	Current index set of those variables set to one.
x(S)	Current solution vector, i.e. x(S)(j)=1 if j ε S and 0 otherwise.
F(S)	Index set of current candidate variables to be assigned, i.e. F(S) = {j|j ε J-S, A(j) ≤ ONE-b(S)}.
A	Coefficient matrix of (P), A=(a(i,j)), i ε I, j ε J
A(j)	Column vector j of A
ONE	m-vector of one's.
b(S)	Current right hand side vector, i.e. b(S) = A * x(S)
z(S)	Current objective function value, i.e. z(S) = c * x(S)
PU(S)	Penalty vector of dimension m.
<e,f>	Denotes the scalar product of two vectors e, f.
‖e‖	Denotes the euclidean norm of a vector e.

0. (Initialization). Create (P) from (P') by dividing each row by b(i). Set S = ∅, F(S) = J, b(S) = 0, z(S) = 0, ONE(i) = 1, for i ε I. Let F(S) ={j|j ε J, A(j) ≤ ONE}. If F(S) = ∅, then stop. Compute G(j) = (c(j)*/m)/<A(j),ONE> for j ε F(S). Find the smallest index jm such that G(jm) = max{G(j)|j ε F(S)}.
1. (Set x(jm) to one). Set S = S ∪ {jm}, F(S) = F(S)-{jm}, z(S) = z(S) + c(jm), b(S) = b(S) + A(jm). If F(S) = ∅ then stop.
2. (Fix variables to zero which would violate feasibility). Set F(S) = {j|j ε F(S), A(j) ≤ ONE-b(S)}. If F(S) = ∅ stop.
3. (Calculate penalty vector). Let r = 0.5*max{b(S)(i)|i ε I}. Calculate PU(S):PU(S)(i) = b(S)(i)-r if b(S)(i) > r and PU(S)(i) = 0 otherwise.
4. (Select a candidate vector). Calculate for j ε F(S)<A(j),PU(S)>. If a scalar product is zero, let jm be the smallest index of such columns and go to 1. Otherwise calculate G(j) = c(j)/<a(j), PU(S)/‖PU(S)‖> for j ε F(S) and let jm be the smallest index with G(jm) = max{G(j)|j ε F(S)}. Go to 1.

3. Implementation

Our objective was to develop a portable FORTRAN implementation of AIII which is able to 'solve' large sparse problems in main memory. The implementation is based on a conventional machine architecture with no apparent parallelism. An implementation of AIII for a processor with vector instructions would certainly look totally different (see for example [7]).

In all implementations of AIII we use the substitution b'(S) = ONE-b(S), to streamline the computations. Program A is a straightforward FORTRAN implementation where the coefficient matrix is stored as a two dimensional array and F(S) is represented as a subset of a boolean array of order n. Program A could be implemented with a few APL statements. Such simple implementation makes sense if one is only interested in comparing different algorithms from a mathematical point of view without paying any attention to program performance.

The most time consuming part of program A is step 2. The computational time complexity of an iteration depends on the size |F(S)| of the current set F(S), how many components of a column A(j) with $j \epsilon F(S)$ have to be examined to exclude it from F(S) and on the size |F(S')| of the new set F(S'). On the average we have to inspect about m/2 components of b(S) to exclude a column. Thus, step 2 has an average asymptotic time complexity of 0(n,|F(S')|*m, (|F(S)|-|F(S')|*m/2) operations. Step 1 and 3 can be combined and are of order 0(m). Step 4 is of order 0(n,|F(S')|*m). Static storage requirements are of order 0(3n,2m,n*m) computer words.

Program A can be improved by exploiting the sparsity of (P). If nc denotes the average number of nonzeros in a column, step 4 would be of order 0(n+|F(S')|*nc) if nothing else is changed. For problems with several thousand rows it is not uncommon that nc is about 10. In such a case we would have a tremendous speedup of the program. We used a column oriented sparse matrix scheme where the nonzeros are stored together with their row indices. A pointer vector defines the starting point of each column. A nonzero and its row index are stored in consecutive full words. Storing a nonzero and its row index together increases data locality, resulting in lower cache miss ratio and working set size. In FORTRAN this can only be achieved by an artificial construction which takes advantage of the language implementation. Instead of representing the sparse matrix as DIMENSION AC(NZ), IA(NZ), where AC is a floating point array for storing the nonzeros and IA is an integer array for storing the corresponding row indices, the following representation was used: DIMENSION AC(2,NZ), IA(2,NZ) EQUIVA-

LENCE (AC(1,1), IA(1,1)). Because in FORTRAN arrays are stored column wise, resulting in the following storage layout:

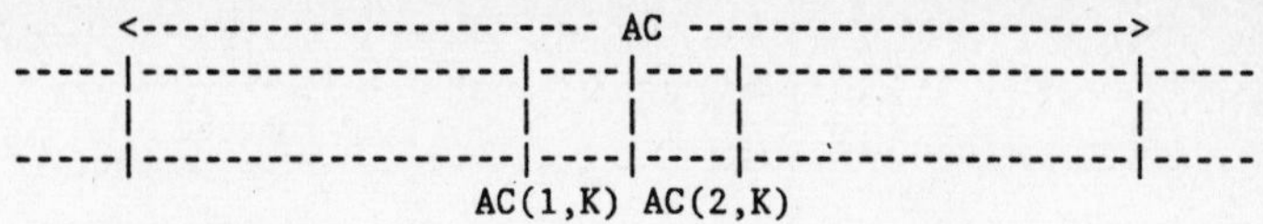

AC(1,K) was used for storing the floating point values and AC(2,k) for storing the corresponding row indices. AC(1,K) has then to be used for referencing the nonzeros and IA(2,K) for referencing the row indices to inform the compiler about the proper data types. Since the first subscript is a constant, a good optimizing compiler will generate efficient code to access elements of AC. The enhanced FORTHX compiler available on IBM machines generated better code for this representation (despite the EQUIVALENCE statement) than for the one where the data is stored in two arrays. This representation restricts portability to those machines where fixed and floating point representations exist which have the same length in storage.

Another improvement can be achieved by a proper representation of the set F(S). Identifying this set by means of a boolean array is very costly because it is of order 0(n), regardless of the size of |F(S)| and results in executing conditional branch instructions which limit instruction overlap on highly concurrent machines. We use an array of dimension n for representing S and F(S):

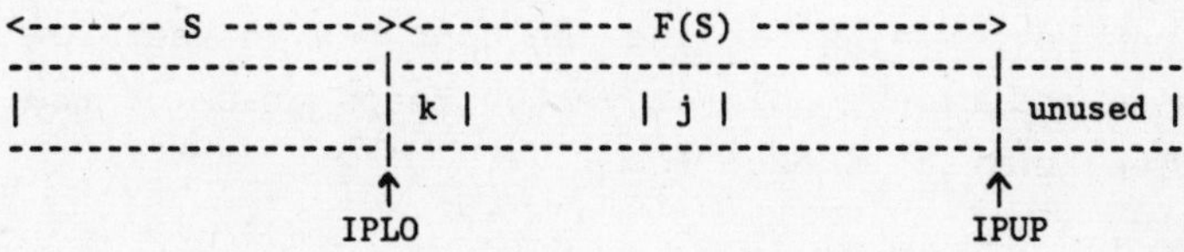

IPLO acts as stack pointer for S and is also the lower end of the stack F(S). IPUP is the stack pointer for F(S). This data representation allows fast access to all elements of F(S). If variable j has to be fixed to one, it changes the position with k and IPLO is incremented by one. All free variables are then tested if they can be fixed to zero, i.e. deleted from F(S). The new reduced stack is created in one pass by overwriting its predecessor:

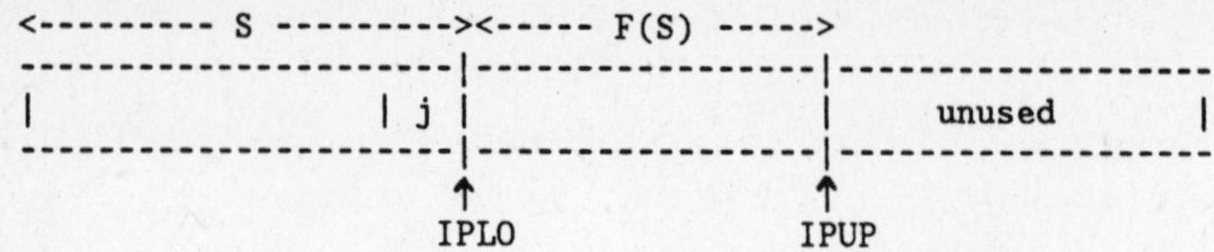

Thus, during each iteration of the algorithm IPLO is increased and IPUP is decreased. F(S) is empty if IPLO = IPUP. Note, that any initial ordering of F(S) cannot be preserved with this data representation during execution of the program. As a consequence, if in step 4 a scalar product is zero, we still have to scan the remaining free variables for the one with the smallest index. We used this array also for storing a full (0-1) solution vector generated in one pass after the algorithm is finished. Static storage requirements are now of the order 0(2nz,2m,2n) computer words, where nz denotes the number of nonzero elements of the coefficient matrix. A simple analysis shows that the average time complexity of step 2 is of order 0(nc*(|F(S)|+|F(S')|)). The average time complexity of step 4 is of order 0(|F(S')|*nc). Program B is based on these data structures.

We can still improve the average asymptotic time complexity of program B. Whenever a variable x(jm) is fixed to one we test all free variables if they can be fixed to zero, even those which are not affected by fixing x(jm). Instead, we need to look only at those variables which have nonzeros in rows where A(jm) has nonzeros. This requires an additional row wise storage of the coefficient matrix. We can do even better by arranging the nonzeros in each row in decreasing order. An additional pointer vector keeps track of the address of the first free nonzero in each row:

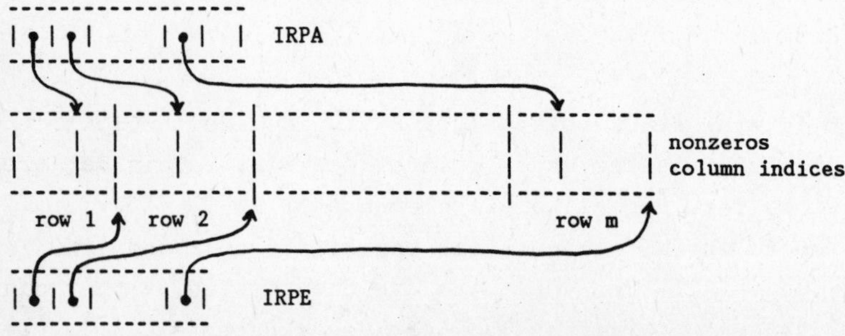

We have here an example of m stacks represented sequentially in an array. The lower end of stack i is defined by IRPA(i) and changes as more variables are fixed. The upper end of stack i is defined by IRPE(i) and remains constant. The nonzeros in each row are sorted in decreasing order. When a variable jm is fixed to one the update of b(S) is accomplished by using the column wise representation. After b(S)(i) was updated for a nonzero a(i,jm) the free variables of row i are scanned by using the row wise data representation. An n-vector IX is used for storing the solution and status record: IX(j) is set to one if j is set to one, to 0 if j is free and to -j if j is fixed to zero. After row i has been processed we compress it so that the free nonzeros of this row are stored contiguously in the next iteration. Compression of those rows which have already been processed is performed in future iterations. This data representation results in a reduction of the average asymptotic time complexity of step 2: We look at most |F('S)|*nc nonzeros, where F('S) is the set of free variables before the previous jm was fixed and updated. However, this is a conservative bound, because we ignored the sorting effect and that for a sparse problem many variables of |F('S)| have no nonzeros in common with A(jm). Clearly this is highly data dependent.

The substitution b'(S)=ONE-b(S) results in determining the minimum of all components of b'(S) in step 3. At first glance this operation seems to be of order O(m). Since a subset of the components of b'(S) is always decreased and the others are unchanged we only have to look at those components of b'(S) where A(jm) has nonzeros. Thus, combining step 1 and step 3 allows us to determine the minimum in O(nc) operations.

For storing F(S) we used in this case a double linked circular list. We gain some speed in those cases in which in step 4 <A(j),PU(S)> is zero, since the list allows us to maintain F(S) as an ordered set.

Storage requirements are now of order O(5n,4m,4nz) computer words which is for large problems about twice as much as for program B. Program C is an implementation based on these ideas. Some preprocessing is required to execute program C: We have a routine for building the column wise sparse matrix structure, a routine for generating the row wise sparse matrix structure with a fast transpose algorithm and a routine for sorting the nonzeros in each row.

4. Testing the Programs

The computational time complexity analysis has shown that program C should execute the fastest on the average. It is not clear, however, how much better program C will be compared to program B. We used a set of test problems to study the influence of problem size and compiler optimization on CP-time and working set size.

The test problems were randomly generated as square problems. The objective function coefficients are equally distributed in the interval (1,100). For a given expected value of the density of the coefficient matrix the locations of the nonzeros were randomly generated. The size of nonzeros were equally distributed in the interval (0.1,0.9). The expected density of the problems decreases linearly with problem size, i.e. for a problem of size (n2xn2) and a problem of size (n1xn1) with $n1 < n2$, the density for the first problem is $d2=d1*n1/n2$. The expected density of the smallest problem (200x200) is 10%, whereas the expected density of a problem (2000x2000) is 1%. Note that the number of nonzeros increases linearly with problem size. A parameter t, with $t \in \{0.2,0.4,0.6,0.8\}$ was used to determine the tightness of the right hand side vector b. For $i \in I$ b(i) was defined as the maximum of t times the sum of the nonzeros of row i and the maximum nonzero in row i. Ten problems were always generated and solved for a given t. The numbers reported in the tables represent the mean of the measured test values.

The programs were tested on an IBM 3033 under VM/CMS. The FORTRAN H compiler (FORTHX) and its enhanced version (FORTHQ) were used to compile the programs. The virtual central processor time, i.e. the CPU time to execute instructions within the virtual machine was measured with a system clock which has a resolution of 13 microseconds. Since during a cache line fault the running task is not interrupted, there is still some variability in the measured times. We ran identical problems under different work load and observed a maximum deviation of 20 milliseconds for a task which requires on the average a virtual CP-time of about 50 seconds on the IBM 3033. Note, that all reported CP-times are the mean of a sample of ten solved problems.

Like on most other machines the number of page faults under VM is workload dependent. For this reason we report only working set sizes as an indicator of the dynamic memory requirements of a task.

5. Test Results

Static program size of a compiled program is a good indicator for the quality of the generated code. This is true because code optimization consists partly of removing instructions and manipulating as much data in registers as possible. The following tables show a comparison of the three programs compiled with the FORTHX and the FORTHQ compiler using different optimization levels. Note, that routines for preprocessing the input data are not included in table 1. However, table 2 contains the time for preprocessing the data. Several conclusions can be derived from tables 1,2:

--full 'optimization' resulted in a code reduction of about 50% compared to unoptimized code.
--program A compiled with FORTHQ opt. 3 ran about four times faster than compiled at level 0. For the more complex programs B, C the improvements were still about a factor of two.
--the worst implementation, i.e. program A compiled with FORTHQ opt. 0 is for the smallest solved problem about 40 times slower than the best implementation, i.e. program C compiled with FORTHQ opt. 3. As one can see from table 4 the speed gap between these implementations increases rapidly to arbitrarily large numbers if problem size grows.
--code optimization is based on heuristics. A higher optimization level does not always produce better code as can be seen by program A. The programmer has some influence on code optimization by experimenting with different syntactical constructions [6]. Solution times for program C were further reduced by about 10% using the following trick: The innermost code segment S representing step 2 was replaced by a segment which contains an artificial DO-loop around S which is executed exactly once. This produced much tighter code for S. This construction takes advantage of the fact that an optimizing compiler pays considerable attention to produce good code for loops, especially for the inner most ones. This trick worked also for the FTN compiler on a CYBER 175. Note that portability is not affected.
--improvements based on complexity considerations had a more significant impact on CP-times than pure code optimization. Inspection of the assembly code showed that program A cannot be improved significantly by hand coding. Program B and especially program C were not optimal. It is hard to estimate, but we felt that at most 20% time reduction could be achieved by hand coding.

Opt. level	Program A		Program B		Program C	
	FORTHX	FORTHQ	FORTHX	FORTHQ	FORTHX	FORTHQ
opt 0	2228	2240	2680	2692	3342	3358
opt 1	1362	1344	1872	1872	2300	2296
opt 2	1292	1126	1906	1652	2170	1974
opt 3	n.a.	1212	n.a.	1616	n.a.	1946

TABLE 1. Static Program Size in Bytes without Arrays on an IBM 3033

Opt. level	Program A FORTHX	Program A FORTHQ	Program B FORTHX	Program B FORTHQ	Program C FORTHX	Program C FORTHQ
opt 0	20408	21768	2155	2217	1242	1256
opt 1	14157	14339	1761	1765	977	898
opt 2	6344	5334	1285	1134	701	602
opt 3	n.a.	5339	n.a.	1132	n.a.	583

Problemsize: N=200, M=200, NZ=4093, density=10%, t=0.8
FORTHX: Extented FORTRAN H compiler
FORTHQ: Enhanced version of FORTHX compiler
n.a. : not available

TABLE 2. The Influence of Compiler Optimization on Virtual CP-Time Measured in Milliseconds on an IBM 3033

Table 3 shows virtual CP-time for a given set of 10 problems of size (400x400) with an average of 8397 nonzeros as a function of parameter t. The programs were compiled with the FORTHQ compiler opt. 3. As can also be seen from complexity considerations, problems with a 'loose' right hand side, i.e. a large t are much harder to solve, because we have to perform more iterations.

t	Program A	Program B	Program C
0.2	4180	583	228
0.4	21967	2179	1356
0.6	33648	3270	2030
0.8	41555	4069	2502

TABLE 3. Virtual CP-Time in Milliseconds as a Function of Parameter t

The next table shows the average virtual CP-time as a function of problem size. For all test problems the parameter t = 0.8 was used. The programs were compiled with FORTHQ opt. 3.

Problem size n	m	nz	Program A	Program B	Program C
200	200	4093	5339	958	583
300	300	6275	18250	2218	1443
400	400	8397	41555	4069	2502
500	500	10491	78799	6261	3565
800	800	16865	X	16963	9633
1200	1200	25263	X	37726	21050
1600	1600	33755	X	67954	38337
2000	2000	41988	X	105725	58083
2400	2400	50460	X	152803	83816
2800	2800	59047	X	209501	113502
3200	3200	67479	X	271061	134530

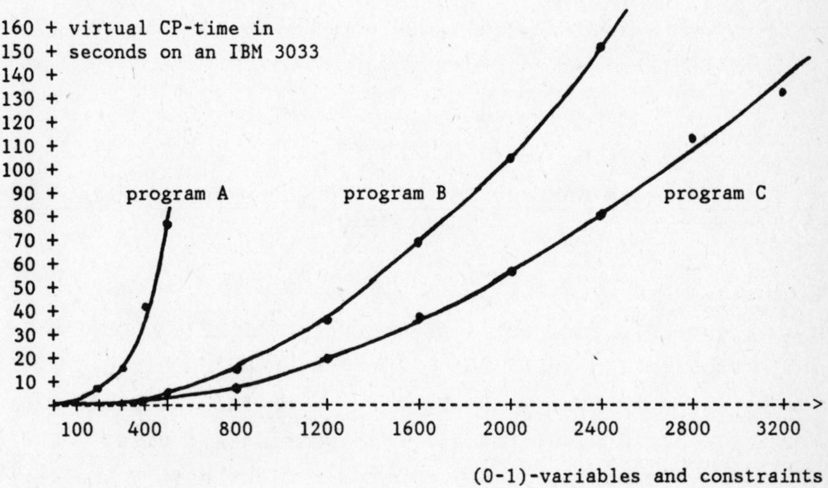

TABLE 4. Virtual CP-Time in Milliseconds as a Function of Problem Size

The working set sizes of the programs are mainly determined by the amount of array space needed during program execution and are only minimally influenced by static program sizes. Unfortunately the estimation of the working set size of a task under VM is workload dependent. Thus these figures provide only a rough comparison of the dynamic memory requirements of the programs executing on a 3033.

Problem size			Program A	Program B	Program C
n	m	nz			
400	400	8397	192	X	X
500	500	10491	285	X	X
2000	2000	41988	X	163	227
2400	2400	50460	X	187	262
2800	2800	59047	X	209	293
3200	3200	67479	X	228	329

TABLE 5. Average Working Set Size in Pages as a Function of Problem Size

6. Conclusions

Which program is the best implementation of AIII? Obviously not program A. Program C outperforms program B based on virtual CP-time. For large problems program C is about twice as fast as program B. However, program C has a significant larger working set size. As a consequence, if real memory is tight, program B might have a smaller execution time than program C.

For solving even larger problems or in environments where main memory is severely limited it might be necessary to store the coefficient matrix on an external device. Program B has the potential for an efficient out-of-core implementation, because the sparse columns can be processed sequentially. Therefore, we can highly overlap processing of data by the CPU and data transfer of blocks of sparse columns between main memory and device. Program B does not allow a significant overlap between CPU and I/O processor, since the update of b(S) requires column access to specific sparse rows of the coefficient matrix, which may be randomly distributed over the row file. Out-of-core implementations of AIII would have an even greater impact on program performance since the design of the I/O part would be crucial.

Some conclusions can be drawn from this case study which seem to be quite general:

--Program performance indices, in particular time and space are frequently conflicting goals. It is, therefore, important to decide which one to give preference.

--Structural decisions during program design and development have a more significant impact on program performance than coding and compiler optimization aspects. It is a common error to believe that a program coded in assembly language is more efficient than an equivalent one in a high level language. In our case neither program A nor program B coded in assembly language would reach the performance of program C. Should the inner loop of program C be coded in assembly language? We do not think so. The next improvement step is to think about a 'better' algorithm, for example one which avoids the costly inner products. Only if this fails, should assembly language be used in the innermost loop.

--It is in general only possible to compare different algorithms based on machine independent measures such as the number of key comparisons, the number of iterations, inspected extreme points or the number of nodes developed in a branch-and-bound algorithm. The actual performance achieved with a program based on a given algorithm depends very much on the implementation.

In summary one can conclude that a formal and an empirical program performance evaluation seem to be necessary for the design and development of efficient software. Both approaches are complementary and cover performance aspects not otherwise obtainable.

7. References

[1] Aho, A. V., Hopcroft, J. E. and Ullman, J. D., The Design and Analysis of Computer Algorithms, Addison-Wesley, Reading, Mass., 1976.

[2] Crowder, H. P., Dembo, R. S. and Mulvey, J. M., On Reporting Computational Experiments with Mathematical Software, ACM Trans. on Math. Software, 5, 2 (1979), 193-203.

[3] Ferrari, D., Computer Systems Performance Evaluation, Prentice-Hall, Englewood Cliffs, N.J., 1978.

[4] Horowitz, E. and Shahni, S., Fundamentals of Data Structures, Computer Science Press, Inc., Woodland Hills, California, 1976.

[5] Horowitz, E. and Sahni, S., Fundamentals of Computer Algorithms, Computer Science Press, Inc., Woodland Hills, California, 1977.

[6] Larson, C., The Efficient Use of FORTRAN, Datamation 17, 8 (1971), 24-31.

[7] Owens, J. L., The Influence of Machine Organization on Algorithms, in Complexity of Sequential and Parallel Numerical Algorithms, Traub, J. F. (ed.), Academic Press, New York and London, 1973, 111-130.

[8] Tarjan, R. E., Complexity of Combinatorial Algorithms, SIAM REVIEW 20, 3 (1978), 457-491.

[9] Toyoda, Y., A Simplified Algorithm for Obtaining Approximate Solutions to Zero-One Programming Problems, Man. Science, 21, 12 (1975), 1417-1420.

WHICH OPTIONS PROVIDE THE QUICKEST SOLUTIONS

William J. Riley
Texas A and M

Robert L. Sielken, Jr.
Texas A and M

1. Introduction

Frequently algorithm users can select their solution strategy by choosing from among various options for each of several algorithm factors. If the algorithm will always eventually find a solution, the important question is which combination of options is likely to be "best". A general statistical approach to answering this question is illustrated in the context of a new integer linear programming algorithm where "best" is quickest.

In this algorithm, named SLIP (Solves Linear Integer Problems), there are 14,400 different combinations of options available to the user, in addition to three covariates which may be set at any level desired by the user. Exhaustive testing of all options is impractical due to time and cost. Two questions, nevertheless, are of interest. First, what is the average usefulness of an option? The answer to this question is important to the direction of future research. Secondly, what is the fastest combination of options in solving a problem? The large number of options and the impracticality of testing all their possible combinations point toward a statistical experimental design as an approach to answering the preceding questions. Of course the answers, if found, are technically applicable only to SLIP, programmed in Fortran and run on an Amdahl 470 V/6 computer, and to the population of problems from which the test problems were randomly selected. The method of analysis, however, is applicable to any algorithm, not just to integer programming algorithms and not just to algorithms in which time is the function to be minimized.

2. The Algorithm

The algorithm SLIP is a hybrid of implicit enumeration, surrogate constraints, and cutting plane techniques for solving the general linear integer programming problem. The general problem is converted by SLIP into an all 0-1 integer minimization problem. Glover's implicit enumeration scheme, as modified by Geoffrion (1967) is used to enumerate and examine partial solutions. It is assumed that any pre-processing of the problem has been accomplished before SLIP is used. Of course, highly structured problems can probably be more efficiently solved using special purpose algorithms.

The SLIP solution process can be summarized as follows: an attempt to fathom the current partial solution is made by using the binary fathoming techniques developed by Balas (1965) along with several modifications developed by the authors. If binary fathoming techniques are unsuccessful, SLIP follows one of two branches. One, a variable may be augmented to the current partial solution. Two, fathoming attempts are continued by relaxing the integrality requirements and solving the resulting linear programming problem. If the first branch is chosen, an attempt to fathom the new partial solution is made by using binary fathoming techniques. If the second branch, relaxed fathoming, is attempted, the solution to the relaxed problem is examined for integrality. If the solution is all-integer, the current partial solution is fathomed. If the solution is not all-integer, it is rounded and the rounded solution is examined for feasibility and, if feasible, is examined to see if the incumbent (best solution found so far) has been improved. If so, the incumbent is updated. If the rounded solution becomes the new incumbent and the objective function value beats the objective function associated with the current partial solution, the current partial solution is fathomed. If the current partial solution is not fathomed, a penalty for an all-integer completion is calculated. If the penalty plus the objective function value associated with the relaxed problem exceeds the objective function value of the incumbent, the current partial solution is fathomed. If not fathomed, up and down penalties are calculated for each fractional-valued integer variable in the solution to the relaxed problem. If up or down penalties indicate that a variable must be added to the current partial solution at the one or zero level, that variable is augmented and the new partial solution is returned to be examined by binary fathoming techniques. If the current partial solution still has not been fathomed, a Gomory (1960) cut may be added to the relaxed problem and relaxed fathoming attempts begin anew. The process is repeated until the partial solution is fathomed or until the maximum number of cuts specified by the user have been added. If the maximum number of cuts have been added and the partial solution is still not fathomed, a surrogate constraint may be added to the problem at the option of the user. If still not fathomed, a variable is augmented to the current partial solution and binary fathoming attempts are begun anew. Any time a partial solution is fathomed, the partial solution is backtracked and binary fathoming procedures are begun again. The entire process is repeated until enumeration is complete.

There are a number of options available to the user. The options can be categorized into four types: (1) master options, (2) binary fathoming options, (3) relaxed fathoming options, (4) augmentation and backtracking options.

MASTER OPTIONS: The user specifies how often, FREQ, relaxed fathoming is to be attempted. The user also indicates one of two methods used to calculate the number of iterations before relaxed fathoming is attempted. One, relaxed fathoming is attempted after FREQ attempts at binary fathoming. Two, relaxed fathoming is attempted after FREQ consecutive unsuccessful attempts at binary fathoming.

BINARY FATHOMING OPTIONS: There are five options and several sub-options here. (1) The first option is to perform the basic Balas zero-one fathoming tests with a slight modification. Those variables which, if augmented at the one level, would cause the current best value of the objective function to be exceeded are augmented to the current partial solution at the zero level and underlined (fixed). Underlining or fixing means that completions with the alternative binary value, here one, cannot be better than the current incumbent. A test for binary infeasibility is also performed. (2) Perform the basic Balas checks, including binary infeasibility. If the partial solution is not fathomed, check to see if each constraint can achieve feasibility before the current best objective fuction value is exceeded. This check involves the ratio of the improvement in infeasibility to the increased cost in the objective function. If any variables are augmented to the current partial solution as a result of these ratio checks, RC, then fathoming attempts continue by returning to the beginning of the binary fathoming routine. As a sub-option, if the number of free variables is less than some maximum number specified by the user, all possible completions, APC, of the current partial solution are evaluated. If APC is performed, the current partial solution is automatically fathomed. (3) Perform the basic Balas checks without the binary infeasibility check. Perform RC and/or APC. (4) Perform the basic Balas checks along with the binary infeasibility check. If the current partial solution is not fathomed, attempt to fix free variables by applying several additional binary infeasibility criteria. In particular, if all the free variables with negative coefficients in a given constraint need to be augmented to the current partial solution in order for that constraint to achieve feasibility, those variables are augmented to the current partial solution at the one level and underlined. Also an additional criterion proposed by Geoffrion (1969) for indicating whether variables need to be in the completion of any partial solution at either the zero or the one level is used to fix such variables. If any variables are fixed at the one level, binary fathoming attempts are begun again. The preceding procedure is referred to as FIXEM. (5) Balas checks, a binary infeasibility check, RC, APC, and FIXEM are all applied to the current partial solution.

RELAXED FATHOMING OPTIONS: (1) Solve the relaxed problem. Check for feasibility in the relaxed problem and see if the optimal relaxed objective function value exceeds the current best solution found so far. Check the rounded solution. (2) In addition to the checks in (1) calculate a penalty for an integer completion (Tomlin (1970)). Also calculate individual variable up and down penalties. If any variables are augmented as a result of up and down penalties, perform binary fathoming checks. (3) Same as (2), except if problem is still not fathomed, add a surrogate constraint, up to some maximum number of surrogate constraints specified by the user. (4) Same as (2), except if problem has not been fathomed, add a cut and repeat the relaxed fathoming process up to some maximum number of cuts specified by the user. (5) Same as (4), except that if cuts have not fathomed the problem, add a surrogate constraint.

AUGMENTATION AND BACKTRACKING OPTIONS: Each of these procedures has six options. (1) Augment the first variable in the set of variables eligible to be augmented, or backtrack the last variable in the current partial solution eligible to be backtracked. (2) Backtrack or augment that variable which leaves the current partial solution with the least total infeasibility. (3) Augment or backtrack that variable which results in the least increase in the value of the objective function associated with the current partial solution. (4) Use the ratio of cost gap to total infeasibility in order to augment or backtrack. (5) Augment or backtrack that variable whose worst infeasibility constraintwise is the best. (6) Same as (4), except use constraintwise infeasibility instead of the sum of the infeasibilities.

Complete details of all options are available in the documentation for SLIP, Riley (1981a).

3. Initial Experimental Design

The analysis of the empirical behavior of the algorithm developed sequentially. In order to help guide future software analysts, the authors describe the evolution of their analysis and the decisions that had to be made along the way.

The options in SLIP suggested a factorial design to test the effectiveness of the various choices available to the user. The design considered was a $2^4 \times 6^2 \times 5^2$ with three covariates (number of cuts, number of surrogate constraints, and number of free variables before APC is invoked). There are at least two problems with this design. First, the covariates are nested within certain levels of given factors; the analysis of such a design would be extremely complicated. More importantly, perhaps, the number of runs of SLIP required to estimate all effects, interactions, and error would be very expensive. Based on some preliminary test runs, an estimate was made of the number of runs of SLIP which would keep the study within the authors' budget. Further designs were considered with this budget constraint in mind.

The covariate problem was resolved by considering each covariate to be either absent or present at some preset level. Either no or a maximum of 10 surrogate constraints were added. Either no or a maximum of six cuts were added. Either APC was not used, or it was used with a maximum of four free variables. The fact that the covariates could be considered to be at either a low level (not used)or a high level (used) suggested a further simplification of the design. The remaining factors could also be thought of as either invoked by the user or not invoked. That is, the design would consist of a number of factors each of which could be at 2 levels, either on or off, at a low level or at a high level. The next question was: How many effects and interactions could be estimated within budgetary constraints?

An assumption made in many factorial experimental designs is that higher-order interactions are negligible, and therefore are not worth the additional experimental units needed to estimate them. If that assumption is made, and additional experimental units are costly, a fractional factorial experimental design is desirable. Consequently, fractional factorial designs with all factors at two levels were examined and considered for the experiment (see for example Kempthorne (1979), Hicks (1973), and National Bureau of Standards (1957)).

The initial estimate of an upper bound for the number of problems of intermediate difficulty that could be tested by SLIP within the budget was 75. Designs which used less than 75 experimental units were considered. In such designs, there is a trade-off between the number of main effects the experimenter wishes to estimate and the number of two-factor interactions which are estimable. For example, in a 1/256 replication of 14 factors, requiring 64 experimental units, of the 91 two factor interactions only 25 are estimable. The largest number of main effects estimable in a fractional factorial design in which all two-factor interactions are estimable and which requires less than 75 experimental units is a 1/4 replication of 8 factors.

The eight factors of primary interest to the authors were: RC,

APC, FIXEM, CUTS, SURROGATES, FREQ, and whether COST or INFEASIBILITY or both were the best criterion for augmentation and backtracking. For the purposes of the experiment, the criterion used in backtracking was also used in augmentation. All other options in SLIP were fixed prior to the experiment. The above eight factors were either at the high level or the low level. The high level meant that the option was used, the low level not used, except in the case of FREQ, where the high level was to perform relaxed fathoming after every two unsuccessful attempts at binary fathoming, the low level to perform relaxed fathoming after every six unsuccessful attempts.

Finally, some experimental units had to be chosen. A set of test problems assembled by Haldi (1964) was used. This set of problems was further narrowed by considering only problems of small to intermediate difficulty, as reported by Haldi and various other sources in the literature. The motivation for picking problems of small to intermediate difficulty was the fear that inefficient options in the experiment would cost too much to test on difficult problems. A possible disadvantage in the selection procedure for experimental units is that the population to which the results of the experiment apply is smaller.

Another decision regarding experimental units was whether to test one problem with each of the 64 treatment combinations or to test several problems each on an equal number of treatment combinations. Here "treatment combination" means a combination of factor levels (high or low and on or off). The latter procedure was chosen for two reasons: The generality of the results is enhanced, and by "blocking" on problems the average effect of the problem is eliminated. A disadvantage of blocking is that by blocking, some of the two-factor interactions are no longer estimable. Eight problems were chosen and became the blocks to which 8 treatment combinations were assigned and run. Two two-factor interactions were no longer estimable. The experimenter is free to choose which interactions are to be non-estimable. In this experiment, the authors felt that the CUT-INFEASIBILITY and SURROGATE-COST interactions were not worthy of estimation.

A timer was inserted into SLIP so that a precise estimate of each actual solution time was obtained. The timer was started after the problem had been converted to an all 0-1 format. Thus the run times considered below do not include non-treatment related functions such as program compilation, reading the problem, etc.

4. Analysis of Initial Experiment

The analysis of a statistical experimental design assumes that the observed dependent variable is the sum of its mean, which is a linear function of the factor levels, and a random "experimental error" component. The experimental error here is due to a particular experimental unit not necessarily being the average unit in the population and also a very small timing error. The usual assumption is that the errors are independent, normally distributed, with zero mean and common variance. To the extent that the experimental units were chosen at random from the population of interest and assigned randomly to blocks, the errors are independent. The normality assumption can be accommodated to some extent by invoking the Central Limit Theorem. The F-tests involved in the Analysis of Variance (ANOVA) can be quite sensitive, however, to the equal-variance assumption. In this experiment, the block (problem) variances were found to be too different to justify the equal variances assumption.

To more nearly justify the equal-variance assumption, two transformations were considered, a natural logarithm transformation and a rank transformation. Both transformations yielded block variances that were at least of the same order of magnitude. The logarithm transformation also linearizes a multiplicative model where the factor levels have percentage effects as opposed to additive constant effects. All F-tests of factor significance were performed using the logarithms of the data.

The results of the initial experiment are presented in modified ANOVA format in the table. Significant effects were found to be FREQ, FIXEM and INFEASIBILITY. Apart from the two-factor interactions, the interpretation of significance is as follows: On the average, over the population of problems considered, a faster solution should be obtained by attempting relaxed fathoming every 2 iterations, as opposed to every 6 iterations. Similarly, FIXEM should be effective in reducing the solution time, and INFEASIBILITY should be the criterion used in augmentation and backtracking.

By examining significant two-factor interactions, some additional insight can be gained. For example, there is a significant RC-FIXEM interaction. When FIXEM is off, having RC on results in a faster solution time. With FIXEM on, having RC on actually increases the solution time. The complete results, including ANOVA information on all the interactions is in Riley (1981b).

5. Follow-up Experiment

An implicit assumption in the analysis of a design using blocks is that there is no block-factor interaction. Here, a block is a problem and it is good that the average effect of the problem is removed. However, the existence and impact of a block-factor interaction where the relative usefulness of a factor would vary considerably from problem-to-problem was of concern.

Since the cost of initial experiment was much less than estimated, a follow-up experiment was conducted to investigate the problem-factor interaction. Each of the eight problems in the initial experiment was run on the remaining 56 treatment combinations. The corresponding ANOVA results are also in the table.

From the table it is apparent that there is a strong problem factor interaction. That is, even in this group of relatively homogeneous problems, which solution options are best is problem dependent. By blocking on problems, the fact that cuts, surrogate constraints, RC, and APC are significant in some problems was masked. Nevertheless, the initial experimental information on the average effect of certain solution options over the larger class of problems is useful.

6. The Fastest Combination of Options

For each problem, the fastest solution time differed from the slowest solution time by at least a factor of 10. This suggests that there is a fastest combination of options which shortens the solution time by a significant amount. Although the question of which options, on the average, are effective has been answered through ANOVA, the fastest combination of options has not necessarily been found.

To gain some insight, the five fastest and five slowest solution times among the 64 runs for each problem were identified. In each category, there were several multiple winners. The most frequent winners in the fastest-time category were:

1. RC on, appearing 19 out of 21 times.
2. FIXEM on, appearing 13 out of 21 times.
3. INFEASIBILITY on, appearing 10 our of 21 times.

The most frequent winners in the slowest-time category were:

1. CUTS on, appearing 21 out of 27 times.
2. FREQ = 6, appearing 20 out of 27 times.
3. COST on, appearing 18 out of 27 times.
4. SURROGATES on, appearing 17 out of 27 times.

However, CUTS on and SURROGATES on also appeared in several of the fastest combinations.

The ANOVA and the ranking procedure suggest that, to maximize the probability of obtaining the fastest solution process, set

1. FREQ = 2,
2. RC on,

3. FIXEM on,
4. INFEASIBILITY on, and
5. COST off.

These were the only clear-cut decisions. CUTS and SURROGATES appeared in winning and losing combinations; so their contribution to fast solution times is not clear.

One possible explanation of the effect of CUTS and SURROGATES is that, when they work, they work well. When they do not, they greatly lengthen the solution time. The next step is perhaps to consider an experiment on larger problems where the use of cuts and surrogate constraints may be more uniformly beneficial. If the factor levels which were most beneficial in the class of problems previously considered maintain their relative usefulness in larger problems, then the experiments analyzed herein will have eliminated the need to consider several factors further. Thus, the experiments herein will have made more economically feasible a study of the value of cuts and surrogate constraints in larger problems.

7. Conclusions

The empirical behavior of a new integer linear programming algorithm on a class of moderately difficult problems has been considered. The study suggested that on the average it is beneficial to (1) perform the binary fathoming based on the procedures RC and FIXEM, (2) use infeasibility as opposed to objective function cost as the criteria for augmenting and backtracking partial solutions, and (3) attempt relaxed fathoming frequently. The use of cuts and surrogate constraints sometimes shortens the solution time but not always and not on the average.

A possible sequel to this paper should be entitled "What Makes Problems Hard." The problems selected for testing had approximately the same number of variables, number of constraints, and sparseness. Yet there were significant differences in solution times and significant problem - factor interactions. The selection of the best combination of options for a particular problem (as opposed to a selection for a class of problems) is unsolved and needs more research. Perhaps, if we could diagnose a problem's difficulty, we could prescribe the best combination of options.

The primary objective of this paper has been to illustrate the benefits of statistical experimental designs and analysis-of-variance techniques in determining the average usefulness of algorithm options and combinations of options. While this analysis cannot necessarily identify the best combination of options for a particular problem, it can make economically possible the identification of superior combinations of options for broad classes of problems. Furthermore the average usefulness of options can direct and focus research. The authors recommend the use of these general statistical techniques to those developing, testing, and validating software.

References

Balas, E., "An Additive Algorithm for Solving Linear Programs with Zero-One Variables," Oper. Res., Vol. 13, (1965), pp. 517-546.

Dembo, R.S., "A Set of Geometric Programming Test Problems and Their Solutions," Math Programming, Vol. 10, (1976), pp. 192-213.

Geoffrion, A.M., "Integer Programming by Implicit Enumeration and Balas' Method," SIAM (Soc. Ind. Appl. Math.) Rev. 7, (1967), pp. 178-190.

Geoffrion, A.M., "An Improved Implicit Enumeration Approach for Integer Programming," Oper. Res., Vol. 17, (1969), pp. 437-454.

Geoffrion, A.M., and Marsten, R.E., "Integer Programming: A Framework and Statement-of-the-Art Survey," Management Sci., Vol. 18, (1972), pp. 465-491.

Geoffrion, A.M., A Guided Tour of Recent Practical Advances in Integer Linear Programming, Working Paper No. 220, Western Management Science Institute, University of California, Los Angeles (1974).

Gomory, R.E., An Algorithm for the Mixed Integer Problem, RM-2597, Rand Corp., Santa Monica, California (1960).

Haldi, J., 25 Integer Programming Test Problems, Working Paper No. 43, Graduate School of Business, Stanford University (1964).

Hicks, C.R., Fundamental Concepts in the Design of Experiments, Holt, Rinehart and Winston, New York (1973).

Kempthorne, O., The Design and Analysis of Experiments, Robert E. Krieger Publishing Company, Huntington, NY (1979).

Riley, W.J. and R.L. Sielken, Jr., The User's Guide to SLIP, Texas A&M University, College Station, Texas (1981a).

Riley, W.J., Dissertation, Texas A&M University, College Station, Texas, (Forthcoming), (1981b).

Statistical Engineering Laboratory, National Bureau of Standards, Fractional Factorial Experimental Designs for Factors at Two Levels, National Bureau of Standards Applied Math. Series 48, (1957).

Taha, H.A., Integer Programming, Academic Press, New York, (1975).

Tomlin, J.A., "Branch and Bound Methods for Integer and Non-Convex Programming," In Integer and Nonlinear Programming (J. Abadie, ed.), Amer. Elsevier, New York, pp. 437-450 (1970).

Trauth, C.A., Jr. and R.E. Woolsey, "Integer Linear Programming: A Study in Computational Efficiency," Management Science, Vol. 15, pp. 481-493 (1969).

Wahi, P.N. and G.H. Bradley, Integer Programming Test Problems, Report No. 28, Administrative Sciences, Yale University, New Haven, CT (1969).

Summary of Experimental Results(1)

		Initial Experiment	PROB 1	PROB 2	PROB 3	PROB 4	PROB 5	PROB 6	PROB 7	PROB 8	
FREQ	6 Iter.	20.57 *	3.71	4.47 *	7.64	4.72 *	47.40 *	34.62	16.34 *	41.24 *	RELAXED FATHOMING OPTIONS
	2 Iter.	16.10	3.43	2.45	7.20	3.11	24.92	41.15	19.33	24.00	
		(5.85)	(0.36)	(154.99)	(2.60)	(6.44)	(203.17)	(0.57)	(7.44)	(505.24)	
SURROGATES	On	18.97	3.65	3.51	7.67	3.70	35.78	37.67	19.13 *	32.89	
	Off	17.70	3.49	3.40	7.18	4.13	36.55	38.09	16.55	32.34	
		(0.67)	(0.72)	(0.24)	(0.97)	(0.02)	(0.02)	(2.05)	(6.79)	(1.45)	
CUTS	On	21.81	3.84	3.53	7.22 *	4.22	43.56	51.59 *	21.71 *	44.83 *	
	Off	14.87	3.31	3.39	7.63	3.61	28.76	24.18	13.96	20.40	
		(0.49)	(3.84)	(0.03)	(5.85)	(0.05)	(2.80)	(129.65)	(20.71)	(1050.59)	
RC	On	16.73	3.14 *	3.01 *	6.56 *	3.46 *	29.77	36.17 *	8.78 *	35.72 *	BINARY FATHOMING OPTIONS
	Off	19.94	4.00	3.90	8.28	4.36	42.56	39.60	26.90	29.52	
		(0.36)	(14.35)	(18.91)	(7.97)	(9.45)	(3.46)	(12.14)	(422.53)	(90.37)	
APC	On	17.74	3.20 *	3.22	7.40	3.60	36.01	37.88	17.64	32.53	
	Off	18.93	3.94	3.70	7.44	4.23	36.32	37.89	18.04	32.70	
		(0.24)	(10.87)	(3.24)	(0.07)	(1.43)	(0.05)	(0.88)	(1.04)	(0.108)	
FIXEM	On	16.09 *	2.27 *	1.94 *	3.75 *	2.01 *	35.81	37.64	17.13 *	22.49 *	
	Off	20.58	4.87	4.98	11.09	5.82	36.52	38.12	18.55	42.74	
		(14.47)	(240.31)	(425.25)	(153.18)	(180.51)	(0.001)	(0.052)	(6.48)	(560.42)	
COST	On	18.92	3.70	3.35	8.60	4.14	44.25	38.00 *	20.45 *	32.93	AUGMENTATION AND BACKTRACKING OPTIONS
	Off	17.75	3.44	3.56	6.24	3.69	28.08	37.76	15.23	32.31	
		(0.17)	(3.36)	(4.45)	(2.71)	(1.42)	(3.85)	(8.77)	(7.03)	(0.42)	
INFEASIBILITY	On	9.27 *	3.48	3.17	4.46 *	2.71 *	24.10 *	16.97 *	2.83 *	32.03	
	Off	27.40	3.66	3.75	10.39	5.12	48.23	58.80	32.85	33.21	
		(40.12)	(0.26)	(1.29)	(95.59)	(21.71)	(9.32)	(577.32)	(2624.72)	(0.56)	
TOTAL TIME		36.67	7.14	6.92	14.84	7.83	72.33	75.77	35.68	65.24	

(1) Values in body of table are the total items (in seconds) to solve 32 problems with the option on and 32 with the option off. The problems are from Haldi (1964) with problems 1-8 being his fixed charged problems 1-4 and IBM test problems 1-4 respectively. The * indicates that the F-test indicated a significant effect at the 5% significance level when the logarithm transformation of the data was used. The value in parentheses is the F-value in the F-test. For the initial experiment: $P(F>2.95) = 0.10$, $P(F>4.30) = 0.05$, $P(F>7.95) = 0.01$, and $P(F>14.38) = 0.001$. For the follow-up experiment: $P(F>2.90) = 0.10$, $P(F>4.21) = 0.05$, $P(F>7.68) = 0.01$, and $P(F>13.61) = 0.001$.

AN INTEGER PROGRAMMING TEST PROBLEM GENERATOR

Michael G. Chang
University of Pennsylvania

Fred Shepardson
Stanford University

In this paper we present a methodology for evaluating new integer programming algorithms -- an area of mathematical programming which is not yet well understood (see [Lin and Rardin, 1979]). We present our procedure not so much as a recommendation for others to follow but as a basis for discussion and further work.

The work reported here was part of a project to develop a new integer programming algorithm for the general class of 0-1 integer programming problems:

$$\text{(IP)} \qquad \text{Max} \sum_{j=1}^{n} C_j X_j$$

$$\text{st} \quad (1) \quad \sum_{j=1}^{n} a_{ij} X_j \leq b_i \qquad i = 1, \ldots, m$$

$$(2) \quad X_j = 0,1 \qquad j = 1, \ldots, n$$

where C_j and a_{ij} are integers unrestricted in sign. As we turned from development to testing we were confronted with the problem of performance evaluation. The first question we encountered was whether the evaluation should be based on theoretical analysis of convergence properties or on empirical results. While the former option is more elegant and intellectually appealing, the state of the art in integer programming precludes its use in most practical applications. The 0-1 integer programming problem (IP) is known to be NP-complete; hence the computation times of the best optimization techniques known to date are exponentially bounded. For heuristic procedures, worst case behavior is of little interest to most practitioners and expected convergence properties are often derived only for limiting cases.

For these reasons we turned to empirical testing. However, this approach is also less than satisfactory. The primary difficulty is that there does not exist any generally accepted set of canonical test problems. The reasons for this range from the proprietary nature of some test data to the lack of representativeness that any small set of problems will exhibit. In particular, it is one of the features of integer programming problems that they lack uniformity. One problem may be very easily solved while a problem derived from it merely by perturbing a few coefficients may be very difficult. It also appears that the difficulty of integer programming problems is often related to the actual application setting. For instance, it has been our experience that bus crew scheduling problems tend to be very difficult to solve while airline crew scheduling problems, which share the same set partitioning structure, tend to be easier [Marsten and Shepardson].

Once we had decided to proceed with empirical testing two questions remained:

1. What problems do we use to evaluate the algorithm?
2. To what algorithms do we compare our own?

These two problems are not completely independent, for the choice of problems will often influence the choice of algorithms for comparison.

We chose to address the first problem in the belief that if a good set of easily accessible test problems were to exist, it would eventually be used routinely in testing all algorithms. In this way, choosing comparative algorithms would become unnecessary, simply because by solving the standard test problems one could compare a new algorithm to many others. Consequently, our problem was reduced to putting together a reasonably good set of test problems.

1. Defining the Test Problems

Our strategy was first to identify those properties which determine whether an integer programming problem will be difficult or easy to solve. Test problems should reflect these properties. Based on conventional wisdom, practical experience, and intuition, we identified four such properties.

Property 1. The SIZE of the problem as measured by the number of rows and columns.

Property 2. The DENSITY of the constraint matrix $[a_{ij}]$, defined as the ratio of the number of nonzero entries to the total number of entries in the matrix.

Property 3. The duality GAP as measured by the difference between the continuous linear programming solution and the discrete integer programming solution expressed as a percentage of the integer programming solution.

Property 4. The LATITUDE of the problem as measured by the magnitude of the epsilon set where the epsilon set is the set of all feasible solutions of the integer programming problem whose objective function value is within epsilon of the optimal objective function value and the magnitude is the number of distinct solutions in that set. Epsilon is some small positive number, on the order of one percent of the value of the optimal integer programming solution.

It is generally agreed that Property 1, the size of the problem, is very important in determing the computational difficulty of a problem. To date, all known algorithms for solving the integer programming problem have solution times which are bounded by an exponential function of the amount of data required to represent the problem. Neglecting the structural constraints (1) in (IP), there are 2^n possible vectors in the unit hypercube defined by the integrality constraints (2). Most integer programming algorithms include some sort of branch and bound procedure which is at least implicitly enumerating these 2^n vectors.

The number of rows in a problem also has a complicating effect. For many algorithms, the number of computations is an increasing function of the number of rows in the problem. A secondary effect is that the inclusion of many constraints may reduce the size of the feasible set to the extent that no, or very few, discrete solutions may exist while solutions to the continuous relaxation:

$$\text{(LP)} \quad \text{Max} \quad \sum_{j=1}^{n} C_j X_j$$

$$\text{st} \quad \sum_{j=1}^{n} a_{ij} X_j \leq b_i \qquad i = 1,\ldots,m$$

$$X_j \leq 1 \qquad j = 1,\ldots,n$$

$$X_j \geq 0 \qquad j = 1,\ldots,n$$

may be found. In this case it may be very difficult to find a feasible integer solution, not to mention the optimal answer.

Property 2, the density of the problem, is important in that many integer programming algorithms take advantage of sparsity to reduce computation time and storage requirements. In addition, the greater the density, the more data required to represent the problem and as mentioned previously computation time is an increasing function of this amount.

The duality gap, Property 3, is important for all algorithms utilizing continuous relaxations. This class includes branch and bound procedures depending on bounds derived from continuous linear programming relaxations, as well as Lagrangian relaxations which have the integrality property [Geoffrion, 1974]. In general, the larger the duality gap, the looser the bounds that will be generated by such an algorithm and the more extensive the enumeration process required.

Primarily, the duality gap is a measure of the degree to which the feasible region of the continuous linear programming relaxation (LP) of (IP) approximates the convex hull of the integer extreme points of the (IP) feasible region in the vicinity of the optimal (IP) answer. Consequently, one way of creating a problem with a small duality gap is to start with a problem with zero duality gap, for example a network problem, and perturb the constraints slightly so that the feasible region of the perturbed problem closely approximates the feasible region of the original problem. This is an approach we have taken in generating test problems.

In 0-1 integer programming problems, large duality gaps may be a result of tightly constrained problems. For instance, one can envision a feasible region which eliminates all of the extreme points of the unit hypercube. In such a case the integer programming problem may be infeasible while the continuous linear programming relaxation is not.

Another example of the sort of conditions which may cause large duality gaps and make an integer programming problem very difficult to solve is a constraint typically used to share the payment of a fixed cost among a number of variables:

$$\sum_j Y_j \leq M Z$$

where each Y_j and Z is constrained to be either 0 or 1, Z is assumed to have a negative objective function coefficient, and M is a large positive number. This constraint assures that none of the Y_j's is used without first incurring the fixed cost of turning on Z. Once Z is turned on (set to 1), the constraint, by design, becomes nonbinding. The problem is that in the continuous linear programming relaxation, Z will usually take on a fractional value and the constraint will be binding. Consequently, by the very formulation of the problem the (LP) and (IP) feasible regions will not coincide in the region of the (IP) optimal answer. This situation almost inevitably results in significant duality gaps. Hence, it is generally accepted that well formulated integer programming problems are much easier to solve than equivalent more parsimonious formulations. A better formulation in the above example is:

$$Y_j \leq Z \qquad j = 1, \ldots, n.$$

The latitude of the problem, Property 4, is a characteristic which affects optimization procedures differently from heuristic techniques. For a problem with a large epsilon set, a good answer should be relatively easy to find for both heuristic and optimization procedures. However, even after the optimal answer has been found, optimality may be very difficult to ascertain, since only the tightest of bounding techniques is able to avoid explicitly enumerating the epsilon set.

A number of factors may influence the latitude of a problem. The size of the feasible region is certainly one. The distribution of the coefficients in the objective function is another, in the sense that if one variable has a very large objective coefficient relative to the others, it is likely to be in the optimal answer if any feasible answer contains this variable. Consequently, the optimal answer will tend to dominate the alternative feasible answers and the epsilon set will tend to be small. Our tendency is to think that if the distribution of coefficients in the objective function is not widely dispersed, then the latitude of the problem will be relatively large and the problem will be relatively difficult to solve.

In addition to reflecting the above four properties, test problems should be readily accessible by practitioners. An adequate supply of test problems should also exist to allow users to do extensive testing. For each of the four properties outlined above there is a range of possibilities. To consider all possible combinations of these four properties requires a very large body of test problems, particularly since it is advisable to do a number of replications in each case. For this reason we turned to developing an easily reproducible method for generating test problems as needed.

Ideally, such a problem generator should directly control each of these four suggested properties. In this way, one could study the conditions believed to influence the computational difficulty of integer programming problems. The intent is that the test problems will adequately reflect the structure of problems found in the real world. However, this is a stringent and ill defined goal. By controlling the four properties described above, are we adequately characterizing real world problems or are there other important properties, such as the structure of the constraint matrix, which must also be considered?

2. Developing the Problem Generator

In this section, the problem generator is described according to its inputs. By duplicating these parameters anybody with a copy of the code, which has been written to be machine independent, can duplicate test problems developed elsewhere using this program. First, the user directly controls problem size by inputting the number of rows and columns. A seed value for the random number generator is then specified, along with a distribution type for the objective function coefficients. Either a uniform distribution, specifying upper and lower bounds, or a normal distribution, specifying mean and standard deviation, may be selected. Hopefully, this control over the distribution will allow the user to directly influence the size of the epsilon set in the generated problem.

The user then indicates the number of columns to be used in determining the right hand side vector for the problem constraints. The specified number of columns are randomly selected from among the generated vectors and the right hand side values are obtained by summing the coefficients in these columns. This procedure directly influences the number of variables in the optimal solution to the problem and assures that the problem is feasible and nontrivial. (It also appears that increasing the number of columns determining the right hand side increases the number of feasible solutions and decreases the duality gap.) The user also specifies the density of the generated coefficient matrix; the precise method by which the generator enforces this condition will be explained below.

The user then indicates the number of right hand sides to be perturbed. This perturbation slightly increases the duality gap, although unperturbed problems rarely have a zero duality gap due to the manner in which the constraint matrix is generated. The user may also specify how many matrix coefficients to perturb. This process serves two purposes. First, it tends to increase the size of the duality gap. Second, as will be discussed later, the generated matrices tend to have a very special structure and perturbation of the entries helps to eliminate this structure.

3. How the Problem Generator Works

Test problem generation is a six stage operation; most of the work, however, is directed to the development of the constraints:

1. A random transshipment problem is generated with the user specified number of rows (m) and columns (n). Each column contains precisely one +1 and one -1 entry, with no duplicate columns permitted. In addition, no two columns are allowed to be the negative of each other.
2. A random set of columns from this transshipment matrix N is chosen and summed to generate the right hand side. The cardinality of this set is determined by the user.
3. A square (mxm) transformation matrix T of appropriate density is generated.
4. The (IP) problem is produced by premultiplying both the N matrix and the right hand side by T.
5. The user specified number of right hand sides are perturbed by random amounts in the range 1 to 3.
6. At the user's option, random coefficient matrix elements are perturbed by amounts in the range of -3 to +3. The corresponding right hand side values may also be modified depending

on whether the given column is in or out of the optimal solution to the transshipment problem. This latter adjustment helps minimize the change in duality gap resulting from coefficient perturbation and assures the existence of a feasible solution.

The first stage of this process allows us to begin with a unimodular coefficient matrix which implies no duality gap. The right hand side generation method insures a feasible, nontrivial problem. The transformation matrix T is then chosen to achieve three objectives:

1. to yield the proper density for the final coefficient matrix TxN, since the density of TxN is roughly twice that of T due to the +1/-1 combinations in matrix N;
2. to eliminate coefficient repetition and disguise the underlying transshipment problem structure;
3. to yield a minimal duality gap in the final problem.

If the original transshipment problem consisted of equality constraints, then the transformed problem would have no duality gap if T were invertible. We assumed that choosing T to be nonsingular may help to minimize the duality gap.

Unfortunately, there is a tradeoff among these three goals. Two methods of generating the transformation matrix T yielding different results with respect to these objectives have been explored by the authors. Method A begins with a nonsingular diagonal matrix. A denser matrix is then produced using elementary row operations in such a way as to maintain a lower triangular structure. The resulting transformation matrix is nonsingular and the generated problems tend to have small duality gaps (<10%) even at densities as high as 40% when a small number of right hand sides (1-3) is perturbed. However, there is considerable coefficient repetition within rows for the resulting constraints. Not only is this condition esthetically unappealing, it may also contribute to the relative ease with which these problems are solved, above and beyond the influence of the small duality gaps.

Method B adds coefficients to a nonsingular diagonal matrix in symmetric positions to reduce coefficient repetition in the final constraints. Random elements are then added to the transformation matrix to reach the correct density in the (IP) coefficient matrix. While these T matrices are not necessarily invertible, coefficient repetition is greatly reduced in comparison with Method A. However, the duality gaps also tend to be considerably larger (30-1000%) at similar coefficient matrix densities (10-40%). The resulting problems are more difficult to solve and possibly unrealistic.

At this time, we are unable to pinpoint whether it is the invertibility of T, the coefficient repetition, or the smaller duality gaps, which makes Method A problems so much easier to solve. It is also uncertain to what extent these three issues are interrelated. Low density problems generated by Method B also appear to have small duality gaps although coefficient repetition is somewhat reduced. The crux of the problem is adequate control of the T matrix density, and hence the resulting coefficient matrix density, as well as T-matrix invertibility and coefficient repetition in the generated constraints.

A number of possible shortcomings are apparent in problems generated using these methods. Most obviously, the multiplicative process by which the coefficient matrix is derived offers the possibility of quickly solving these problems by inverting the T matrix. While optional perturbation of random matrix elements addresses this concern, we feel that as long as algorithms are not tailored to solving these specific types of problems, this limitation is not significant. More importantly, the underlying transshipment network forces a lower bound on the coefficient matrix density of 2/m where m is the number of rows in the problem.

4. Our Evaluation Procedure

Once a working version of the test problem generator was completed, an evaluation procedure was needed to test the prototype of the proposed IP algorithm. Two approaches were considered:

1. generate a random set of problems ignoring the size of the resultant duality gaps and then analyze the algorithm's performance with respect to this range of gap sizes;
2. select an ideal set of problems with specific gap sizes, if necessary by generating and discarding test problems until sufficient numbers in each category were found.

While in principle the use of an ideal test set is appealing, the former method appears more defensible. First, we know of no a priori range of duality gaps for evaluating or comparing algorithms, although a case might be made for gaps in the range 0 to 20% based on our experience with real world problems. Second, and more significantly, since the problem generator cannot control the resulting duality gap as precisely as desired (for instance, there is some variation due to the choice of the initial random seed value), we might expect to generate and solve many problems before completing the desired test set. Consequently, both computer time and a certain amount of performance data would be lost as a result of this approach. In addition, since duality gap size alone does not determine difficulty -- i.e. not all problems with the same gap size are of equal difficulty -- selection among problems with equal duality gaps is also complicated.

5. Implementation

While it is easy to assess the generator's control over problem characteristics such as density and duality gap, it is more difficult to determine whether unusual structure in the test problems or some other unrecognized property of our prototype algorithm is the cause of particular solution times. Hence, in this section, we provide a summary of several sets of test problems under various conditions with only nominal emphasis on problem difficulty as reflected by the solution times. These particular results were generated using a prototype of our IP algorithm coded in standard FORTRAN IV and run on a DEC System 20/60 under TOPS-20 using the OPT and NODEBUG options.

Two different sizes of problems were examined using the test problem generator. A 40 row by 100 column format appeared to represent a reasonable size for applied problems. Since problems of this size generated by Method A could generally be solved in a reasonable amount of time, all Method A tests were performed using this format. However, similar problems generated under Method B were considerably harder or unsolvable within several hours of computing; hence, a smaller 20x50 format was used in Method B tests.

In order to test the generator's degree of control over three of the hypothesized dimensions of problem difficulty -- density, duality gap, and latitude -- a number of experiments were performed. First, as Tables 1 through 4 indicate, two densities -- 15% and 30% -- were employed. Second, the duality gap was adjusted by controlling the number of columns forming the right hand side. Tests using 3 and 5 columns were performed; it was conjectured that increasing the number of columns used would decrease the duality gap due to an increase in the number of feasible solutions. Finally, two experiments were performed by varying the distribution of objective function coefficients to examine problem latitude. The first experiment compared a uniform distribution over the interval [-100,100] to a normal distribution with mean 0 and standard deviation 25; the former having greater variance and hopefully lower latitude -- i.e., fewer nearly optimal solutions (see Tables 1 and 2). The second experiment compared two normal distributions with mean 1000 and standard deviations 50 and 200 (Tables 3 and 4, respectively).

6. Results

Ideally we would like to be able to answer two questions: to what extent does our problem generator control the four properties listed above and to what extent does each of these properties influence computational difficulty? Of course, the answer to the second question is in turn very much affected by the particular algorithm being used. Actually it is the algorithm we would like to be investigating, not the problem generator.

By definition, the problem generator controls problem size absolutely. We also found that problem density was controllable within 2-3%. Examination of the results in Tables 1-4 suggests that some control over duality gap is obtained by varying the number of columns used in creating the right hand side. Without explicitly generating the epsilon set, it is difficult to measure the latitude of each problem.

We found, of course, that problem size had an overwhelming effect on computational complexity as measured by CPU time to solve a problem. The results in Tables 1-4 indicate that increased density also leads to increased computation time, although the results are much clearer for Method B than Method A. The tables also indicate that as the size of the duality gap increases, solution times generally increase. Although we could not measure latitude directly, tighter distribution of objective function coefficients appears to be associated with increased solution times.

Assuming a multiplicative effect for these dimensions of problems difficulty, two simple regression models were tested using solution time as the dependent variable. The first model examined the direct control of the three generator parameters -- problem density, number of right hand sides adjusted, and variance in the objective function coefficient distribution -- over solution time. As Table 5 shows, this model has considerably more explanatory power for problems generated using Method B than Method A. As expected, the regression coefficients indicate that higher density increases solution time while using additional columns to generate the right hand side reduced difficulty. Latitude has almost no consistent effect on solution time, at least for this set of problems.

The second model studied the influence of the problem's characteristics on the solution time. In addition to the three 0-1 coded independent variables in the previous model, a variable measuring the actual duality gap was used. Although this model improves the fit for both methods, the results are again more compelling for Method B. For these test problems, the size of the duality gap appears to strongly influence solution time -- a finding which is quite consistent with the branch and bound procedure of the test algorithm.

The relatively low overall explanatory power of these two models may be attributed to several factors in the experimental setup. Notably, the use of one specific solution algorithm may be confounding the analysis due to idiosyncrasies of the algorithm itself -- such as program coding or presumed problem structure. Also, since the number of adjusted right hand sides is only loosely correlated with the duality gap, it is possible to improve the quality of the regression fit by adding a variable representing size of duality gap to the model. Unfortunately, in practice, duality gap size is not directly controllable and is never known prior to solving the problem.

7. Conclusions and Observations

While this test data is not complete enough to draw any definitive conclusions, several general observations are possible:

1. The overall solution time for a problem is not dictated solely by the size of the duality gap, thereby lending credence to the use of randomly generated test problems in investigating algorithm performance as suggested in Section 4.
2. Method A problems are generally easier to solve than Method B ones; as mentioned previously, however, it is an open question as to whether it is coefficient repetition, transformation matrix invertibility, or duality gap that creates this condition.
3. It is possible to at least loosely control the four problem dimensions and in turn influence although not control overall solution time.

Ideally, this same set of problems should be solved with other IP codes to help separate the effects of generated problem structure from properties of our solution method. However, during the test period, all of the alternative integer programming codes available to us were discovered to have bugs and hence provided no reliable basis for comparison.

8. Concluding Remarks

A machine independent FORTRAN code has been prepared by the authors for generating integer programming test problems. Copies of this code are available for a nominal distribution charge from Michael Chang, Wharton School of Business, University of Pennsylvania. The generator allows users to develop test problems by merely inputting a set of parameters. This procedure simplifies reporting results, as well as replicating test problems reported by others.

The test problem generator is designed to control certain properties of the problems generated, including size, density, duality gap, and latitude. Since these properties appear to influence the difficulty of an integer programming problem, this allows users to reflect real world problem structure with randomly generated data. More work is required to determine whether these four properties are indeed solely responsible for the difficulty of real world integer programs. Additional work is also required to ascertain whether the problem generator adequately controls these four properties without introducing other characteristics that influence the resulting test problems.

Results reported here demonstrate that the problem generator does adequately control problem size and density. The results also support the contention that the generator can influence the size of the problem's duality gap and hence its computational difficulty.

	15% Dense				30% Dense			
	3 RHS		5 RHS		3 RHS		5 RHS	
	Unif	Norm	Unif	Norm	Unif	Norm	Unif	Norm
(1)	64	12	5	9	6	2	6	7
	95	170	92	298	543	40	84	525
(2)	8	11	5	8	10	12	2	10
	39	1533	34	344	1400	2311	327	2253
(3)	7	8	5	8	2	2	2	3
	91	98	863	1084	122	52	63	73
(4)	9	5	4	5	6	5	8	6
	379	268	88	72	223	133	245	299

Table 1. Method A: 40 x 100 Format Problems

for: Uniform Distribution [-100,100]
Normal Distribution: Mean 0, Std. dev. 25

where values are:

Gap %	
	Solution Time (in seconds)

	15% Dense				30% Dense			
	3 RHS		5 RHS		3 RHS		5 RHS	
	Unif	Norm	Unif	Norm	Unif	Norm	Unif	Norm
(1)	4	3	1	6	23	191	15	22
	6	11	7	7	231	689	29	18
(2)	1	2	11	1	28	13	75	19
	8	8	12	15	40	31	161	72
(3)	<1	8	7	6	9	22	4	6
	7	9	10	8	16	21	11	14
(4)	1	0	0	0	39	18	11	2
	12	9	6	3	859	82	29	10

Table 2. Method B: 20 x 50 Format Problems

for: Uniform Distribution [-100,100]
Normal Distribution: Mean 0, Std. dev. 25

where values are:

Gap %	
	Solution Time (in seconds)

	15% Dense				30% Dense			
	3 RHS		5 RHS		3 RHS		5 RHS	
	s=50	s=200	s=50	s=200	s=50	s=200	s=50	s=200
(1)	7	6	3	3	2	1	2	2
	92	47	168	170	71	44	26	28
(2)	5	5	5	5	*	4	3	3
	88	110	364	189	422	291	1342	1119
(3)	6	5		<7	1	1	2	2
	257	105	(>7200)	(>7200)	130	66	205	232
(4)	2	2	3	3	2	2	3	3
	498	427	220	527	228	196	3146	1013

Table 3. Method A: 40 x 100 Format Problems

for: Normal Distribution: Mean 1000, Std. dev. 50
Normal Distribution: Mean 1000, Std. dev. 200

where values are:

Gap %	
	Solution Time (in seconds)

and '*' indicates that no LP optimum is available

		15% Dense				30% Dense			
		3 RHS		5 RHS		3 RHS		5 RHS	
		s=50	s=200	s=50	s=200	s=50	s=200	s=50	s=200
(1)	Gap %	<1	1	3	3	24	25	14	14
	Time	12	14	9	7	393	463	196	170
(2)	Gap %	<1	<1	7	8	13	13	11	11
	Time	12	8	14	14	97	47	21	21
(3)	Gap %	2	2	4	4	11	11	3	4
	Time	10	7	8	8	193	153	10	13
(4)	Gap %	2	2	<1	1	29	25	7	4
	Time	10	17	6	6	2925	1755	22	18

Table 4. Method B: 20 x 50 Format Problems

for: Normal Distribution: Mean 1000, Std. dev. 50
Normal Distribution: Mean 1000, Std. dev. 200

where values are:

Gap %
Solution Time
(in seconds)

Direct Control Analysis *

	Method A ** Table 1	Method A ** Table 3	Method A ** Combined	Method B ** Table 2	Method B ** Table 4	Method B ** Combined
adj R-sq	.02	.0	.01	.6	.5	.5
regr. coef.						
density	.3(.6)	.04(.1)	.2(.6)	1.8(5.1)	2.3(6.0)	2.1(7.8)
RHS	.04(.1)	.8(1.8)	.4(1.3)	-.7(-1.9)	-1.3(-3.4)	-1.0(-3.7)
latitude	.5(1.2)	.2(.5)	.4(1.2)	-.2(-0.6)	.1(.3)	-.05(-0.2)

Influence of Problem Characteristics *

	Method A ** Table 1	Method A ** Table 3	Method A ** Combined	Method B ** Table 2	Method B ** Table 4	Method B ** Combined
adj R-sq	.02	.2	.1	.8	.6	.6
regr. coef.						
density	.8(1.7)	.4(.7)	.5(1.7)	.9(2.1)	.5(1.2)	1.1(3.4)
RHS	.3(.7)	.7(1.7)	.4(1.5)	-.5(-1.6)	-1.2(-4.6)	-8.8(-3.7)
latitude	.5(1.1)	.2(.5)	.3(1.2)	-.2(-0.6)	.009(.3)	-3.8(-0.2)
gap	.9(2.7)	.5(.9)	.6(2.7)	.4(3.2)	1.1(5.4)	4.7(4.2)

* computational time is the dependent variable
** t-statistics listed in parentheses

Table 5. Regression Models

References

[1] Geoffrion, A., "Lagrangian Relaxation for Integer Programming," Mathematical Programming Study, 2, (1974), 82-114.

[2] Lin, B.W. and Rardin, R.L., "Controlled Experimental Design for Statistical Comparison of Integer Programming Algorithms," Management Science, 25 (12), Dec. 1979, 1258-1271.

[3] Marsten, R.E. and Shepardson, F., "Exact Solution of Crew Scheduling Problems Using the Set Partitioning Model: Recent Successful Applications," to appear in Networks.

COMPARATIVE COMPUTATIONAL STUDIES IN MATHEMATICAL PROGRAMMING

by

Ron S. Dembo*

Abstract

This session is concerned with identifying ideas that will guide future studies on comparisons of algorithms and codes. To do so we have selected three recent computational studies with differing viewpoints. The authors of these studies will make a brief presentation** outlining the methodology they used, their views on the good and bad points in the study, and their suggestions for future studies. Following these presentations members of the panel, who have been selected so as to cover a wide range of opinions on the subject, will comment on the points that have been made. The discussion will then be open to the general audience.

Introduction (R.S. Dembo): Ever since the development of computer implementations of mathematical programming algorithms began, there has been an interest in being able to compare algorithms so as to guide future research. As early as 1953, Hoffman et al. [2] ran a set of comparative tests to evaluate LP codes. Since then there have been numerous studies; for a comprehensive survey see Jackson, Mulvey [4], and Hoffman, Jackson [3]. Periodically, there have been attempts at improving testing methodologies. Major landmarks have been Colville's 1964 study [1], which had a major impact on algorithm development, and Schittkowski's 1979 study [7], which in my view, represents the current state-of-the art in testing methodology.

Unfortunately, there has been no attempt at a formal evaluation of these studies in the literature. The purpose of this session is to provide incentive for periodic evaluations of testing methodologies. To focus the discussion, three recent studies were selected for critical review: They are:

1. Sandgren and Ragsdell's study [6]
2. Schittkowski's study [7]
3. Miele and Gonzalez's study [5].

Schittkowski's study was by far the most comprehensive and expensive of the three. The Sandgren and Ragsdell study was similar in

*School of Organization and Management, Yale University

**These reports are listed in the Appendix, this Proceedings spirit to the Colville study except that all codes were run on a single computer. Miele and Gonzalez addressed the specific issue of CPU time and number of equivalent function evaluations as valid performance measures. It is clear that it has become far more difficult and expensive to draw conclusions regarding algorithm behavior than it was when Hoffman et al. conducted their study in 1953. Algorithm development is at a point where there is no algorithm for NLP that dominates all other competitors. It is hoped that this session will be the first of many attempts at reassessing testing methodology.

References

[1] Colville, A., "A Comparative Study of Nonlinear Programming Codes," IBM Scientific Center Report 320-2949 (June 1968), pp. 487-500.

[2] Hoffman, A., Mannos, M., Sokolowsky, D., and Wiegmann, N., "Computational Experience in Solving Linear Programs," J. SIAM, 1, (1953), pp. 17-33.

[3] Hoffman, K.L., and Jackson, R.H.F., "In Pursuit of a Methodology for Testing Mathematical Programming Software," this proceedings, 1982.

[4] Jackson, R.H.F., Mulvey, J.M., "A Critical Review of Methods of Comparing Mathematical Programming Algorithms and Software: 1953-1977," Journal of Research of the National Bureau of Standards, 83, 6, Nov.-Dec. 1978, 563-584.

[5] Miele, A., and Gonzalez, S., "On the Comparative Evaluation of Algorithms for Mathematical Programming Problems," Nonlinear Programming, 3, Academic Press, NY (1978), 337-359.

[6] Sandgren, E., and Ragsdell, K.M., "The Utility of Nonlinear Programming Algorithms: A Comparative Study -- Parts 1 and 2," ASME Journal of Mechanical Design, 102, 3, (July 1980), 540-551.

[7] Schittkowski, K., Nonlinear Programming Codes - Information, Tests, Performance, Lectures Notes in Economics and Mathematical Systems, No. 183, Springer-Verlag, Berlin, Heidelberg, New York, 1980.

REMARKS ON THE EVALUATION OF NONLINEAR PROGRAMMING ALGORITHMS

D.M. Himmelblau
The University of Texas at Austin

What do we expect from a nonlinear programming code that implements a known mathematical algorithm? Our primary interest seems to lie not in the design of the code and its structures but in the execution of the code since the latter will determine the efficiency, reliability, and robustness of the code. While the details of the algorithm can be read from the literature and the code can be examined from a (hopefully) well-documented listing of the programs, the author of a code should also explain the rationale behind the chosen strategy and the various choices which have been made. It is the choices rather than the decisions that represents the essential information.

A report about the analysis of codes should begin with a clear specification of the design objectives: What is the experimentation intended to achieve? Without such a statement it is impossible to judge the results of the experimentation. Second, the report should explain the design strategy: What approach was selected to solve the design problem and why? If possible, defects in the approaches of previous investigators that have been ameliorated should be pointed out. Third, the results of the experimentation have to be analyzed by sound statistical techniques if the relative merits of various codes are to be evaluated.

All of the authors have focused to some extent on these three features of an evaluation. All agree that the design objective is to determine the best codes (in some sense) for general or specialized use. Schittkowski has provided the most detailed analysis of the performance criteria and test problems used in the design strategy. Evaluation of the results of tests is difficult because of the noncompatible evaluation criteria and the uncertainty that exists with respect to what statistical tests are to be used. It is in this third area that we should focus our attention for the future. Before commenting on these factors, I would like to mention some facets of the evaluation of codes that were not mentioned but only implied by the speakers.

(1) Evidence of the proper programming and functioning of a given code cannot normally be achieved by simply testing the code on problems. It might easily happen that the individual who has implemented the code has made one or two minor errors, or perhaps changes in a code to make

it run on another machine, introduces some minor errors. Test results will demonstrate poor performance (or better perhaps!) than warranted. Indistinguishable test results may arise because part of the software disguises the existing differences or because the battery of test problems is not adequate for exhibiting such differences. What is needed is to ascertain that the input-output relationship which the code is supposed to produce actually occurs. How to accomplish this task other than by tracing each computation segment by hand is not clear at the present time.

(2) All codes are comprised of parts which together generate some synergism. Most codes can be improved by fitting the proper unidimensional searches, or absence thereof, within the proper unconstrained multivariate searches within the proper constraint satisfaction tests using proper matrix manipulations within the overall code. A test of a general NLP code may only demonstrate that one component part is more effective than a similar component part in an analogous code, but the effect of the component is lost by examining only final results. A report about the results of a carefully planned and executed set of experiments will lend more authority to an assertion about the relative merits of a code than could ever be achieved by some arbitrary test runs by the author of the code. But such tests rarely provide much insight into the reasons for the superiority of one code vs. another, nor delve into the merits or shortcomings of various individual parts of the codes.

(3) A widely accepted definition of and comparison of the degree of difficulty of test problems need to be accomplished. While other fields of science and engineering have established ways of judging the merits of problems, this is not the case so far in nonlinear programming. The consequence is that evaluations retain a substantial subjective element. Perhaps this is quite natural and should not be considered unscientific, but it makes the evaluation of experimental results quite diffuse.

Now let me turn to some general comments relative to the presentations that have been made. The detailed outline by Schittkowski is quite thoughtful and he mentions many important points such as what the user needs to know and the difficulty in selecting a suitable set of test problems. It is only with respect to his remark on the location of a global solution that I would differ. A way to test the probability that a local solution is the global solution lies far in the future, if ever. Schittkowski takes into account nearly all of the features that must be considered in an evaluation.

Ragsdell is specific with regard to the tolerance and convergence parameters used in evaluation, but one must keep in mind the user. The time needed by the user to prepare a problem for solution, i.e., to fit the problem to the form called for by the code, and the time needed by the user to debug the problems that arise when the code does not work the first time, are as important (more important?) as the execution time for a code.

I do not believe that the traditional form of testing as described by the speakers necessarily provides a proper means of evaluating the relative merits of alternative strategies of NLP. If we repeatedly run a code, with variations in one or more parameters, against the usual battery of test problems and carry out the typical statistical evaluation of the results, we may be able to eliminate a particularly poor strategy but we cannot hope to recognize more subtle properties of a particular strategy. We cannot trace the produce to its real source. More important, we may not even realize a crucial sensitivity of a strategy to a certain computational requirement. The requirement may not occur within the test problem battery, or if it does, we probably will not recognize it.

Finally, if statistics are to be used in evaluating the results of experimentation via test problems, appropriate statistics should be used. One should not employ tests based on the assumption of a normal distribution for a single variable unless the assumptions underlying the test are verified by theory or experiment. It seems to me that comparisons of means are not necessarily a good way to evaluate test results, and that if they were, some distribution free test based on, say, the Chebychev criterion should be employed. Nonparametric tests (such as Friedman's test for several ranks) or pattern analysis would be the more appropriate approach to decision making.

COMMENTS ON EVALUATING ALGORITHMS AND CODES FOR MATHEMATICAL PROGRAMMING

Robert B. Schnabel
University of Colorado at Boulder

1. Introduction

In this paper we comment on comparative computational studies for mathematical programming, in particular the papers by Miele and Gonzalez [3], Sandgren and Ragsdell [5], and Schittkowski [7], and give some of our views on this subject. We suggest the papers by Gill and Murray [2] and Moré [4], and the book they are contained in (Fosdick [1]), as other important reading on this subject.

The most important distinction we wish to make is that the papers under consideration are concerned with two different types of evaluations:

Type A: Evaluation of mathematical methods (algorithms) for mathematical programming

Type B: Evaluation of software for mathematical programming.

Sandgren and Ragsdell are concerned with A, Miele and Gonzalez with an aspect of A, Schittkowski with B. Clearly, the criteria used in A are a proper subset of those used in B. We believe that a clear distinction should be made between studies of these two types. In Section 2 we comment on the criteria for A that are used in the papers under consideration. In Section 3 we comment on the criteria for B that are used by Schittkowski. In Section 4 we offer a suggestion for organizing comparative evaluations into studies of types A and B.

2. Criteria for Evaluation of Mathematical Methods

There seems to be agreement among the authors that the criteria to be used in evaluating the ability of programs to solve problems in this field are efficiency, reliability and robustness. Sandgren and Ragsdell [5] present nice techniques for measuring reliability and robustness. The measurement of efficiency continues to fascinate researchers. In nonlinear programming, computer storage requirements are usually similar and reasonably small, so computational effort is the interesting measure of efficiency. We agree with Miele and Gonzalez [3] that both CPU time and the number of function and derivative evaluations used on test problems should be reported. Since virtually all test problems involve inexpensive functions, the total CPU times used on test problems may indicate which methods are preferable to a user whose problems involve inexpensive functions. (Such problems do occur in practice, often in cases where one is solving many similar problems, so that the total cost is still large.) The number of function and derivative evaluations used on test problems may indicate which methods are preferable for a user whose problems involve expensive functions. Thus we would prefer just reporting these data, rather than the more complex measure suggested by Miele and Gonzalez [3], which they admit is akin to CPU time.

The results of the evaluation of mathematical methods can be strongly dependent on the selection of test problems, and we are far from having a good idea of what problems to use in this field. Schittkowski [7] gives a nice summary of the available options. We have some misgivings about his randomly generated test problems (see [6]), because all the objective functions are of the same mathematical type, signomials. Some wrong conclusions have been made in this field from tests that involved functions of one type only. Thus we would be more comfortable with Schittkowski's technique if the class of functions generated could be shown to be quite general.

3. Criteria for Evaluation of Mathematical Software

Schittkowski [7] gives a nice summary of the criteria that are important when evaluating a piece of mathematical software. In our experience, the factors that a user considers most strongly are often those that Schittkowski cites in his Section 2: availability, documentation, ease of use, quality of output, portability. The performance of the mathematical method is often a secondary consideration. For this reason, it is crucial that developers of mathematical programming software give considerable attention to documentation, output, ease of use, and portability. Eventually, it would be helpful if some standards were proposed in this area, to aid inexperienced software developers.

4. A Suggestion

In reading the papers of Schittkowski [7] and Sandgren and Ragsdell [5] we were struck by two features: the differences as to what was being evaluated, and the large number of codes tested in both studies. We have tried to address the first point by defining tests of types A and B, which evaluate mathematical methods and mathematical software, respectively. We believe that if studies are properly organized into these two categories, then the number of codes evaluated in any one test can be significantly reduced as well. In conjunction, we believe that tests of type A should be conducted in as controlled an environment as possible. We now propose a framework that supports these suggestions.

Suppose we are testing software for a particular class of mathematical programming problems, say nonlinearly constrained optimization. There are likely to be a number of classes of mathematical methods for this problem, for example reduced gradient, penalty function, augmented Lagrangian, and successive quadratic programming. We suggest that first, studies of type A be done between programs within each class. Ideally, these could be done before all the trappings of mathematical software (e.g., extensive documentation) were added to the programs. Then we suggest that a study of type B be conducted between the dissimilar methods that were the winners in their respective studies of type A. While one might retain more than one program from a particular class, one should be able to eliminate many programs in the first round that are either inferior to or virtually the same as the retained programs.

It may also be possible to conduct the studies of type A in a controlled environment. Ideally, we mean a modular environment such as is discussed by Schnabel [8] and implemented by Weiss [9], where the codes differ only in specific modules and are identical otherwise. In our experience, this environment is very helpful in testing related mathematical methods. If it is impractical to write or rewrite the codes in this manner, some measure of control is still possible. As Sandgren and Ragsdell [5] indicate, the most uncontrolled feature of a comparative study in this field is usually the different stopping criteria of the programs.

As the use of mathematical programs increases, it becomes less possible for there to be a "best" program for a particular problem class. Instead, different programs turn out to be best suited to different applications. Comparative studies can still suggest a group of leading programs for a problem class, and indicate inferior programs that should be discarded. The aim of our suggestions is to reduce the size and cost of these studies through a two-tiered testing system, and to make the data more reliable and understandable by doing controlled testing.

References

[1] L.D. Fosdick, ed., Performance Evaluation of Numerical Software, North-Holland, 1979.

[2] P.E. Gill and W. Murray, "Implementation and Testing of Optimization Software," in Performance Evaluation of Numerical Software, North-Holland, 1979, pp. 221-234.

[3] A. Miele and S. Gonzalez, "On the Comparative Evaluation of Algorithms for Mathematical Programming Problems," in: O.L. Mangasarian, R.R. Meyer and S.M. Robinson, eds., Nonlinear Programming, 3, Academic Press, 1978, pp. 337-359.

[4] J.J. Moré, "Implementation and Testing of Optimization Software," in Performance Evaluation of Numerical Software, North-Holland, 1979, pp. 253-266.

[5] E. Sandgren and K.M. Ragsdell, "The Utility of Nonlinear Programming Algorithms: A Comparative Study -- Parts I and II, ASME Journal of Mechanical Design, 102, 1980, pp. 540-551.

[6] K. Schittkowski, Nonlinear Programming Codes -- Information, Tests, Performance, Lecture Notes in Economics and Mathematical Systems, No. 183, Springer-Verlag, 1980.

[7] K. Schittkowski, "A Model for the Performance Evaluation in Comparative Studies," this proceedings.

[8] R.B. Schnabel, "Developing Modular Software for Unconstrained Optimization," Performance Evaluation of Numerical Software, North-Holland, 1979.

[9] B.E. Weiss, "A Modular Software Package for Solving Unconstrained Nonlinear Optimization Problems," M.S. Thesis, Department of Computer Science, University of Colorado at Boulder, 1980.

SOME COMMENTS ON RECENT COMPUTATIONAL TESTING IN MATHEMATICAL PROGRAMMING

Jacques C.P. Bus
Mathematical Centre, Amsterdam

The scope of these comments is too limited for extensive discussion concerning studies (1), (2) and (3). We shall restrict ourselves to comments concerning (1).

First of all we would like to emphasize that our comments are meant to contribute to the general discussion about testing and validating. Any criticism is meant to be constructive. We express our admiration for the hard and elaborate job that Schittkowski did. With this in mind we shall discuss five aspects of testing which, in our opinion, are of great importance for mathematical programming software. For a more elaborate discussion, refer to (4).

1. Testing versus validating

A clear distinction should be made between testing and validating of algorithms and software. By testing we mean the production of numerical evidence, possibly evaluated by statistical methods and mathematical modelling. One obtains hard figures having some statistical relevance, related to the significance of the mathematical models used, which express the relationship between performance measures and problem characteristics (or problem subclasses). On the other hand, validating is the decision process leading to "good" and "bad" programs or algorithms. This decision process may be defined mathematically (see (5)), but it is based on subjective premises defining the weights of the various performance measures in a certain environment. Clearly, (1) combines testing and validating. Given several performance measures, the test results are reported in tables. The choice of the performance measures related to efficiency and reliability, among others, are subjective. One can imagine other measures leading to different conclusions. Furthermore, the programs are ranked according to the selected performance measures and weights. Distinctions between testing and validating are not very explicit, but upon careful reading some obvious distinctions can be made. Regarding the conclusions, one should realize that many subjective choices and assumptions have led to these conclusions.

2. Programs versus algorithms

The quality of the implementation of an algorithm may influence the test results obtained, as is also noted in (1, p. 105), where programs are in fact compared. Nevertheless, some conclusions there refer to algorithms (see p. 104) and such conclusions are desirable. It would be very interesting to distinguish between algorithms and programs more explicitly during the testing. This can be done by conducting tests in two phases. First, we test different implementations of one algorithm using specific performance criteria. Next we compare the "best" implementations of a number of algorithms. Such testing might more clearly separate performance due to programming expertise and that due to the basic algorithms. This distinction is very attractive from the point of view of software designers. We would like to find a fuller discussion in (1) about the influence of the programming on test results.

3. Test problems

Given algorithms for solving certain problems, one has to choose test problems based on the testing objectives and the problem characteristics which are relevant to the testing objectives. Of course, testing with a number of test problems should lead to conclusions for classes of problems. The validity of such conclusions depends on the choice of the test problems and assumptions about the problem class. The rationale of the choice of the test set should therefore be clearly specified. In (1) the author uses a set of test problems which claims to represent the user-world. This is done by creating a test problem generator which may generate problems with different characteristics (variable sparsity, variable dimensions, upper and lower bounds, degenerate, ill-conditioned, indefinite, etc.). Thus, test problem sets can be designed to some extent to represent the user-world. However, we would like to make some remarks on the test sets used in (1).

- All problems are defined by signomials, but the observation that 'many real world problems are defined by signomials' is subjective. It would have been preferable to present the results for this special class of problems separately and to save the observation about them for the conclusion.
- It is unclear how the choice of parameters of the problem generator is related to the user-world distribution of problems.
- A discussion of the relationship between problem characteristics and relevant performance measures and sufficient characterization of the

problem class by certain parameters would have been welcomed.

4. Efficiency

Efficiency, particularly of nonlinear programming software, may depend highly on the ratio of the overhead time per iteration step and the problem function evaluation times, as well as on the availability of analytical derivatives. For instance, the advantage of a low iteration overhead in variable metric methods ($O(n^2)$) above the overhead of Newton's method ($O(n^3)$) becomes negligible for large order n if the function evaluation time is very high (more than $O(n^3)$). The efficiency measures reported in (1) are (arithmetic means of) total execution time and the total number of problem function and gradient evaluations. However, all test problems are relatively inexpensive to evaluate, which is unrepresentative of real world problems. This means that the efficiency results do not apply to problems with other evaluation costs, unless one can assume that the overhead is negligible so that the number of function calls can be used, or one assumes that the cost of the test function evaluations is negligible with respect to the overhead. In the last case the reported CPU-times can be assumed to be total overhead times.

A second remark concerns the order (dimension) of problems. One expects the solution of a nonlinear programming problem to require about $O(n^3)$ arithmetical operations or some other expression involving the number of constraints and variables. It therefore seems inadequate to give the arithmetic mean of required CPU-times for problems of different dimensions. It would seem more appropriate to model these results according to problem dimensions (variables and constraints).

5. Computer Environment

Many test results reported in (1) are heavily dependent on the computer environment. Some programs are self-contained, while others are software library (at least in the original environment) or even machine language routines (inner product in Powell's routine). An extensive discussion devoted to the influence of the test computer environment on the test results, particularly in relation to expected results in the original environment, would have added significantly to the study.

References

[1] K. Schittkowski, Nonlinear Programming Codes - Information, Tests, Performance, Lecture Notes in Economics and Mathematical Systems, No. 183, Springer-Verlag, 1980.

[2] A. Miele, "Some remarks on the comparative evaluation of algorithms for mathematical programming problem," this volume, 1982.

[3] K. Ragsdell, "The evaluation of optimization software for engineering design," this volume, 1982.

[4] J.C.P. Bus, "A methodological approach to testing of NLP-software," this volume, 1982.

[5] F.A. Lootsma, "Performance evaluation of non-linear programming codes from the viewpoint of a decision maker, in: L.D. Fosdick (ed.), Performance Evaluation of Numerical Software, North Holland, (1979).

REMARKS ON THE COMPARATIVE EXPERIMENTS OF MIELE, SANDGREN, AND SCHITTKOWSKI

by

K.M. Ragsdell, P.E.

Visiting Professor
Aerospace and Mechanical Engineering
The University of Arizona
Tucson, Arizona

I will in this brief discussion offer critical remarks on the recent comparative experiments of Miele [1], Sandgren [2] and Schittkowski [3]. Because of my closeness to the work of my former doctoral student, Eric Sandgren, it is difficult for me to be completely objective in evaluating his work. I will comment on his work in spite of this obvious bias.

I agree with the analysis and conclusions of Miele and Gonzalez. They conclude that "... in spite of its obvious shortcoming, the direct measurement of the CPU time is still the most reliable way of comparing different minimization algorithms." "...it is necessary that the comparison of different algorithms be done on a single computer, with the same programming language, with the same compiler, with the same subroutines, under similar workload conditions of the computer, and by the same programmer." It is useful to have the issues addressed in such an orderly manner and the conclusions so carefully documented. I am certain that both Sandgren and Schittkowski have assumed the result in the formative stages of their work some time ago.

I view Sandgren's work as the first scientific comparison of optimization codes, and consider the design of the experiment itself to be the outstanding characteristic of the work. There are several deficiencies associated with his comparative study. First, too few problems and codes were used. More performance data is needed on practical problems, especially large, nonlinear ones. Several of the latest methods such as Han's [4] were not included, because they were not conveniently available or simply not yet ready for testing. This is truly unfortunate. Second, there has been an incomplete analysis of the Sandgren results; in particular the data could be used in conjunction with various additional ranking schemes. Finally, it would have been useful for other experts in the field to have a more direct involvement in the experiments.

The work of Schittkowski represents the best and most complete comparative experiment to date. The problem set is very large and varied in character. He includes randomly generated problems, some of which are linearly constrained, degenerate, ill-conditioned, or indefinite. The experimental procedure is sound and thoughtfully executed. The very latest codes are included by Schittkowski, which increases the value of the results even more. The ranking scheme is excellent. The use of Saaty's priority theory [5] as outlined by Lootsma [6] is noteworthy and will hopefully encourage others to do likewise in future studies. On the other side, the study does contain imperfections. Before commenting further, let me again say that Schittkowski's work is the best available to date. He does not rate the codes at equal levels of accuracy. Eason [7] has shown this to be dangerous. Furthermore, Eason, Sandgren [8] and Fattler [9] give easy to implement procedures for making equal accuracy comparisons. This problem is mitigated to some extent by the ranking scheme employed. Second, he does not rate or even record the storage capacity required for the various codes to load and execute a problem. This is unfair to those codes which are variable dimensioned, and I suggest that in the future some subset of the problems be used to record this important information for all tested codes. This becomes of increasing importance as we attempt larger and larger problems on smaller and smaller machines. More information is needed on the effect of different starting points. Randomly generated starting points of equal distance from the solution would have been very useful. I feel that Schittkowski should have included some problems from previous studies, and some practical (real world) and larger problems. Finally, it would have been good to have a more direct involvement of others in the conduct of the experiment and analysis of the data. Errors such as those on page 111 (last sentence of page is simply incorrect) might have been avoided. But this problem is not unique to Schittkowski, all studies to date share this problem. We need to work more closely to remove this problem from future experiments.

References

[1] Miele, A., and Gonzalez, S., "On the Comparative Evaluation of Algorithms for Mathematical Programming Problems," Nonlinear Programming, 3, ed. by Mangasarian, Meyer and Robinson, Academic Press, 1978, pp. 337-359.

[2] Sandgren, E., "The Utility of Nonlinear Programming Algorithms," Ph.D. Dissertation, Purdue University, Dec. 1977, available from University Microfilm, 300 North Zeeb Road, Ann Arbor, Michigan 48106, USA, document no. 7813115.

[3] Schittkowski, K., Nonlinear Programming Codes -- Information, Tests, Performance, Lecture Notes in Economics and Mathematical

Systems, 183, Springer-Verlag, Berlin, 1980.

[4] Han, S.P., "Superlinear Convergent Variable Metric Algorithms for General Nonlinear Programming Problems," Mathematical Programming, 11, No. 3, 1976, pp. 263-282.

[5] Saaty, T.L., "A Scaling Method for Priorities in Hierarchical Structures," J. Math. Psych., 15, 1977, pp. 234-281.

[6] Lootsma, F.A., "Ranking of Nonlinear Optimization Codes According to Efficiency and Robustness," in Konstruktive Methoden der finiten nichtlinearen Optimierung, ed. Collatz, Meinardus and Wetterling, Birkhauser, Basel Switzerland, 1980.

[7] Eason, E.D. and Fenton, R.G., "A Comparison of Numerical Optimization Methods for Engineering Design," ASME Journal of Engineering for Industry, Series B, Vol. 96, No. 1, Feb. 1974, pp. 196-200.

[8] Sandgren, E. and Ragsdell, K.M., "The Utility of Nonlinear Programming Algorithms: A Comparative Study - Part 1 and 2," ASME Journal of Mechanical Design, 102, No. 3, pp. 540-551, July 1980.

[9] Fattler, J.E., Sin, Y.T., Root, R.R., Ragsdell, K.M. and Reklaitis, G.V., "On the Computational Utility of Posynomial Geometric Programming Solution Methods," to appear in Mathematical Programming.

IN PURSUIT OF A METHODOLOGY FOR TESTING MATHEMATICAL PROGRAMMING SOFTWARE

K. L. Hoffman and R. H. F. Jackson
Center for Applied Mathematics
National Bureau of Standards
Washington, D. C. 20234

ABSTRACT

To resolve the often conflicting and confusing results of computational testing of mathematical software reported in the literature, the Committee on Algorithms of the Mathematical Programming Society has developed guidelines which present minimum standards to which all papers reporting such efforts should conform. Although guidelines now exist, the development of sound methodologies for this testing is at the embryonic stage. This paper surveys the results to date of computational test efforts in each of the major fields of mathematical programming and reviews to what extent these efforts have advanced the goal of developing methodologies for testing MP software. Directions for future research are presented.

Keywords: Algorithm testing; mathematical programming software; software testing; testing methodologies.

1. Introduction

This paper presents the historical development of methodologies for testing mathematical programming software. Since this paper is concerned with methodologies for testing MP software rather than specific test results, many large-scale test efforts will not be covered by this survey. We should also point out that this paper is concerned with measuring the effectiveness of computer-implemented algorithms (i.e., software) rather than the theoretical complexity of mathematically described algorithms. Thus, no discussion of complexity analysis will be presented in this paper.

We begin this paper by defining what we mean by "testing." The literature on this subject can be divided into three distinct categories. The first we label "feasibility testing." This phase of testing consists of executing a given software package on a small set of test problems in order to insure that the program is an accurate representation of the conceptional solution procedure. The second we label "developmental testing." This phase of testing is concerned with determining the best implementation strategies (e.g., the order in which arithmetic operations are performed, the handling of lists, sorting routines, etc.). To determine which of the implementations is best, modules of alternative implementations are created, a battery of test problems is executed, and, based on certain performance measures, one of the modules is chosen as the permanent implementation. The alternative strategies are not ordinarily distributed as part of the software package. Thus, when the package is finally ready for distribution, the user must accept the package as it stands without the opportunity to choose among implementations. It is at this stage that anyone other than the code developer has the opportunity to test the software. This third phase of testing we label "selectivity testing," i.e., testing to determine which of a number of software packages is most capable of solving specific sets of test problems. It is this third phase of testing which this paper will address (since most of the literature on MP software testing concerns this phase) although, where applicable, developmental testing will be discussed.

Since we consider the task of testing software to be equivalent to that of performing any scientific experiment, we have chosen that framework prescribed by experimental design methodology for reporting the methodological developments in this field. The natural questions one must answer before performing any experiments are:

(1) What is it you want to know when the experiment is completed, i.e., what type of conclusions do you expect?

(2) What do you want to measure?

(3) Is it possible to measure these things and to what precision can such measurements be taken?

(4) What is the test environment, i.e., under what conditions do you require these measurements?

(5) How can one best (most economically) obtain these measurements?

Sections of this paper address the issues raised in the above questions. Section 2 describes the possible performance measures which might be used to test MP software while Section 3 addresses the confounding factors which may affect the measurements taken. Section 4 describes the test problems available, since the test problems chosen for an experiment define the class of conditions for which conclusions can be drawn. Section 5 describes developments in experiment design, and Section 6 describes the analysis and reporting of test results. Finally, Section 7 will present our conclusions and future research directions.

2. Performance Criteria

With respect to the testing of mathematical programming software, performance criteria can naturally be divided into three separate categories. The first category is "efficiency," which encompasses both the computational effort required to obtain the solution and the quality of the solution obtained. The second category we define as "robustness," which measures the reliability of a code (i.e., how large a class of problems can this code solve?). The third category, "ease of use," refers to the amount of effort required in order to use the software.

2.1 Efficiency. Let us begin the discussion with "efficiency." As noted above, this measure encompasses both computational effort and quality of solution. Of these two, computational effort--usually measured by CPU time but often by the number of iterations or number of function evaluations--has been the most widely reported performance measure. In a 1976 literature survey, Jackson and Mulvey [61] reviewed 50 papers and found that over 70 percent of those papers discussed time required to solve a problem as measured in Central Processing Units (CPU), and in over half of those papers conclusions were drawn about the efficiency of codes based upon that one single measure. This fact is rather surprising since the first computational study on mathematical programming software, by Hoffman, et al. [58] in 1953 considered three measures--iteration counts, CPU time, and accuracy of solution--specifically because they believed that a single measure of "efficiency" might not provide sufficient information.

Colville [13] in 1968 observed wide variability in the times reported for the same codes using the same test problems on various computers, and, in an attempt to "normalize" this variability, he constructed a measure which he calls "standard time." Colville developed a test problem package (the inversion of a 40 x 40 matrix, 10 times) which can be used to get a relative measure of the speed of various machines. The time taken to run this problem set was to be reported along with any test results. Colville in that same paper notes, however, that he experienced a 10 percent variation in the CPU times obtained on a single machine for his standard test package. A newer test package is presented in Curnow & Wichmann [18] which in at least one test effort proved more accurate than the Colville Timer (see Eason, 1977 [30]). In 1975 Shanno and Phua [102], in an attempt to reduce this variability, used a measure called "average time." Every problem in the test set was solved repeatedly 10 times on the same code and the average of these 10 replications was reported.

Glover et al. [42] argue that CPU time should not be reported at all, since CPU time is merely a surrogate measure for the cost of performing the computations. Using CPU as a surrogate measure, they argue, is especially misleading when an installation charges as much or more for input/output and core storage as it does for processing time. Hillstrom [53], furthering this notion, prescribed an "overhead measure," which is defined to be CPU time relative to elapsed time. In this same paper he also argues that accuracy measures should be relative to more than one norm and argues persuasively for collecting both absolute and relative measures of accuracy. Finally, Miele and Gonzales [79] present arguments for why, even with all the noted problems, CPU time is still a better measure of computational effort than is the number of function evaluations.

As is obvious from the above discussion, there is no concensus as to how computational effort ought to be measured. In addition to the controversy over how to measure computational effort, there are those who believe that computational effort cannot be separated from measures of accuracy. Ostrowsky [86] in 1960 presented an "efficiency index" which measured accuracy relative to the number of function evaluations required to obtain that accuracy. This is the first occurrence in the literature of a measure which reflected the viewpoint that accuracy and computational effort are not independent measures.

In a similar approach, Eason and Fenton [29] chose a variety of tolerances for stopping criteria and plotted the solution time for each code relative to each tolerance setting. This study proved through the data collected that accuracy and computational effort are not independent. Adding to these ideas, a study by Sandgren and Ragsdell [99] presents a measure which incorporates both the accuracy of the objective function and the satisfaction of the constraint set. Finally, recent testing by Gilsinn et al. [40] attempted to find a composite measure of efficiency (which included computational effort and accuracy in both the objective function and the constraint set); those authors conclude that a single measure did not exist for the specific test experiment they performed. They therefore chose to report data for a number of measures and let the user decide on the "best" code based on those measures deemed most important. Included in this report were two new measures for quality of solution: "consistency" (which measures the consistency among various values output by a code) and "veracity" (which measures how often a code reported truthful auxiliary information--e.g., information about alternative optima, the rank of the matrix, etc.).

This discussion has presented the various views in the literature about measures of efficiency. The diversity of opinion and the conflicting conclusions about "best" codes reflects the fact that there is currently no consensus regarding which performance measures accurately represent the efficiency of code performance.

2.2 Robustness. For the second major classification of performance measures, "robustness," the situation is somewhat better defined. Shanno and Phua [102] define robustness as the number of problems solved with a prescribed tolerance and within a maximum computing time. Many other authors used this concept of robustness prior to this 1975 paper, but this is the earliest paper we found in which the authors concisely define "robustness". An alternative definition for "robustness" is one presented by Bus [8]. He defines robustness as the ability of a code to "exit gracefully from specific difficulties." Moré [81] expands on this definition by stating that robustness is the ability of a code to "exit properly from overflows, underflows, and invalid termination criteria."

These two rather different definitions (that of Shanno and Phua, and Bus) are used by other authors as well, sometimes calling the measure "reliability" and other times "robustness." Since both measures are useful and since no consistent definition has emerged, we recommend that users of the terms robustness and reliability define them before presenting their computational results.

2.3 Ease of Use. As noted earlier, the third major classification of performance measures is that called "ease of use." Although this criteria has received much lipservice in the literature, scientifically sound methodologies for measuring "ease of use" do not exist. As early as 1968, Colville [13] considered a measure he called "set-up time," and defined that to be the amount of time applied to solving a problem, from the moment one obtains a specific problem to the moment one obtains the solution. To measure this time, he sent a set of test problems to each code developer and asked them to report back the time it took to solve each problem. Colville's paper reports these times. Himmelblau [53] also tried the same idea but chose to use his graduate students instead of the code developers, and reports those times. The obvious problem with the approach of these authors to measuring set-up time is that the varying experiences and capabilities of the people involved may bias the results.

Brockelhurst and Dennis [6] take a somewhat different approach. They surveyed users requesting their opinions on how well a selected set of six codes performed and how easy the codes were to use. No quantifiable results are reported but user impressions are presented. Finally, Waren and Lasdon [112] present a survey of available NLP software in which they provide comments about the documentation, I/O requirements, and "user friendliness" of each code. Again, these results are impressionistic but do provide far more information on the subject than has previously existed in the literature.

2.3 Performance Criteria Research. We have, in the above discussion, highlighted those papers which have presented new ideas about performance measures which might best be used in computational testing of MP software. Clearly, there has not emerged one or two measures thought to be most important. For that reason, Crowder and Saunders [17] queried the mathematical programming community as to which performance measures were believed to be most important. The survey was mailed to members of the Mathematical Programming Society, the Special Interest Group on Mathematical Programming of the ACM, and the Computer Science Technical Section of ORSA. Table 1 presents the results of that survey. CPU time and documentation were found to be the two most important measures. These survey results highlight the fact that although methodology for measuring good documentation does not exist, users of MP software consider ease-of-use a very important criteria when selecting among software packages.

As a final note on the topic of performance measures, we would like to direct the interested reader to current research on the development of methodologies both for reporting results when many measures are used and for providing the user with a mechanism for selecting among codes when many results exist. Hiebert [52] should be consulted for information on the first of these two topics and Lootsma [72] and Saaty [96] for the second.

3. Confounding Factors

Let us assume that, based on some rational criteria, an experimenter has chosen, among the large class of feasible performance measures, those measures which will be used in some test effort. The experimenter must next determine how accurately these measurements can be performed. Factors which bias or mask the measurements one wishes to obtain, are called "confounding factors" and when they exist, steps must be taken to overcome the problems which they create.

One such factor in the area of software testing is the variability of the machine environment (e.g., the number of jobs on the computer at one time, the types of jobs competing with this job for processing, the amount of core storage available on the computer, etc.). In the 1953 paper by Hoffman, et al. [58], iteration counts were chosen in lieu of CPU time because the authors could find no way of controlling the variability in CPU time.

Wolfe and Cutler [114] considered using iteration counts and found that measure to be biased as well since specific codes performed different arithmetic operations within a single iteration. They therefore estimated the amount of effort required for one iteration for each code and reported that along with the number of iterations required.

In a paper by Gilsinn, et al. [40], it was shown that even when the environment appears to be quite comparable (same time of day, same number of jobs, etc.), the timing variability caused by machine environment can be more significant than the differences in timing among the codes when timing variability is held constant. They overcame this difficulty by using a dedicated machine (no other jobs being processed by the machine when the testing was done). For other instances of reported problems with timing measurements see Enquist [32], Eason [30], Bell [4], and Eason and Fenton [29].

Eason [28] has indicated that comparable testing across machines creates even more significant problems than does testing on one machine, while both Parlett and Wong [88], Shier [103], and Suhl [107] indicate that even in a dedicated environment when testing on only one computer, the compiler used to translate the FORTRAN version of a code to machine language can drastically affect the conclusions drawn.

Research to date has shown that obtaining conclusions about code performance is a very difficult task. Due to all of the things which confound the testing environment, we must conclude that in order for test results to be taken seriously, the experiments must, at a minimum, indicate the specific circumstances under which the testing was performed. One must not only state the computer and the compiler used, but define the test problems used and describe the entire environment under which the testing took place, preferably with measurements of the variability in the performance measures reported.

4. Test Problems

Assuming one has chosen the performance indicators to be used and has determined how to measure accurately each of these, one must next select the test problems to be used. There are three distinct types of test problems: (1) hand-picked, (2) randomly-generated, and (3) perturbed application problems. The first type--hand-picked--consists of problems that arise from real-world applications and those that are specially constructed to test the limitations of codes. These problems are useful in testing because they represent the types of problems likely to be found in the "real" applications or because they have known structural characteristics; unfortunately there are usually only a very small number of them.

The second type--randomly-generated problems--has the desirable property that one has a sufficient number of similarly structured problems to permit statistical inference-making. One often wonders, however, how closely these problems resemble those encountered when solving real-world problems.

Finally, the third type--perturbed application problems--has emerged as a compromise between the first two types. This class of problems is constructed by taking an application problem of some importance and making small perturbations in the data describing the problem. Thus, from one problem, a large number of problems all with similar structure can be created.

Tables 2-4 present references to the literature for test problems of each of these three types. We have classified the problems within a test problem type by the structure of the test problems (e.g., integer programming, constrained nonlinear programming, etc.). Examination of Table 2 reveals some interesting observations. For example, although linear programming is widely used in industrial settings, very few test problems are available to code developers and testers because of their proprietary nature. In the area of nonlinear programming the situation is very different: Hock and Schittkowski [56] published a test problem set consisting of 120 problems. In integer programming the situation is similar to that of LP. Taken together, there are a rather large number of hand-picked IP problems, but for any specific integer structure (e.g., traveling salesman, set covering, plant location) few test problems exist.

Most of the test problem generators have been developed in the past six years. Table 3 does not present any generators which create problems without structure or known solutions since these generators have been superceded by the recent research which has uncovered a variety of approaches to the generation of problems with user-controllable structure and known global solutions for difficult classes of problems (e.g., integer and nonconvex).

Tables 2-4 taken together reflect that although there is not a large set of mathematical programming test problems, recent efforts toward test problem collection have been productive. In closing the section on test problems, we direct interested readers to a few recent papers which examine the underlying structure of test problems and generators. Papers by Jackson [60] and O'Neill [83] examine the similarities and differences between and among test problems and test problem generators, while research by May and Smith [76] and van Dam and Telgen [109] investigate the "randomness" of certain test problem generators.

5. Experiment Design

As with all scientific experiments, good experiment design, prior to the actual performance of the test, is always recommended. In our search through the mathematical programming literature, the oldest computationally oriented paper we encountered which discusses experiment design is a paper by Moore and Whinston [80], published in 1966. The authors perform a factorial design experiment with each factor considered separately. In 1969, Bourgoin and Heurgon [5] use the technique of latin squares and postulate a linear form for their results. They test this hypothesis and, due to the high correlations among the independent variables, reject the original hypothesis. In 1976 Dembo and Mulvey [23] describe approaches to testing which might allow one to draw statistical inferences. Lin and Rardin [71] present a full-factorial design methodology for testing integer programming codes and describe methods for overcoming the difficulties of censored data and interaction among variables. Recently, a paper by Riley & Sielken [91] expands the notion of factorial design to that of fractional designs and reports the results of such experiments when testing IP codes. Papers by Sandgren [98] and Graves [47] indicate that analysis of variance techniques and hypothesis testing using paired observations are also useful in the evaluation of mathematical programming software.

Although the mathematical programming community has rarely used the available statistical techniques of experiment design for computational studies of software performance, the papers cited above indicate the applicability and usefulness of such techniques. For more discussion of how to use such techniques, see Fosdick [39], Lyness [74], Lyness and Greenwell [73] and Nelder [82].

In the area of reporting results obtained from computational experiments current applied statistics research is again useful. Heibert [52] illustrates the use of Chernoff faces [12] as a meaningful summary technique, while an alternative approach suggested in Pereyra and Russell [89] is the use of star plots [113]. Multi-criteria clustering techniques can also help in analyzing the wealth of data a large test-effort is likely to generate (see [27], [63], [66], and [115]).

6. Major Test Efforts

In the above discussion we have presented the papers which have significantly advanced the development of testing methodologies. There are, however, certain other test efforts which, although having used no new methodological techniques, have indeed advanced the field through their thoroughness and completeness in testing and reporting the results. This section will highlight such efforts, and Table 5 lists the authors, fields, and sizes of the testing efforts.

In the area of network optimization, Glover, Klingman, et al. are continually expanding both the size of the problems which codes can tackle and the breadth of applications which may be solved by such techniques (see [24], [45], and [64] for examples). These developments have taken place because of the careful modular coding and extensive testing being performed. Some of their test efforts required that more than 2,000 problems be tested on more than 10 different codes. Similarly, in the areas of geometric programming and nonlinear programming a number of authors have performed very extensive tests (see Table 5).

A rather recent phenomena is that national laboratories (such as Argonne, Sandia, NBS) have performed extensive testing of a variety of codes because of a need to create software libraries. An interesting aspect of these test efforts is that they are being performed by those not responsible for the development of any of the codes included in the test.

7. Conclusions and Future Directions

This paper has discussed research directed toward the development of methodologies for testing mathematical programming software. Computational testing has changed significantly in the past 10 years: generators of test problems with known solutions and controllable structures have emerged and hand-picked problems are becoming more accessible to a larger audience. We notice that the performance measures most used prior to 1970--CPU time, function evaluations, iteration counts--are being augmented with new measures related to accuracy, ease-of-use, core storage requirements, and input/output requirements. We believe this trend toward more concern for user-time and effort will be even more prevalent in the 1980's, with the ever-increasing storage capacity and computational speed of computers. With this change comes the need for quantitative measures for the terms "ease-of-use" and "set-up-time." Certainly the challenge of finding a consistent measure for computational effort continues to exist as well.

We hope that future papers on mathematical software testing will rely more heavily on the statistical techniques of "experiment design." By using the procedures described in the statistical literature, one can be more assured of the reproducibility and statistical validity of the results obtained. A related statistical field, that of multi-criteria analysis and clustering, can also make analysis easier and the results of this analysis more understandable to potential users of test results. Certainly, anyone publishing results on computational testing should adhere to the guidelines presented in [15].

Finally, we would like to direct those readers interested in performing future computational tests to some software which might make that testing easier. There are currently packages which

(1) test the portability of software (see [35], [39], [84], and [95]),

(2) "tidy" up software so that it is more readible (see [27] and [39]),

(3) perform routine counts of function and derivative evaluation (see [7]),

(4) supply users with large sets of test problems in a standard format (see [7]),

(5) count the number of times each portion of code is executed (see [7] and [75]),

(6) scan and summarize the output of general MP software packages (see [7]).

In addition, researchers at the University of Colorado are developing a package, called TOOLPAC [84], which attempts to incorporate the above described features into one package, thereby simplifying the task. We believe that these packages make large-scale test efforts feasible to a much larger group of researchers.

We would like to conclude this paper with a few observations about testing mathematical programming software. First, testing is extremely costly and time-consuming. Second, one must be very careful in designing the experiment if one hopes to obtain reproducible results (especially with the noted confounding problems attributed to compilers and the computing environment). Third, since the methodologies for testing are still at their embryonic stage, one can expect many alternative approaches to emerge in the future.

On the positive side, there are many researchers--organized under the Committee on Algorithms (COAL) of the Mathematical Programming

Society--who are willing to assist anyone wishing to perform such testing. Test problems and test problem generators have been collected, a survey of available software is currently being taken, guidelines for reporting test results exist, and the Committee is currently investigating the organization of a testing laboratory which would assist in funding large test efforts. Also, the response this Committee has received from the application community has been very positive and supportive. There are many people interested in knowing which of the many software packages available are capable of solving their problems quickly and easily.

	ALL RESPONDENTS		GENERAL NONLINEAR		INTEGER PROGRAMMING		LINEAR PROGRAMMING		LARGE SCALE LP	
No. of Responses	275		58		53		39		38	
	Rank	Score	Rank	Score	Rank	Score	Rank	Score	Rank	Score
CPU Time	1	1.16	1	1.26	1	1.62	4	.85	5	.79
Documentation	2	1.15	3	.98	4	.89	1	1.56	3	1.08
Rel. Dif. Obj. Fn.	3	.93	2	1.05	2	1.28	3	1.03	---	
Cost/Run	4	.89	---		3	1.07	2	1.10	1	1.32
Sensitivity Measures	5	.85	4	.90	---		5	.82	2	1.21
No. Funct. Evals.	---		5	.85	---		---		---	
Max. Rel. Error in Soln. Vec.	---		---		---		---		4	.90
Ease of Modification	---		---			.70	---		---	

TABLE 1

TOP FIVE PERFORMANCE INDICATORS FOR ALL RESPONDENTS AND FOR TOP FOUR MP CLASSES

Source: H. P. Crowder and P. B. Saunders [17].

TABLE 2

TEST PROBLEMS: HAND-PICKED

Title	Number	Year & Reference
Linear Programming:		
IBM, Share Test Problems	5	1963 [114]
McCoy & Tomlin	3	1974 [78]
Staircase LP	1	1978 [55]
Nonlinear Programming:		
Colville	8	1968 [13]
Himmelblau	25	1972 [54]
Bus	16	1978 [10]
Cornwall, et al.	32	1978 [14]
Sandgren	30	1978 [97]
Hock & Schittkowski	120	1979 [56,57]
Geometric Programming:		
Dembo	7	1976 [21]
Rijckaert & Martens	24	1978 [94]
Approximation:		
Nonlinear Least Squares Hiebert	36	1979 [104]
L_1 Linear Approximation Shier, et al.	27	1979 [104]
Nonsmooth Optimization:		
Lemarechal & Mifflin	4	1967 [90]
Integer:		
Capital Budgeting Peterson	7	1967 [90]
General I. P. Trauth & Woolsey	29	1969 [108]
General I. P. Wahi & Bradley	28	1969 [111]
Fixed Charge Gray	8	1974 [50]
General I. P. Haldi	25	1974 [50]
Travelling Salesman Crowder & Padgerg	14	1980 [16]

TABLE 3

TEST PROBLEMS: GENERATORS

(Controlled Characteristics and Known Solutions)

Date	Author	Discipline	Acronym
1974	Klingman, Napier, & Stutz [65]	Network Optimization	NETGEN
1975	Fleisher [36,37]	Plant Location and Capital Budgeting Problems	
1975	Michaels & O'Neill [77]	Linear & Nonlinear Prog.	MPGENR
1976	Schlick & Nazareth [101]	Nonlinear Prog.	
1979	Schittkowski [100]	Signomial (NLP)	
1979	Lin & Rardin [70]	Integer Programming	
1979	Domich, et al. [25]	Linear L_1 Polynomial Approx.	POLY1 & POLY2
1980	Hoffman & Shier [59]	Linear L_1 Approx.	LIGNR
1981	Lidor [69]	Geometric & Signomial	
1981	Chang [11]	Linear L_2 Approximation	
1981	Rosen [92]	Concave Minimization	
1981	Elam & Klingman [31]	Network-based Test Problems	NETGEN II

TABLE 4

TEST PROBLEMS: PERTURBED HAND-PICKED

Date	Author	Title
1976	Scklick & Nazareth [101]	λ-Perturbation of Constrained NLP
1979	Dembo [19]	Perturbed Geometric Problems
1979	More [81]	Perturbed Hand-Picked to Test Starting Points and Scaling
1979	vanderHock & Dijkshoorn [110]	Hand-Picked NLP (Poorly Scaled)

TABLE 5

MAJOR TEST EFFORTS

Networks

Gilsinn and Witzgall [41] (1973)
7 codes; 12 problems

Dembo and Mulvey [23] (1976)
2 codes; 100 problems

Glover and Klingman [24] (1979)
2000 problems; 15 codes

Glover and Klingman [45] (1980)
185 problems; 11 codes

Shier [103] (1980)
220 problems; 5 codes, 3 compilers

Geometric

Dembo [21] (1976)
11 codes; 8 test problems

Rijckaert and Martens [93] (1978)
17 codes; 24 test problems

Fattler, et al. [33] (1981)
20 random starting points; 42 problems; 10 codes with multiple error levels

Nonlinear Programming

Ragsdell and Sandgren [99] (1979)
17 codes; 30 test problems

Schittkowski [100] (1979)
26 codes; 370 random problems; 120 hand-picked problems

van der Hoek [110] (1979)
3 codes; 24 problems

Bus [10] (1980)
18 codes; 24 problems

Approximation

Gilsinn, Hoffman, Jackson, Leyendecker, Saunders, and Shier [40] (1978)
4 codes; 27 hand-picked problems; 1000 randomly generated

Chang [11] (1979)
24 codes; 7000 data sets

Hiebert [52] (1979)
12 codes; 36 problems

8. References

[1] J. Abadie and J. Guigou, "Numerical Experiments with the GRG Method," in F. Lootsma, Numerical Methods for Nonlinear Optimization, Academic Press, New York (1972), pp. 529-536.

[2] E. Balas and E. Zemel, "Al Algorithm for Large Zero-One Knapsack Problems," Operations Research, Vol. 28, No. 5, pp. 1130-1154.

[3] R. Braitsch, "A Computer Comparison of Four Quadratic Programming Algorithms," Management Science, Vol. 15, No. 11 (1972), pp. 631-643.

[4] T. E. Bell, "Computer Performance Variability," Computer Performance Evaluation, National Bureau of Standards Special Publication 406 (August 1975).

[5] M. Bourgoin and E. Heurgon, "Study and Comparison of Algorithms of the Shortest Path Through Planned Experiments," Project Planning by Network Analysis, North-Holland, Amsterdam, Netherlands (1969), pp. 106-118.

[6] E. Brocklehurst and K. Dennis, "A Comparison of Six Algorithms for Dense Linear Programs," NPL Report NAC51, National Physical Laboratory, Teddington, Middlesex, England (June 1974).

[7] A. Buckley, "Testing Minimization Codes," COAL Newsletter #5 (February 1981), pp. 49-52.

[8] J. C. P. Bus, "Performance Measures for Evaluating Mathematical Programming Software," COAL Newsletter (January 1980), p. 8.

[9] J. C. P. Bus, "A Proposal for the Classification and Documentation of Test Problems in the Field of Nonlinear Programming," Design and Implementation of Optimization Software (ed. H. Greenberg), Nato Advance Study Institute Series, Series E--Applied Science, No. 28, pp. 507-518.

[10] J. C. P. Bus, "Numerical Solution of Systems of Nonlinear Equations," Doctoral Thesis presented to the faculty of the University of Amsterdam Mathematics Center, Amsterdam (April 1980).

[11] R. E. Chang, "An Evaluation and Comparison of Curve Fitting Software," SAND80-8727, Sandia National Laboratories, Albuquerque, New Mexico (1981).

[12] H. Chernoff, "The Use of Faces to Represent Points in k-Dimensional Space Graphically," Journal American Statistical Association, Vol. 68 (1973), pp. 361-368.

[13] A. Colville, "A Comparative Study of Nonlinear Programming Codes," IBM Scientific Center Report No. 320-2949 (June 1968), pp. 487-500.

[14] L. W. Cornwall, P. A. Hutchinson, M. Minkoff, and H. K. Schultz, "Test Problems for Constrained Nonlinear Mathematical Programming Algorithms," TM 320, Argonne National Laboratories, Applied Mathematics Division (1978).

[15] H. Crowder, R. S. Dembo, and J. M. Mulvey, "On Reporting Computational Experiments with Mathematical Software," ACM TOMS, Vol. 5 (1979), pp. 193-203.

[16] H. P. Crowder and M. W. Padberg, "Large Scale Symmetric Travelling Salesman Problems," Management Science, Vol. 26, No. 5, pp. 495-509.

[17] H. P. Crowder and P. B. Saunders, "Results of a Survey on MP Performance Indicators," COAL Newsletter (January 1980), pp. 2-6.

[18] H. J. Curnow, and B. A. Wichmann, "A Synthetic Benchmark," The Computer Journal, Vol. 19, No. 1 (February 1976) pp. 43-46.

[19] R. S. Dembo, "Random Generation of Problems with a Given Structure," COAL Newsletter (January 1980), pp. 25-26.

[20] R. S. Dembo, "The Current State-of-the-Art of Algorithms and Computer Software for Geometric Programming," Working Paper 88, School of Organization and Management, Yale University, New Haven, CT (1976).

[21] R. S. Dembo, "A Set of Geometric Programming Test Problems and Their Solutions," Mathematical Programming, Vol. 10 (1976), pp. 192-214.

[22] R. S. Dembo and H. Avriel, "Optimal Design of a Membrane Separation Process Using Signomial Programming," Mathematical Programming, Vol. 15 (1978), pp. 12-25.

[23] R. S. Dembo and J. M. Mulvey, "On the Analysis and Comparison of Mathematical Programming Algorithms and Software," in W. White, Computers and Mathematical Programming, Natl. Bur. Stds. Special Publications 502 (February 1978).

[24] R. Dial, F. Glover, D. Karney, and D. Klingman, "A Computational Analysis of Alternative Algorithms and Labeling Techniques for Finding Shortest Path Trees," Networks, Vol. 9 (1979), pp. 215-248.

[25] P. Domich, J. Lawrence, and D. Shier, "Generators for Discrete Polynomial L_1 Approximation Problems," J. of Res. of the National Bureau of Standards, Vol. 84, No. 6 (1979), pp. 485-487.

[26] J. Dorrenbacher, et al., "POLISH-A FORTRAN Program to Edit FORTRAN Programs," Dept. of Comp. Sci., University of Colorado, Tech. Report #CU-C5-050-76 (Rev.)(May 1976).

[27] J. S. Dyer, "Interactive Goal Programming," Management Science, Vol. 19 (1972), pp. 62-70.

[28] E. D. Eason, "Evidence of Fundamental Difficulties in Nonlinear Optimization Code Comparisons," to appear in the Proceedings of the Conference on Testing and Validating of Algorithms and Software, held in Boulder, CO (January 1981).

[29] E. D. Eason and R. G. Fenton, "A Comparison of Numerical Optimization Methods for Engineering Design," J. of Engineering for Industry, ASME, Vol. 96, No. 1 (1974), pp. 196-200.

[30] E. D. Eason, "Validity of Colville's Time Standardization for Comparing Optimization Codes," ASME Paper No. 77-DET-116 (September 1977).

[31] J. J. Elam and D. Klingman, "Netgen II: A System for Generating Structured Network-based Mathematical Programming Test Problems," to appear in The Proceedings of the Conference on Testing and Validating Algorithms and Software, held in Boulder, CO (January 1981).

[32] M. Enquist, "A Successive Shortest Path Algorithm for the Assignment Problem," Research Report CCS 375, Center for Cybernetic Studies, The University of Texas, Austin, TX 78712.

[33] J. E. Fattler, Y. T. Sin, R. R. Root, K. M. Ragsdell, and G. V. Reklaitis, "On the Computational Utility of Posynomial Geometric Programming Solution Methods," to appear in Mathematical Programming (1981).

[34] D. Fayard and G. Plateau, "Resolution of the Zero-One Knapsack Problem: Comparison of Methods," Mathematical Programming, Vol. 8 (1975), pp. 272-307.

[35] J. Feiber, R. N. Taylor, and L. J. Osterweil, "NEWTON-A Dynamic Testing System for Fortran 77 Programs; Preliminary Report," University of Colorado, Dept. of Comp. Sci., Tech. Note (November 1980).

[36] J. M. Fleisher, "Constructure of Plant Location Test Problems with Known Optimal Solutions," Computer Sciences Technical Report #263, University of Wisconsin, Madison, WI (December 1975).

[37] J. M. Fleisher, "Construction of Generalized Capital Budgeting Test Problems with Known Optimal Solutions," Computer Sciences Technical Report #260, University of Wisconsin, Madison, WI (August 1975).

[38] J. M. Fleisher and R. R. Meyer, "New Sufficient Optimality Condition for Integer Programming and Their Application," Communications of the ACM, Vol. 21, No. 5 (1978), pp. 412-418.

[39] L. D. Fosdick, (ed.) 1979, Performance and Evaluation of Numerical Software, North-Holland Press.

[40] J. Gilsinn, K. Hoffman, R. H. F. Jackson, E. Leyendecker, P. Saunders, and D. Shier, "Methodology and Analysis for Comparison Discrete Linear L_1 Approximation Codes," Communications in Statistics, Simulation, and Computations, Vol. B6, No. 4 (1977), pp. 399-413.

[41] J. Gilsinn and C. Witzgall, "A Performance Comparison of Labeling Algorithms for Calculating Shortest Path Trees," Technical Note 772 (May 1973) Natl. Bur. of Stds., Washington, D.C.

[42] F. Glover, J. Hultz, and D. Klingman, "Improved Computer-Based Planning Techniques, Part 1," Interfaces, Vol. 8, No. 4 (August 1976), p. 16.

[43] F. Glover, D. Karney, D. Klingman, and A. Napier, "A Comparison of Computation Times for Various Starting Procedures, Basis Solution Criteria, and Solution Algorithms for Distribution Problems," Research Report CS#4, Center for Cybernetic Studies, Graduate School of Business, University of Texas (May 1971).

[44] F. Glover, D. Karney, D. Klingman, and A. Napier, "A Comparison Study of Start Procedures, Basis Charge Criteria, and Solution Algorithms for Transportation Problems," Research Report CS#93,

Center for Cybernetic Studies, Graduate School of Business, University of Texas (September 1972).

[45] F. Glover, D. Klingman, J. Mote, and D. Whitman, "Comprehensive Computer Evaluation and Enhancement of Max Flow Algorithms," Research Report CC5356, Center for Cybernetic Studies, Graduate School of Business, University of Texas (1979).

[46] B. Golden, "Shortest Path Algorithms: A Comparison," Operations Research, Vol. 24, No. 6 (Nov.-Dec. 1976), pp. 1164-1168.

[47] J. S. Graves, "An Improved Branch and Bound Algorithm for Integer Programs with Set Partitioning Constants," Ph.D. Thesis presented to the faculty of Carnegie Mellon University, Pittsburgh, PA (1976).

[48] J. S. Graves, "An Approach to Design of Experiments for Computational Testing of Math Programming Algorithms," to appear in the Proceedings of the Conference on Testing & Validating Algorithms and Software, held in Boulder, CO. (January 1981).

[49] P. Gray, "Exact Solution of the Fixed Charge Transportation Problem," Operations Research, Vol. 19, No. 6 (1971), pp. 1529-1537.

[50] J. Haldi, "25 Integer Programming Problems," Working Paper No. 43, Graduate School of Business, Stanford University (1964).

[51] K. L. Hiebert, "An Evaluation of Mathematical Software Which Solves Nonlinear Least Squares Problems," Sandia Report, SAND80-0479 (1980), Sandia National Laboratories.

[52] K. L. Hiebert, "A Comparison of Nonlinear Least Squares Software," Sandia Technical Report, SAND79-0483 (1979), Sandia National Laboratories.

[53] K. Hillstrom, "A Simulation Test Approach to the Evaluation and Comparison of Unconstrained Nonlinear Optimization Algorithms," Argonne National Laboratory Report No. ANL-76-20 (February 1976).

[54] D. M. Himmelblau, Applied Nonlinear Programming, McGraw-Hill, New York (1972).

[55] J. K. Ho and B. Loute, "A Set of Staircase Linear Programming Problems," Applied Mathematics Dept., Brookhaven National Laboratory Upton, New York, 11973 (1979).

[56] W. Hock and K. Schittkowski, "Test Examples for the Solution of Nonlinear Programming Problems, Part 1," Preprint No. 44, Institut fur Angewandte Mathematik and Statistik, Universitat Wurzburg, W. Germany (1979).

[57] W. Hock and K. Schittkowski, "Test Examples for the Solution of Nonlinear Programming Problems, Part 2," Preprint No. 45, Institut fur Angewandte Mathematik and Statistik, Universitat Wurzburg, W. Germany (1979).

[58] A. Hoffman, M. Mannos, D. Sokolowsky, and N. Wiegmann, "Computational Experience in Solving Linear Programs," J. SIAM, Vol. 1 (1953), pp. 17-33.

[59] K. L. Hoffman and D. R. Shier, "A Test Problem Generator for Discrete Linear L_1 Approximation Problems," ACM TOMS Vol. 6, No. 4 (1980), pp. 587-593.

[60] R. H. F. Jackson, "Measures of Similarities for Comparing Test Problems," presented at the 10th International Conference on Mathematical Programming, Montreal, Canada (1979).

[61] R. H. F. Jackson and J. M. Mulvey, "A Critical Review of Comparisons of Mathematical Programming Algorithms and Software (1953-1977)" J. Research of the National Bureau of Standards, Vol. 83, No. 6 (November 1978).

[62] D. Karney and D. Klingman, "Implementation and Computational Study of In-Core, Out-of-Core Primal Network Code," JORSA, Vol. 24, No. 6 (November-December 1976), pp. 1056-1077.

[63] R. Keeney and H. Raiffa, Decision Making with Multiple Objectives: Value and Preference Tradeoffs, John Wiley and Sons (1976).

[64] D. Klingman, A. Napier, and G. Ross, "A Computational Study of the Effects of Problem Dimensions and Solution Times for Transportation Problems," Research Report #CS135, Center for Cybernetic Studies, Graduate School of Business, University of Texas (October 1973).

[65] D. Klingman, A. Napier, and J. Stutz, "NETGEN: A Program for Generalizing Large-Scale Assignment, Transportation, and Minimum Cost-Flow Network Problems," Management Science, Vol. 20 (1974), pp. 814-821.

[66] J. S. H. Kornbluth, "A Survey of Goal Programming," Omega, Vol. 1 (1976), pp. 193-205.

[67] L. Lasdon and D. Waren, "Survey of Nonlinear Programming Applications," Operations Research, Vol. 28, No. 5 (September 1980), pp. 1029-1074.

[68] C. Lemarechal and R. Mifflin, Nonsmooth Optimization, Pergamon Press, New York, 1978.

[69] G. Lidor, "Construction of Nonlinear Programming Test Problems with Known Solution Characteristics," to appear in the Proceedings of the Conference on Testing and Validating Algorithms and Software, Boulder, CO (1981).

[70] B. W. Lin and R. L. Rardin, "The RIP Random Integer Programming Test Problem Generator," Georgia Institute of Technology, Atlanta, Ga. (June 1980).

[71] B. W. Lin and R. L. Rardin, "Controlled Experimental Design for Statistical Comparison of Integer Programming Algorithms," Mgt. Science, Vol. 25, No. 12 (1979), pp. 1258-1271.

[72] F. L. Lootsma, "Performance Evaluation of Nonlinear Programming Codes from the Viewpoint of a Decision Maker," in L. D. Fosdick (ed.) Performance Evaluation of Numerical Software, North Holland, Amsterdam (1979), pp. 285-297.

[73] J. N. Lyness and C. Greenwell, "A Pilot Scheme for Minimization Software Evaluation," Argonne Natl. Lab., Tech. Report II, TM-323 (1977).

[74] J. N. Lyness, "Performance Profiles and Software Evaluation," in Fosdick, L. D. (ed.), Performance and Evaluation of Numerical Software, North-Holland Press (1979).

[75] G. Lyon and R. B. Stillman, "A FORTRAN Analysis," NBS Technical Note 849 (1974), National Bureau of Standards, Washington, D.C. 20234.

[76] J. May and R. Smith, "The Definition and Generation of Geometrically Random Linear Constraint Sets," Tech. Report No. 80-6, Dept. of Industrial and Operations Engineering, College of Engineering, University of Michigan, Ann Arbor, MI (October 1980).

[77] W. M. Michaels and R. P. O'Neill, "A Mathematical Programming Generator: MPGENR," ACM Transactions of Mathematical Software, Vol. 6 (1980), pp. 31-40.

[78] P. F. McCoy and J. A. Tomlin, "Some Experiments on the Accuracy of Three Methods of Updating the Inverse in the Simplex Method," Technical Report No. SOL 72-2, Stanford University, Stanford, California (December 1974).

[79] A Miele and S. Gonzalez, "On the Comparative Evaluation of Algorithms for Mathematical Programming Problems," Nonlinear Programming 3, Ed., O. L. Mangasarian, R. R. Meyer, and S. M. Robinson, Academic Press, New York (1978), pp. 337-359.

[80] J. H. Moore and A. B. Whinston, "Experimental Methods in Quadratic Programming," Management Science, Vol. 13, No. 1 (September 1966).

[81] J. J. More, "Implementation and Testing of Optimization Software," in Fosdick, L. D. (ed.) Performance Evaluation and Testing of Numerical Software (1979) North-Holland Press.

[82] J. A. Nelder, "Experimental Design and Statistical Evaluation," in Fosdick, L. D. (ed.) Performance and Evaluation of Numerical Software, North-Holland Press (1979).

[83] R. P. O'Neill, "A Comparison of Real-World Linear Programs and Their Randomly-Generated Analogs," to appear in the Proceedings of the Conference on Testing and Validating Algorithms and Software, sponsored by the Committee on Algorithms of the Mathematical Programming Society, Boulder, CO (January 1981).

[84] L. J. Osterweil, "TOOLPACK An Integrated System of Tools for Mathematical Software Development," to appear in the Proceedings of the Conference on Testing and Validating Algorithms and Software, held in Boulder, CO (January 1981).

[85] L. J. Osterweil and L. D. Fosdick, "DAVE--A Validation, Error, Detection, and Documentation System for FORTRAN Programs" Software Practice & Experience, Vol. 6 (September 1976), pp. 473-506.

[86] A. M. Ostrowski, Solution of Equations and Systems of Equations, Academic Press, New York (1960).

[87] U. Pape, "Implementation and Efficiency of Moore-Algorithms for the Shortest Route Problem," Mathematical Programming, Vol. 7 (1974), pp. 212-222.

[88] B. N. Parlett and Y. Wong, "The Influence of the Compiler on the Cost of Mathematical Software," ACM TOMS, Vol. 1 (1975), pp. 35-36.

[89] V. Pereyra and R. D. Russell, "Difficulties of Comparing Complex Mathematical Software: General Comments and the BVODE Case," Simon Fraser University, Dept. of Mathematics, Burnaby B. C., Canada.

[90] C. C. Peterson, "Computational Experience with Variants of the Balas Algorithm Applied to the Selection of R&D Projects," Management Science, Vol. 13 (1967), pp. 736-750.

[91] W. J. Riley and R. L. Sielken, "Which Options Provide the Quickest Solution?" to appear in the Proceedings of the Conference on Testing & Validating Algorithms & Software, held in Boulder, CO (1981).

[92] J. B. Rosen, "Global Minimization of a Linearly Constrained Concave Function by Partition of the Feasible Region," (Appendix B), Technical Report 81-11, Computer Science Dept., University of Minnesota, Minneapolis, MN 55455 (1981).

[93] M. J. Rijckaert, "Computational Aspects of Geometric Programming," in Design and Implementation of Optimization Software, ed. H. J. Greenberg, NATO Advanced Study Institute Series, Series E: Applied Science, No. 28, pp. 481-506.

[94] M. J. Rijckaert and X. M. Martens, "Comparison of Generalized Geometric Programming Algorithms," JOTA, Vol. 26, No. 2 (October 1978).

[95] B. G. Ryder, "The FORTRAN Verifier: User's Guide," Bell Laboratories, Computing Science Technical Report #12 (May 1973).

[96] T. L. Saaty, "A Scaling Method for Priorities in Hierarchical Structures," J. Math. Psych., Vol. 15 (1977), pp. 234-281.

[97] E. Sandgren, "The Utility of Nonlinear Programming Algorithms," Ph.D. Dissertation, Purdue University, Lafayette, Indiana (December 1977).

[98] E. Sandgren, "A Statistical Review of the Sandgren-Ragsdell Comparison Study," to appear in the Proceedings of the Conference on Testing & Validating Algorithms & Software, held in Boulder, CO (1981).

[99] E. Sandgren and K. M. Ragsdell, "The Utility of Nonlinear Programming Algorithms: A Comparative Study--Parts 1 & 2," ASME Journal of Mechanical Design, Vol. 102, No. 3 (July 1980), pp. 540-551.

[100] K. Schittkowski, (1978) "A Numerical Comparison of Optimization Software Using Randomly Generated Test Problems," Rechenzentrum Preprint No. 43, University of Wurgburg, West Germany.

[101] F. Schlick and L. Nazareth, "A Performance Profile Study of Three Unconstrained Optimization Routines," in W. White, Computers and Mathematical Programming, National Bureau of Standards, Special Publication 502 (February 1978).

[102]D. Shanno and K. Phua, "Effective Comparison of Unconstrained Optimization Techniques," Management Science, Vol. 22, No. 3 (November 1975), pp. 321-330.

[103]D. R. Shier, "A Computational Study of Floyd's Algorithm," to appear in Computers and Operations Research, 1981.

[104]D. R. Shier, S. J. Neupauer, P. B. Saunders, "A Collection of Test Problems for Discrete Linear L_1 Data Fitting," Natl. Bur. Stds. Tech. Report NBSIR 79-1920, Washington, D.C. (July 1979).

[105]R. L. Staha, "Constrained Optimization Via Moving Exterior Truncations," Ph.D. Thesis, University of Texas, Austin, Texas (1973).

[106]W. R. Stewart, "A Computational Comparison of Five Heuristic Algorithms for the Euclidean Traveling Salesman Problem," to appear in the Proceedings of the Conference on Testing & Validating Algorithms & Software, held in Boulder, CO (1981).

[107]U. Suhl, "Implementation of an Algorithm: Performance Considerations and a Case Study," to appear in the Proceedings of the Conference on Testing & Validating Algorithms & Software, sponsored by the Committee on Algorithms of the Mathematical Programming Society, Boulder, CO (January 1981).

[108]C. A. Trauth and R. E. Woolsey, "Integer Linear Programming: A Study in Computational Efficiency," Management Science, Vol. 15, No. 11 (1969), pp. 481-493.

[109]W. B. Van Dam and J. Telgen, "Randomly Generated Polytopes for Testing Mathematical Programming Algorithms," Report 7929/0, Erasmus University, P.O. Box 1738, Rotterdam, The Netherlands.

[110]G. Van der Hoek and M. W. Dijkshoorn, A Numerical Comparison of Self-Scaling Variable Metric Algorithms," Report 7910/0, Erasmus University, P.O. Box 1738, Rotterdam, The Netherlands.

[111]P. N. Wahi and G. H. Bradley, "Integer Programming Test Problems," Report #28, Administrative Sciences, Yale University, New Haven, Connecticut (1969).

[112]A. D. Waren and L. S. Lasdon, "The Status of Nonlinear Programming Software," Operations Research, Vol. 27, No. 3 (1979), pp. 431-456.

[113]K. L. Weldon, "One-to-One Graphical Strategies in Multiwarrate Data Analysis," Manuscript (1981).

[114]P. Wolfe and L. Cutler, "Experiments in Linear Programming," in R. Graves and P. Wolfe (eds.), Recent Advances in Mathematical Programming, McGraw-Hill, New York (1963), pp. 177-200.

[115]S. Zionts and J. Wallenius, "An Interactive Programming Method for Solving Multiple Criteria Problem," Management Science, Vol. 22 (1976), pp. 652-663.

NONLINEAR PROGRAMMING METHODS WITH LINEAR LEAST SQUARES SUBPROBLEMS

by

Klaus Schittkowski
Stanford University

Abstract

This paper presents the results of an extensive comparative study of nonlinear optimization algorithms, cf. [8]. This study indicates that quadratic approximation methods, which are characterized by solving a sequence of quadratic subproblems recursively, belong to the most efficient and reliable nonlinear programming algorithms available at present. The purpose of the paper is to investigate their numerical performance in more detail. In particular, the dependence of the overall performance on alternative quadratic subproblem strategies is tested. The paper indicates how the efficiency of quadratic approximate methods can be improved.

1. Introduction

The underlying mathematical model is the constrained nonlinear programming problem,

$$x \in \mathbb{R}^n: \quad \begin{array}{ll} \min & f(x) \\ g_j(x) = 0 \;, & j=1,\ldots,m_e \;, \\ g_j(x) \geq 0 \;, & j=m_e+1,\ldots,m \;, \\ x_l \leq x \leq x_u \;, & \end{array} \qquad (1)$$

with continuously differentiable functions f and g_j, $j=1,\ldots,m$. Without loss of generality, upper and lower bounds, x_l and x_u respectively, are allowed.

The extensive numerical tests of a comparative study of optimization codes for (1) prove the outstanding efficiency of the quadratic approximation methods and, in particular, of the method of Wilson, Han, and Powell, cf. [8]. In these algorithms, sometimes denoted as recursive quadratic programming, variable metric, or projected Lagrangian methods, a solution of (1) is obtained by solving a sequence of quadratic programs recursively, in which the Lagrange function of (1) is approximated quadratically subject to the linearized constraints. To get more insight into their numerical behavior, the dependence of the quadratic subproblem on efficiency and reliability of the solution method shall be investigated. A further intention is to compare more precisely the algorithm of Bartholomew-Biggs (XROP) [1] to that of Powell (VF02AD) [6] in terms of efficiency. Although XROP requires more than twice as many function evaluations as VF02AD to solve the test problems of [8], the program is much faster than VF02AD, cf. [8, Section 2 of Chapter V]. For this reason, some modifications of the method of Wilson, Han, and Powell are presented with the intention of reducing the calculation time of the algorithm and to study the effect of different formulation attempts and solution procedures for the quadratic subproblems.

Section 2 describes the algorithm in more detail. It is shown how the quadratic subproblem using the method of Wilson, Han, and Powell can be replaced by a linear least squares problem which takes advantage of the fact that successive subproblems are distinguished only by the constraints and a rank-two-correction of the objective function. The least squares problem can be solved, for example by the methods of Schittkowski, Stoer [9], or of Lawson, Hanson [5]. The implementation of four different versions is described in Section 3. In one of these versions, the derivatives are calculated numerically by forward differences. The resulting programs are tested according to the framework presented in [8]. Each implementation had to solve the same test problems and the results are evaluated in the same fashion as that presented in [8]. The corresponding efficiency and reliability scores are presented in Section 4 together with some more detailed discussions which present reasons for outcomes obtained. Conclusions are summarized in Section 5.

2. Description of the Algorithm

The method of Wilson, Han, and Powell defines a quadratic subproblem which, in each iteration step, is the minimization of approximation of the Lagrange function subject to the linearized constraints of (1). Given a current iterate x, a correction $\bar{d}$ is obtained by solving

$$\begin{aligned} & \min \ 1/2\, d^T B d + \nabla_x f(x)^T d \\ d \in \mathbb{R}^n: \quad & g_j(x) + \nabla_x g_j(x)^T d = 0 \ , \quad j=1,\ldots,m_e \ , \\ & g_j(x) + \nabla_x g_j(x)^T d \geq 0 \ , \quad j=m_e+1,\ldots,m \ . \end{aligned} \tag{2}$$

The method was first studied by Wilson [10] with

$$B:=\nabla_x^{\,2} L(x,u) \ , \tag{3}$$

where $L(x,u)$ denotes the Lagrange function of (1). The optimal Lagrange multipliers of (2) are used as guesses for the multipliers of the original problem for defining the matrix B in the successive subproblem. Han [4] replaced the Hessian of $L(x,u)$ by a quasi-Newton update procedure and proved the local superlinear convergence. Powell [6] implemented the method in an efficient way. In particular, he updated B by the BFGS-formula and introduced a line-search to determine the new iterate in the form $\bar{x} := x + \bar{\alpha}\bar{d}$, where $\bar{d}$ is the solution of (2) and $\bar{\alpha}$ the corresponding steplength parameter. Some local convergence properties of this method are given in Powell [7].

The fundamental difference of the algorithm proposed in this paper to the method of Wilson, Han, and Powell is that the quadratic subproblem (2) is replaced by a linear least squares problem which can be solved by the methods of Schittkowski, Stoer [9], or Lawson, Hanson [5]. Efficient solution to the subproblem is possible, only if the approximation matrix B of the Hessian of the Lagrange function is factorized in an appropriate way. Using the LDL-decomposition

$$B = LDL^T \tag{4}$$

with a lower triangular matrix L and a diagonal matrix D, it is easy to see that (2) is equivalent to the least squares problem

$$\begin{array}{ll} & \min \; \| D^{1/2}L^T d + D^{-1/2}L^{-1}\nabla_x f(x) \| \\ d \in \mathbb{R}^n: & g_j(x) + \nabla_x g_j(x)^T d = 0 \quad , \quad j=1,\ldots,m_e \quad , \\ & g_j(x) + \nabla_x g_j(x)^T d \quad 0 \quad , \quad j=m_e+1,\ldots,m \quad , \\ & x_l - x \le d \le x_u - x \quad , \end{array} \tag{5}$$

where $D^{1/2} := \text{diag}\,(\sqrt{\delta_1}\,, \ldots , \sqrt{\delta_n})$ and δ_i is the i-th diagonal element

the matrix $D^{1/2}$ is well-defined since it is guaranteed that B is always positive definite. In contrast to (2), the upper and lower bounds x_l and x_u, respectively, are treated separately and are not defined in the form of general constraints as required in Powell's method, cf. [6].

The usage of a subproblem of the form (2) or (5) is restricted by the fact that the linear constraints can become inconsistent. In a way similar to Powell [6], this situation is avoided by introducing an additional variable η leading to the following extended least squares program:

$$\min \quad \left\| \begin{pmatrix} D^{1/2}L^T & : & 0 \\ 0 & : & \rho \end{pmatrix} \begin{pmatrix} d \\ \eta \end{pmatrix} + \begin{pmatrix} D^{-1/2}L^{-1}\nabla_x f(x) \\ 0 \end{pmatrix} \right\|$$

$$\begin{pmatrix} d \\ \eta \end{pmatrix} \in \mathbb{R}^{n+1}: \quad \begin{array}{llll} (\nabla_x g_j(x) \; : \; -g_j(x))\begin{pmatrix} d \\ \eta \end{pmatrix} + g_j(x) & = & 0 \;, & j=1,\ldots,m_e \;, \\ (\nabla_x g_j(x) \; : \; \sigma_j g_j(x))\begin{pmatrix} d \\ \eta \end{pmatrix} + g_j(x) & \geq & 0 \;, & j=m_e+1,\ldots,m \;, \end{array} \tag{6}$$

$$x_l - x \leq d \leq x_u - x \;.$$

The coefficients in the inequality constraints are defined by the instruction

$$\sigma_j := \begin{cases} -1 \;, & \text{if} \;\; g_j(x) \quad 0 \;, \\ 0 \;, & \text{otherwise} \;, \end{cases} \qquad j=m_e+1,\ldots,m \;.$$

Obviously, the point

$$d_o := \begin{pmatrix} 0 \\ 1 \end{pmatrix} \tag{7}$$

satisfies the linear constraints in (6) and, in addition, d_o can be used as a feasible starting point for a least squares algorithm. The square of the objective function in (6) can be written in the form

$$||D^{1/2}L^T d + D^{-1/2}L^{-1}\nabla_x f(x)||^2 + \rho^2\eta^2 \tag{8}$$

together with a constant term and shows that ρ plays the role of a penalty parameter forcing η to be as small as possible.

Subsequently, the new iterate is formed with

$$\bar{x}: x + \bar{\alpha}\bar{d} \tag{9}$$

and a steplength parameter $\bar{\alpha} > 0$. Since numerical experiments indicate that in most cases, the steplength $\bar{\alpha} = 1$ is chosen, the line-search procedure of Powell [6] is adopted without any alterations. The choice of $\bar{\alpha}$ is based on minimizing the nondifferentiable L_1-penalty function by a quadratic interpolation.

For the construction of the matrix B, the BFGS-formula is used together with the modifications proposed by Powell [6]. But instead of calculating the new matrix $\bar{B}$ in the form

$$\bar{B} = B + vv^T - ww^T \;, \tag{10}$$

the updating process of Gill, Murray, and Saunders [3] is applied to get the new LDL-factors of $\bar{B}$, i.e. of

$$\bar{B} = \bar{L}\bar{D}\bar{L}^T \;. \tag{11}$$

3. Implementation of Two Different Least Squares Algorithms

Besides the interchanging of the subproblem and the handling of upper and lower bounds separately, the proposed algorithm differs in further numerical details from Powell's program VF02AD. In particular, we describe the implementation of two different least squares codes: the projection method of Schittkowski, Stoer [9] and the dual method of Lawson, Hanson [5].

VF02AD requires the evaluation of the objective function, the constraints and all of the gradients within one block; this leads to some unnecessary gradient calls whenever a line-search is performed with $\bar{\alpha} \neq 1$. For this reason, we have implemented the algorithm described in the last section, so that the objective function and the constraints are processed by a call to one subroutine and their gradients with a call to a separate subroutine. In addition, an 'active set' strategy is included in the following sense: Since we suppose that in the neighborhood of an optimal solution, we can disregard the non-active constraints, only those gradients of the restrictions are recalculated in (6) for which $u_j \neq 0$ or $g_j(x) < 0$.

The implementation of the least squares code of Lawson and Hanson [5] requires additional modifications to the subproblems (5) or (6), respectively. For simplicity, we define the problem as

$$d \in \mathbb{R}^{n'}: \quad \begin{aligned} &\min \|Ud + U^{-T}a\| \\ &A_1 d + b_1 = 0 \quad , \\ &A_2 d + b_2 \geq 0 \quad , \\ &d_l \leq d \leq d_u \end{aligned} \qquad (12)$$

with an upper triangular (n',n') matrix U, a (m_e,n') matrix A_1, a $(m-m_e,n')$ matrix A_2, and some vectors $a \in \mathbb{R}^{n'}$, $b_1 \in \mathbb{R}^{m_e}$, $b_2 \in \mathbb{R}^{m-m_e}$, d_l $\mathbb{R}^{n'}$, d_u $\mathbb{R}^{n'}$. The dimension n' is either equal to n or $n+1$, respectively.

The first step in the method of Lawson and Hanson consists of reducing (12) to an equivalent least distance problem with equality and inequality constraints. By defining

$$z := Ud + U^{-T}a,$$

(12) can be written in the desired format:

$$z \in \mathbb{R}^{n'}: \quad \begin{aligned} &\min \|z\| \\ &A_1U^{-1}z - A_1U^{-1}U^{-T}a + b_1 = 0 , \\ &A_2U^{-1}z - A_2U^{-1}U^{-T}a + b_2 \geq 0 , \\ &U^{-1}z - U^{-1}U^{-T}a - d_l \geq 0 , \\ &-U^{-1}z + U^{-1}U^{-T}a + d_u \geq 0 . \end{aligned} \qquad (13)$$

Householder transformations are subsequently used to eliminate the equality constraints. The dual problem of the resulting least distance problem with inequality constraints is then constructed. It is defined in the form of a linear least squares problem restricted only by lower bounds. This dual problem is solved by subroutine NNLS of Lawson and Hanson [5].

The formulation of subproblem (13) requires the inversion of triangular matrix U leading to the objection that numerical instabilities could occur. To avoid this situation, one proceeds by producing an alternative approximation to the general nonlinear programming problem (1). Instead of approximating the Hessian matrix of the Lagrangian, it is possible to update its inverse by the formula

$$\begin{aligned}\bar{H} &:= H + \left(1 + \frac{q^T H q}{s^T q}\right)\frac{1}{s^T q}\, s s^T - \frac{1}{s^T q}(s q^T H + H q s^T) \\ &= H + \varkappa \left(\frac{1}{\varkappa s^T q}\, s - Hq\right)\left(\frac{1}{\varkappa s^T q}\, s - Hq\right) - \varkappa H q q^T H\end{aligned} \tag{14}$$

with $\varkappa := \dfrac{1}{s^T q + q^T H q}$. It is easy to see that $\bar{H} = \bar{B}^{-1}$, whenever $H = B^{-1}$.

To guarantee that $s^T q > 0$, Powell's [6] modification of the BFGS-formula is adopted to the inverse case. Again, the matrix H is stored in factorized form

$$H = SES^T \tag{15}$$

with a lower triangular matrix S and a diagonal matrix E. Instead of computing the inverse BFGS-formula (14), the update technique of Gill, Murray, and Saunders [3] is again used to determine the new decomposition,

$$H = \bar{S}\bar{E}\bar{S}^T \quad .$$

The quadratic subproblem (2) is equivalent to

$$\begin{aligned} & \min \ \| E^{-1/2} S^{-1} d + E^{1/2} S^T \nabla_x f(x) \| \\ d \in \mathbb{R}^n: \quad & g_j(x) + \nabla_x g_j(x)^T d = 0 \ , \quad j=1,\dots,m_e , \\ & g_j(x) + \nabla_x g_j(x)^T d \geq 0 \ , \quad j=m_e+1,\dots,m , \\ & x_l - x \leq d \leq x_u - x \ . \end{aligned} \tag{16}$$

If we write (16) or the corresponding extended problem for avoiding inconsistency in the form,

$$\min \; \|T^{-1}d + T^{T}a\|$$

$$d \in \mathbb{R}^{n'}: \quad \begin{aligned} A_1 d + b_1 &= 0 , \\ A_2 d + b_2 &\geq 0 , \\ d_l \leq d &\leq d_u , \end{aligned} \tag{17}$$

with a lower triangular matrix T, then it is possible to construct an equivalent least distance problem without inverting a matrix:

$$\min \; \|z\|$$

$$z \in \mathbb{R}^{n'}: \quad \begin{aligned} A_1 Tz - A_1 TT^{T}a + b_1 &= 0 , \\ A_2 Tz - A_2 TT^{T}a + b_2 &\geq 0 , \\ Tz - TT^{T}a - d_l &\geq 0 , \\ -Tz + TT^{T}a + d_u &\geq 0 . \end{aligned} \tag{18}$$

Note that the introduction of an additional variable η replaces the matrix T^{-1} in (17) by the extended matrix

$$\begin{pmatrix} T^{-1} & : & 0 \\ 0 & : & \rho \end{pmatrix} ,$$

cf. (6), and that the least distance problem (18) has to be formulated with

$$\begin{pmatrix} T & : & 0 \\ 0 & : & 1/\rho \end{pmatrix}$$

instead of T.

So far, three different modifications of Powell's optimization method have been described. A fourth modification is obtained by calculating the first derivatives numerically. To sum up, the resulting four programs are designated in the following way:

NLPLSA: The extended least squares subproblem (6) is formulated and solved by LCLSQ, a FORTRAN program of the method of Schittkowski, Stoer [9]. LCLSQ has been implemented by R.L. Crane, K.E. Hillstrom, and M. Minkoff of the Argonne National Laboratory. A feasible starting point $d_o = (0 : 1)^T$ can always be provided.

NLPLSB: The program is identical with NLPLSA with the exception that all gradients are computed numerically by a forward difference formula of the kind

$$\frac{\partial}{\partial x_i} f(x) \approx \frac{1}{h}(f(x + he_i) - f(x)) \qquad (19)$$

with $h := 10^{-6}$, and e_i denotes the i-th axis vector, $i=1,\ldots,n$.

NLPLSC: This program uses the BFGS-update formula to construct a least distance subproblem of the form (13). The additional variable η is introduced if and only if the first attempt to solve (5) failed because of inconsistency. The dual problem is defined and solved by subroutine NNLS of Lawson and Hanson [5]. Since initial experimentation showed numerical instabilities with this approach, a restart is introduced whenever a solution process has been terminated by a failure, e.g. uphill search direction or division by zero in (10). In addition, it turned out that the objective function and constraints had to be scaled to improve the reliability of the code. The inversion of the triangular matrix U in (13) is performed in double precision arithmetic which is also introduced for some critical phases of the least squares algorithm.

NLPLSD: The code is similar to NLPLSC, but uses the inverse BFGS-formula (14) to construct subproblems of the kind (16) or (18), respectively. These subproblems are solved by the method of Lawson and Hanson [5] too. The avoidance of inconsistency and restart capabilities are taken from NLPLSC.

4. Numerical results

The four programs realizing the least squares-based concept to solve nonlinear optimization problems are tested in the framework developed for the comparative study [8], where 26 nonlinear programming codes were investigated. In particular, they had to solve the same randomly generated test problems with predetermined solutions and the performance is evaluated in the same manner, i.e. by the same nine performance criteria. It is outside the scope of this paper to repeat the organizational and computational details of [8], and the reader is referred to that report to get more information about the codes and the evaluation of the performance criteria.

Table 1 shows the efficiency results from [8] for all optimization programs under consideration and Table 2 the corresponding reliability scores. The following abbreviations are used:

ET: Average execution time in seconds.

NF: Average number of objective function calls.

NG: Average number of restriction function calls (each constraint counted).

Code	ET	NF	NG	NDF	NDG
OPRQP	22.6	58	599	40	418
XROP	13.9	40	357	29	270
VF02AD	31.5	16	179	16	179
NLPLSA	17.6	23	219	19	67
NLPLSB	21.9	223	916	0	0
NLPLSC	16.8	25	275	19	83
NLPLSD	13.3	23	244	18	73
GRGA	37.7	204	2946	67	378
OPT	62.5	742	7528	0	321
GRG2(1)	52.6	297	3368	38	423
GRG2(2)	75.0	757	7677	0	0
VF01A	42.2	158	1595	158	603
LPNLP	57.4	252	2518	101	1014
SALQDR	41.6	120	1096	117	1068
SALQDF	88.2	1228	10731	0	0
SALMNF	132.0	78	4610	420	4610
CONMIN	99.9	955	9550	62	618
BIAS(1)	66.7	533	6621	65	729
BIAS(2)	148.1	1805	20901	0	0
FUNMIN	98.8	519	5023	112	1097
GAPFPR	73.1	147	1414	147	1414
GAPFQL	64.3	149	1439	149	1439
ACDPAC	70.5	222	4094	146	748
FMIN(1)	118.8	737	7027	158	1300
FMIN(2)	158.5	1168	18405	0	0
FMIN(3)	169.1	319	3242	620	6211
NLP	88.1	1043	8635	111	957
SUMT	270.1	2335	24046	99	1053
DFP	124.2	782	8214	107	1107
FCDPAK	23.1	125	480	63	251

Table 1: Efficiency.

Code	PNS	FFV	FVC	F
OPRQP	23.2	.27E-7	.18E+1	1
XROP	31.2	.15E-6	.14E+2	0
VF02AD	6.2	.41E-1	.84E-4	5
NLPLSA	13.6	.82E-9	.39E-2	9
NLPLSB	16.0	.52E-9	.36E-2	9
NLPLSC	12.5	.43E-7	.18E-3	0
NLPLSD	15.0	.12E-3	.20E-3	0
GRGA	12.1	.12E-2	.22E-7	3
OPT	44.7	.41E-1	.76E-7	7
GRG2(1)	10.1	.99E-1	.35E-4	0
GRG2(2)	18.3	.53E-6	.75E-3	9
VF01AD	23.9	.46E-6	.49E+1	13
LPNLP	26.7	.65E-7	.18E+3	10
SALQDR	50.7	.25E-3	.35E-4	13
SALQDF	48.7	.69E-3	.29E-4	15
SALMNF	15.6	.34E-5	.67E-3	18
CONMIN	57.6	.53E-4	.43E-3	14
BIAS(1)	25.0	.26E-4	.52E-6	4
BIAS(2)	30.8	.33E-4	.36E-5	13
FUNMIN	32.9	.25E-2	.98E-1	9
GAPFPR	21.4	.45E-2	.23E-7	24
GAPFQL	16.0	.14E-6	.49	30
ACDPAC	10.6	.96E-4	.74E-4	8
FMIN(1)	35.0	.35E-7	.16E-1	0
FMIN(2)	52.5	.52E-6	.85E-2	0
FMIN(3)	22.1	.19E-5	.87	15
NLP	15.6	.28E-7	.29E+1	31
SUMT	69.9	.11E-2	.40E-6	19
DFP	31.9	.27E+6	.80E+2	34
FCDPAK	23.9	.45E-4	.43E+1	2

Table 2: Reliability.

NDF: Average number of gradient calls for objective function.

NDG: Average number of gradient calls for restriction functions (each constraint counted).

PNS: Percentage of non-successful test runs.

FFV: Average objective function value for all non-successful test runs.

FVC: Average sum of constraint violations for all non-successful test runs.

F: Number of failures, i.e. number of problems which could not be solved due to overflow, exceeding calculation time, etc..

Since the programs NLPLSA, ..., NLPLSD use the same stopping criteria as VF02AD, the final accuracy is identical. It is therefore not necessary to relate the efficiency scores to the achieved accuracy in the scope of this report. The results for the remaining seven performance criteria are omitted since in most cases, they are very similar to the results obtained by VF02AD. If significant differences are observed, they are summarized at the end of this section.

Initial experimentation with different modifications of the method of Wilson, Han, and Powell showed that the performance of a quadratic approximation algorithm is very sensitive to the solution method used in the quadratic or least squares subproblem. This observation lead to the additional implementation details such as restart, scaling, or mixed precision capabilities. Nevertheless, the tested versions required a higher number of function or gradient evaluations and are not as reliable as Powell's VF02AD in terms of the percentage of non-successful solutions. One reason for the better performance of Powell's code may be the usage of double precision arithmetic for evaluating inner products in subroutine VE02AD which implements Fletcher's [2] quadratic programming method. On the other side, NLPLSA, NLPLSC, and NLPLSD are much faster than VF02AD, and if one of these codes fails, we can expect a better approximation of the optimal objective function value, than that provided by the starting point.

To investigate these observations in more detail, let us first consider the numerical results of the 5-th test problem class 5A, where problems with 8 variables, 10 constraints, and 4 active constraints are gathered, cf. [8]. Table 3 lists the average accuracy scores defined by

FV: Average objective function value (geometric mean),

VC: Average sum of constraint violations (geometric mean),

Code	VF02AD	NLPLSA	NLPLSB	NLPLSC	NLPLSD	XROP
FV	.60E-8	.49E-8	.45E-8	.99E-8	.91E-8	.18E-7
VC	.10E-11	.22E-11	.13E-10	.13E-11	.18E-11	.79E-8
KT	.11E-3	.10E-3	.97E-4	.16E-3	.15E-3	.48E-4
ED	5.30	5.34	5.33	5.16	5.17	5.48
ET	31.0	12.4	13.6	7.7	6.7	12.2
NF	19	19	188	20	20	24
NG	191	198	531	204	204	244
NDF	19	18	0	19	19	23
NDG	191	37	0	38	38	238

Table 3: Numerical results for test problem class 5A.

KT: Average norm of the Kuhn-Tucker vector (geometric mean),

ED: Average number of exact digits,

and the average efficiency scores ET, NF, NG, NDF, NDG, as described above. For the sake of completeness, the results for XROP are also included. All problems were solved successfully. Since all of the different realizations of the method of Wilson, Han, and Powell required about the same number of iterations, we are able to compare the calculation time more precisely: ET could be reduced from 31.0 sec for VF02AD to 6.7 sec for NLPLSD. But this drastic reduction of the calculation time could only be observed for problems where numerical difficulties in the quadratic subproblem do not occur and where therefore the same iteration process is performed in each version.

The program NLPLSA using the least squares code LCLSQ of Crane, Hillstrom, and Minkoff is slower than the implementations based on the Lawson, Hanson method. One reason is that NLPLSA and NLPLSB must always solve the extended subproblem (6) because of unreliabilities in the feasible point generator. In most cases, the subproblem is therefore a higher dimensional one than that used in NLPLSC and NLPLSD. Furthermore, we can observe that the usage of the inverse BFGS-formula in NLPLSD did reduce the calculation time, but did not improve the reliability of NLPLSC. The reason for both attributes is the fact that the inversion of the triangular matrix U in (13) is performed in double precision arithmetic.

The evaluation of the remaining seven performance criteria did not lead to significant differences between the quadratic approximation methods and is therefore omitted. There is only one exception: The programs VF02AD, XROP, NLPLSA, and NLPLSB are much more sensitive to the position of the starting point than NLPLSC and NLPLSD which do not show any significant alteration of their efficiency scores if a starting point is chosen far away from the optimal solution. This effect can be explained by the additional scaling of the problem functions in the beginning of the solution process.

5. Conclusions

This paper represents the results of computational tests and comparisons of different attempts to implement the quadratic approximation method of Wilson, Han, and Powell. Four modifications of this method are implemented by the author which are characterized by the formulation of a least squares subproblem in each iteration step. Different algorithms for the solution of the subproblem are investigated. All programs are tested in the framework developed for the comparative study [8] and compared with the results presented there. The main conclusions are the following ones:

1. All realizations of the method of Wilson, Han, and Powell are very efficient, in particular with respect to the number of function calls. The least squares based modifications are faster than Powell's VF02AD.
2. The least squares based methods have some numerical difficulties in solving ill-behaved problems compared with VF02AD which uses a robust subroutine for the quadratic subprogram (when it is implemented with double precision inner products).
3. For well-behaved problems, we can expect the same iteration process for all modifications under consideration, but the calculation times of NLPLSA, ..., NLPLSD are very low compared with that of VF02AD.
4. All tested versions of the method of Wilson, Han, and Powell are very reliable compared with the other programs of [8].
5. The method of Wilson, Han, and Powell is very robust with respect to numerical differentiation of the problem functions in contrast to most of the other methods tested in [8] using numerical derivatives.

References

[1] M.C. Bartholomew-Biggs,"An improved implementation of the recursive quadratic programming method for constrained minimization," Technical Report No. 105, Numerical Optimisation Centre, The Hatfield Polytechnic, Hatfield, England, 1979.

[2] R. Fletcher, "A general quadratic programming algorithm," Journal of the Institute of Mathematics and its Applications, 7, 1971, pp. 76-91.

[3] P.E. Gill, W. Murray, M.A. Saunders, "Methods for computing and modifying the LDL factors of a matrix," Mathematics of Computation, 29, 1975, pp. 1051-1077.

[4] S.P. Han, "Superlinearly convergent variable metric algorithms for general nonlinear programming problems," Mathematical Programming, 11, No. 3, 1976, pp. 263-282.

[5] C.L. Lawson, R.J. Hanson, Solving Least Squares Problems, Prentice Hall, Englewood Cliffs, New Jersey, 1974.

[6] M.J.D. Powell, A Fast Algorithm for Nonlinearly Constrained Optimization Calculations, in: Proceedings of the 1977 Dundee Conference on Numerical Analysis, Lecture Notes in Mathematics, Springer-Verlag, Berlin, Heidelberg, New York, 1978.

[7] M.J.D. Powell, "The convergence of variable metric methods for nonlinearly constrained optimization calculations," in: Nonlinear Programming, 3, O.L. Mangasarian, R.R. Meyer, S.M. Robinson, eds., Academic Press, New York, San Francisco, London, 1978.

[8] K. Schittkowski, Nonlinear Programming Codes - Information, Tests, Performance, Lecture Notes in Economics and Mathematical Systems, No. 183, Springer-Verlag, Berlin, Heidelberg, New York, 1980.

[9] K. Schittkowski, J. Stoer, "A factorization method for the solution of constrained linear least squares problems allowing subsequent data changes," Numerische Mathematik, 31, Fasc. 4, 1979, pp. 431-463.

[10] R.B. Wilson, A Simplicial Algorithm for Concave Programming, Ph.D. Dissertation, Graduate School of Business Administration, Harvard University, Boston, 1963.

AN OUTLINE FOR COMPARISON TESTING OF MATHEMATICAL SOFTWARE -- ILLUSTRATED BY COMPARISON TESTINGS OF SOFTWARE WHICH SOLVES SYSTEMS OF NONLINEAR EQUATIONS *

K.L. Hiebert

Sandia National Laboratories +
Albuquerque, New Mexico 87185

ABSTRACT

This paper presents an outline for doing comparison testing of mathematical software. The outline is illustrated with examples from two comparison testings. One comparison tested software that solves nonlinear least squares problems, while the other tested software that solves square systems of equations.

Introduction

This paper is the accumulation of experience gained in doing two comparison testings of mathematical software which solve systems of nonlinear equations. One comparison tested software which solves nonlinear least squares problems. The other tested software which solves only square systems of equations, i.e. finds a zero for all the functions. The comparisons are used as examples. For specific details of the testing and the results, see [2], [3] and [5].

Much of what is said has been generally accepted in testing and has generally been done. Points are made that someone who has not been involved with testing software may have never considered. And the paper emphasizes aspects of comparison testing that have been neglected.

The paper presents an outline for comparison testing of mathematical software that is illustrated with examples from the two comparisons. It is hoped that the outline and the examples will make the job of testing mathematical software easier.

---

* This article sponsored by the U.S. Department of Energy under Contract DE-AC04-76DP00789.

+ A.U.S. Department of Energy Facility

Outline

I. The Purpose of the Testing
   A. Decide on the purpose of the testing.
   B. Decide on which codes to test.
   C. Establish specific goals.

II. The Design of the Testing
   A. Design the tests to demonstrate the specific goals.
   B. Implement the tests in an equitable manner (note limitations in the testing).

III. The Analysis of the Results
   A. Collecting the data.
   B. Displaying the data.
   C. Modifying results because of limitations in the testing.

I. The Purpose of the Testing

The most common purpose for testing mathematical software is to determine the "best" code from a group of codes being tested. There are many difficulties with this kind of testing. Probably the most important difficulty is defining exactly what is meant by the "best." The criteria for "best" depend on many different factors, e.g., the intended problem set, the intended user-group, even the intended machine. The criteria for determining the "best" must be established before the testing begins, because the criteria should play a major part in determining the whole testing procedure. The criteria for "best" should also be clearly stated in the report of the comparison. By giving the reader the criteria for "best," the reader may disagree with the criteria but should not be able to disagree with the results of the report.

Generally there is a specific reason for finding the "best" code. This specific reason can (should) affect the criteria for determining the "best." For example, the reason for the two comparison testings we made was to be able to advise a library committee on which codes should be included in their library. The library is a general purpose library, something that is available to both experienced and novice users. The criteria for the "best" code could be quite different if the purpose of the testing was to find a special purpose code to be part of a much larger code, where once installed solves a very special type of problem and is never tampered with again.

The criteria for "best" may actually determine which codes are to be included in the testing. Again, as an example, for both comparisons the codes needed to be general purpose, high quality, user-oriented, easily available and portable FORTRAN subprograms. Twelve such codes that solve nonlinear least squares problems (NLLS) and eight codes that solve nonlinear square systems of equations (NLSS) were collected. Some of the codes could not be added to the library (because of proprietary rights), but were included in the testing because they met the other specifications. In this case, these specifications were not used for determining the "best," but rather as minimum conditions which the codes had to meet to be considered. The names of the codes collected, as well as where they were developed and the method they implement are listed in Table 1. As another example, the criteria for "best" may limit the appropriate set of codes to only one code. Of the 12 NLLS codes only one code, NS03A, took advantage of the sparsity of the problem. If the criterion for "best" was that the code must be able to handle very large problems, which are probably sparse, then the best code based on that criterion would be NS03A because it is the only code that satisfied that criterion.

Establishing the criteria for the "best" determines goals for the testing. In our testing the criteria for "best" were based on 1) robustness -- the ability to find solutions, 2) reliability -- the ability to correctly identify solutions and 3) efficiency -- execution time and the number of function evaluations. Establishing these criteria determines that the goals of the testing should be to measure the robustness, reliability and efficiency of these codes.

Originally the purpose of testing these codes was to determine the "best" code for a user's library, but because of the codes selected, the criteria used for "best" and the manner in which the testing was done and the data recorded, additional analysis was possible, e.g., comparing the basic methods implemented by the codes or analyzing special features of some of the codes. This aspect of comparison testing is important because it can influence the development of future codes. Also, the information from the testing is complete enough that others can take this data, apply different criteria for "best" and come up with their own conclusions.

II. The Design of the Testing

The design of the testing is very important to the testing procedure. First, the tests must be designed to determine how well the codes perform with respect to the criteria for the "best." Considerations for the design may include the general class of problems that the codes are intended to handle, extreme conditions that the codes may not claim to handle but that do exist, basic requirements for the codes and special features or capabilities of some of the codes. Second, the tests must be implemented in such a manner that all the codes have equivalent problems to solve.

In our testing, the basic design was to measure the performance of the codes, i.e., the robustness, reliability and efficiency, on a large variety of problems and under different adverse conditions. The design was reasonable because of the intended problem set and user-group.

From the literature, a set of "standard" problems was collected as the nominal set. (Another difficulty is what constitutes "standard" problems.) There were 36 problems for the NLLS codes and 23 problems for the NLSS codes.

Both the NLLS and NLSS codes require an initial starting value for the variables. Most of the algorithms implemented by the codes are based on Newton's method. Therefore, only local convergence is guaranteed, i.e., if the starting value is "close enough" to the solution, the algorithm will converge to the solution. Recent developments have tried to improve the global convergence of the algorithms, i.e., convergence to a solution for all starting values. Therefore it seemed reasonable to ask how the codes would perform with different starting values. To test the robustness of the codes with respect to the starting values for some of the problems three different starting values, x, the "standard" starting value, 10 * x, and 100 * x were used. (For problems involving rapidly increasing functions, such as exponentials, only the standard starting value was used to prevent overflow conditions.) In this manner, a set of 72 NLLS problems and a set of 57 NLSS problems were defined.

It is a common opinion that these types of codes will perform better if the functions and/or variables are appropriately scaled, i.e., they are approximately of the same magnitude. In fact, many of the codes in the testing suggest (or warn) that the user scale the functions and/or variables before running the problem. On the other hand, some of the codes provide automatic scaling or an option for user-provided scaling. Therefore, as tests on the robustness of the codes with respect to poor scaling, two new sets of 72 (57) problems were defined for the NLLS (NLSS) codes. One set has poorly scaled variables, i.e.,

$$F(x) \equiv F(\Sigma x)$$

and the other set has poorly scaled functions, i.e.,

$$F(x) \equiv \Sigma F(x) \quad .$$

In both cases, $F(x)$ is the "standard" set of 72(57) problems and Σ is a diagonal matrix with entries ranging in size from 10^{-5} to 10^{5}.

In the case of the NLSS problems, the codes are trying to find a value for the variables such that all the functions are zero. A reasonable question to ask about these codes is, how do they perform when the functions cannot be evaluated very accurately? (None of the codes address this problem.) To test this aspect of robustness of these codes, a new set of 57 problems was defined as

$$\bar{F}(x) = F(x) + \xi$$

where ξ is uniform random noise in the range (0,1.E-5).

Just as important as the actual tests is the implementation of the tests. This aspect of the testing has been neglected. For the results to be unbiased, it is very important to ask all the codes to perform the same task. However, it is usually <u>impossible</u> to implement the tests in a totally equivalent manner because of the different return criteria the codes use. There are two ways to handle this problem: 1) change the codes so that the return criteria are the same for all the codes, or 2) run the codes as they are, set available parameters to minimize the differences and note the differences that still exist. The latter was chosen in our tests for two reasons. First, it was at least impractical if not infeasible to try to modify so many codes. Second, it is still an open question which criteria should be used in deciding when an answer is found.

For most of the codes in our testing, there are two reasons to return to the user: 1) a solution has been found or 2) something has gone wrong and a solution was not returned, e.g., improper input, too much work. A very few codes have a third type of return criteria which is "not all the conditions for a solution has been met, but no better approximation can be found." In other words, the code is not sure if it has found a solution or not, and so leaves the decision to the user. As an example of why it is so difficult to make the return criteria equivalent, Table 2 lists the different types of return criteria for the codes we tested and Table 3 demonstrates how different a particular type of return criterion can be.

In setting the parameters, the easiest parameter to set equivalently was the parameter for the return because of "too much work," i.e., either too many function evaluations or too many iterations, and the most difficult was the tolerances used in accepting a solution. For details in how the parameters were set, see [2] and [4].

An additional comment about the tolerances used is needed. A very stringent tolerance, essentially machine precision, was used in our testing. Such a stringent tolerance is not unreasonable for these types of problems. This tolerance was used to try to detect any difficulties in the code when operating at machine precision. Another very useful test would be to use a less stringent tolerance, say the square root of machine precision. By using this tolerance, a special measure of efficiency could be recorded, i.e., is the code doing too much work for the accuracy required.

One aspect of implementing the tests pertinent to the NLLS codes is that some of the codes require the user to supply the Jacobian, others approximate the Jacobian themselves, and some of the codes have the option to do either. It would seem that by supplying an approximation of the Jacobian to the codes which require the user to supply the Jacobian, that all the codes would be equivalent in this regard. This was not the case. In particular, the codes doing their own approximation had different differencing schemes and information from the problem to determine the step size. Therefore, the codes were divided into two groups, those using the analytical Jacobian and those using an approximation to the Jacobian. (Some codes were in both groups.) Most comparisons made between the groups noted this difference in implementation. Because some codes belonged to both groups, very useful data on the importance of the analytical Jacobian versus an approximation could be collected.

III. The Analysis of the Results

Important aspects of analyzing the results of the testing are which data to collect and how to display the data. Again the criteria used in defining the "best" determines the type of data that should be collected. In our case, the criterion for "best" consisted of robustness, reliability and efficiency. Our definition of robustness is the ability to find solutions. Therefore both the value of the variables returned and the value of the functions at this point were collected to determine if solutions had been returned. Our definition of reliability is the ability to correctly identify solutions. Therefore, why the code returned for each problem was collected. As our measures of efficiency, both the number of function evaluations and the CPU time were collected for each problem.

As can be imagined the amount of data collected was overwhelming. The usual format for displaying data is numerical tables. For the function evaluation count and the CPU time this type of display was reasonable. The rest of the data, i.e., why the codes returned, the value of the variables returned and the function values at the returned point, did not as easily lend itself to this type of display. This data was used to determine how the codes performed on the problems. It seems a simple enough task to record how a code performed on a problem, but eight categories of performance for the NLLS codes and nine categories for the NLSS codes were defined to do this. As an example, the nine categories of performance for the NLSS codes were:

1. The code claimed a solution and it returned a solution.
2. The code did not claim a solution but returned a solution.
3. The code claimed a solution but did not return a solution.

The code returned before finding a solution because:

4. warning of slow convergence;
5. too many function evaluations;
6. warning of possible divergence;
7. singular Jacobian or nearby stationary point.

The run was aborted before the code could return because:

8. an overflow or indefinite condition;
9. time limit.

Because of the way the categories were defined, another aspect of robustness, namely the ability to take reasonable action when a problem is not suitable for the code, could be measured. For example, NLSS problems do not always have solutions. In fact, there were problems in the test set that have no solution. It would be very useful to know if a code can detect this condition and return before it gets into trouble (e.g., overflow condition) or too much work.

One difficulty, in assigning the performance categories to the codes, was determining if a certain value for the variables is a solution. Many of the problems had published solutions, otherwise, the decision was based on the value of the variables, the function values and if another code had found that solution. Another difficulty is that this set of categories did not distinguish the condition that for some of the problems, the codes returned different solutions. Because of the global properties of many of the codes, one solution was considered as good as another. However this information is important especially with respect to the efficiency measures.

Once the performance categories were assigned, they could also be displayed in numerical tables. Table 4 is one such table of the performance. As is the usual case, these tables are at best awkward to read. There is so much information it is hard to discover patterns in the data. An alternative method for displaying this type of data is Chernoff faces [1]. Chernoff faces is a statistical tool used to display multivariant data. In this technique, the features of the face represent the different variables. As an example, the following explains how the nine performance categories for NLSS codes define the Chernoff faces. (Each face represents the performance of one code.)

A. The size of the face depends on the number of solutions.

1. The height of the face depends on the number of "claimed

solutions," category #1.

2. The width of the face depends on the total number of solutions, #1 plus #2.

B. The frown depends on the number of "erroneous solutions," #3.

C. The size of the ears depends on the number of times the job was aborted because of overflow or indefinite conditions, #8.

D. The width of the nose depends on the number of returns because of too many function evaluations, #5.

E. The eyes and eyebrows depend on the number of returns without a solution.

1. The downward slant of the eyes and eyebrows and the ellipsoid shape of the eyes depends on the number of times a code could not find a solution, #4, #6, and #7.

2. The height of the eyebrow over the eye depends only on the number of singular Jacobians or stationary points reported, #7.

Figure 1 displays the same information as Table 4. A quick examination of the Chernoff faces easily identifies patterns, similarities and differences in the performance of the codes. A very strong warning about using Chernoff faces is in order. One is not to pick out the "prettiest" face and conclude it is the "best" code. The intended use of the faces is for comparison. For example, the QN face and the ZONE face are the smallest faces and each have very downward slanted eyes, very narrow nose and small ears. This indicates that the codes performed very similarly, in particular, when a solution could not be found, the code returned with a warning and not because of too many function evaluations or an indefinite or overflow condition. The main difference in the two faces is the height of the eyebrow over the eyes. The height of the eyebrow is higher on the QN face which implies, QN more often did not find a solution because of a singular Jacobian. Once this pattern is discovered it is easy to confirm it by using the numerical tables.

After all is said and done, conclusions must be stated. The most common conclusion is that there is no overall "best" code. While criteria for the "best" have been established and certain measures of these criteria have been obtained, it is still very difficult to come up with conclusions. For example, if one code is more robust than another (found more solutions) but not as reliable (found more "erroneous" solutions), which code is "best?" Lootsma in [6] has addressed this problem of priorities and weighting.

Perhaps more significant conclusions (maybe the only ones possible) are the qualified conclusions, e.g., all the NLLS codes, except one, performed quite well on the nominal set of "standard" problems or the different starting values had very little effect on the performance of the NLSS codes. Conclusions must be qualified by the differences in implementing the tests. Conclusions can also be qualified by the different aspects of the testing design, e.g., poor variable scaling. Because of the differences in the codes, the differences in implementing the tests and the differences in the test design, much insight can be gained about the areas that the software can handle well and areas which need improving.

In summary, testing mathematical software can be an unrewarding task for several reasons. It is very difficult, if not impossible, to make definite conclusions from the testing. The results are soon out of date because a newer version of a code in the test or a new code that should have been in the test has been written. Then there is the problem of justifying your results to the author of a code that didn't do well. But the testing is important and needs to be done. First, the information gathered about the codes can direct users to a particular code, applicable for their situation. Second, the testing can establish algorithms, techniques, convergence tests that work well (and those that do not). And third, it can influence the development of the next generation of software.

References

[1] Chernoff, H., "The use of faces to represent points in K-dimensional space graphically." Journal American Statistical Association, 68 (1973), pp. 361-368.

[2] Hiebert, Kathie, "A comparison of nonlinear least squares software," Sandia Technical Report, SAND79-0483, Sandia National Labs, Albuquerque, NM.

[3] Hiebert, Kathie, "An evaluation of mathematical software which solves nonlinear least squares problems," TOMS, Vol. 7, No. 1, March, 1981, pp. 1-16.

[4] Hiebert, Kathie, "A comparison of software which solves systems of nonlinear equations," Sandia Technical Report, SAND80-0181, Sandia National Labs, Albuquerque, NM.

[5] Hiebert, Kathie, "An evaluation of mathematical software which solves systems of nonlinear equations," to appear in TOMS.

[6] Lootsma, F.A., "Performance evaluation of nonlinear programming codes from the viewpoint of a decision makes," Performance Evaluation of Numerical Software, L.D. Fosdick, ed., North-Holland (1979), pp. 285-297.

NLLS Codes						NLSS Codes		
Group 1			Group 2					
Name	Source	Method	Name	Source	Method	Name	Source	Method
EO4GAF	NAG	LM	EO4FAF	NAG	GN	BRENT	MINPACK	BT
LMDER	MINPACK	LM	LMDIF	MINPACK	LM	SOSNLE	Sandia	BN
LSQFDN	NPL	GN+FD	LSQNDN	NPL	GN+FD	ZSYSTM	IMSL	BN
LSQFDQ	NPL	GN+QN	NL2SOL	NBER	GN+QN	QN	Sandia	QN
NL2SOL	NBER	GN+QN	NSO3A	Harwell	LM	CO5NAF	NAG	PH
N5O3A	Harwell	LM	TJMAR1	Sandia	LM	HYBRD	MINPACK	PH
SNWT	JPL	LM	ZXSSQ	IMSL	LM	NSO1A	Harwell	PH
TJMAR1	Sandia	LM				ZONE	PORT	PH

LM - Levenberg-Marquardt GN - Gauss Newton
GN+FD - augmented Gauss Newton using finite differencing
GN+QN - augmented Gauss Newton using quasi-Newton

BT - Brent's method
BN - Brown's method
QN - quasi-Newton
PH - Powell's hybrid method

Table 1. Codes Used in the Comparisons

Group	$\|F\|$ Rel.	$\|F\|$ Abs.	FSUMSQ Rel.	FSUMSQ Abs.	Δx Rel.	Δx Abs.	Gradient	MAXITR	MAXFEV
Group 1						E04GAF		E04GAF	
	LMDER				LMDER		LMDER		LMDER
			LSQFDN	LSQFDN*	LSQFDN		LSQFDN*		LSQFDN
			LSQFDQ	LSQFDQ*	LSQFDQ		LSQFDQ*		LSQFDQ
				NL2SOL	NL2SOL		NL2SOL	NL2SOL	NL2SOL
				NS03A	NS03A*	NS03A			NS03A
	SNWT*	SNWT				SNWT		SNWT	
				TJMAR1	TJMAR1	TJMAR1	TJMAR1*	TJMAR1	
Group 2			E04FAF*	E04FAF		E04FAF		E04FAF	
	LMDIF				LMDIF		LMDIF		LMDIF
			LSQNDN	LSQNDN*	LSQNDN		LSQNDN*		LSQNDN
			ZXSSQ		ZXSSQ		ZXSSQ	ZXSSQ	

*Parameters set by code, the user cannot change them.

Table 2: Summary of Return Criteria for NLLS Codes

Use of the Gradient to Determine a Solution

did not use	codes sets parameters		user sets parameters	
E04GAF	TJMAR1	$\|J^T f\| \le 0$	ZXSSQ	$\|J^T f\| \le$ GTOL
E04FAF	LSQFDN	$\|J^T f\| \le \xi \|f\|^{1/2}$ in combination with other tests	LMDER	$\|J^T f\| \le$ GTOL and max cosine between f and any column of J $\le$ GRDMIN
NS03A	LSQFDQ		LMDIF	
SNWT	LSQNDN		NL2SOL	

f the vector of functions

J the Jacobian matrix

ξ square root of machine precision

Table 3: NLLS Codes

PROBLEM SET VARIABLE SCALING

TOTALS OF PERFORMANCE

	BRENT	SOSNLE	ZSYSTM	QN	CO5NAF	HYBRD1	HYBRD2	NS01A	ZONE
1	15	21	9	7	8	22	30	7	3
2	0	3	10	0	0	0	0	0	0
3	0	0	0	0	0	2	0	0	0
4	9	0	0	0	0	20	24	0	0
5	3	7	9	2	16	3	1	17	0
6	30	12	0	14	5	0	0	5	47
7	0	12	14	34	28	0	0	1	7
8	0	2	15	0	0	10	2	27	0
9	0	0	0	0	0	0	0	0	0

****

1 The code claimed a solution and the code returned with a solution.
2 The code did not claim a solution but returned with a solution.
3 The code claimed a solution but did not return with a solution.

THE CODE STOPPED BEFORE FINDING A SOLUTION BECAUSE
4 Too many function evaluations (with a warning of slow convergence).

5 Too many function evaluations.
6 Not able to improve or diverging.
7 Singular Jacobian or nearby stationary point.

8 Overflow or indefinite condition.
9 Time limit

Table 4: NLSS Codes

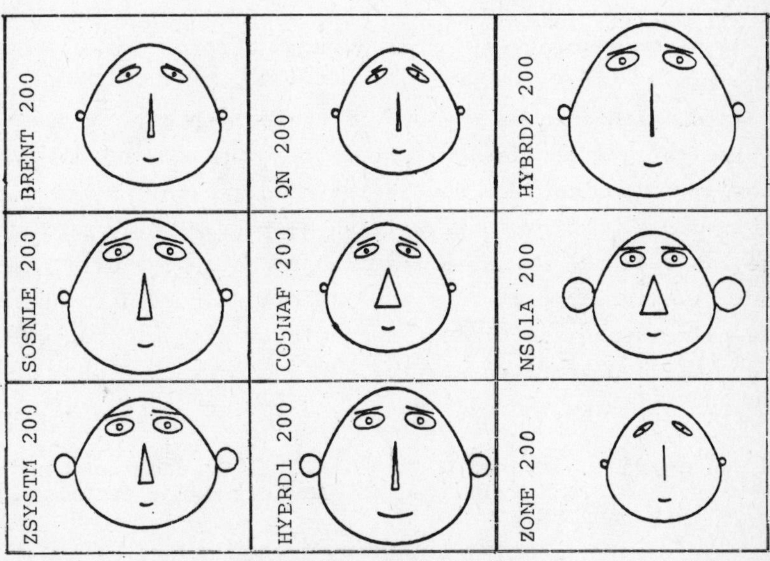

Figure 1: NLSS codes
Chernoff faces based on performance on all variable scaled problems.

A PORTABLE PACKAGE FOR TESTING MINIMIZATION ALGORITHMS*

A. Buckley
Mathematics Department
Concordia University
7141 Sherbrooke Street, W.
Montreal, Canada H4B 1R6

ABSTRACT

A package of routines designed to aid in the process of testing minimization (or other types of) algorithms will be discussed. The intention is to provide a vehicle for evaluating codes which is portable, versatile and simple to use. Ease of use should encourage algorithm designers to use the package, versatility will ensure that it is applicable to testing many different routines, and portability will mean that it is available to anyone. This last point should contribute to consistency in the testing of algorithms. The emphasis is not on actual testing, but is on the presentation of some ideas of how one can design a useful testing mechanism.

1. Introduction

Testing of minimization software generally takes place on one of two levels. In one case, the author of a new or modified algorithm wishes to test his code to see how well it performs. Often a primary intent here is to produce enough results to be convincing when a paper is submitted for publication. The second situation occurs when some major group undertakes extensive comparative analysis of existing software. The emphasis here is generally on algorithms already developed and available.

There are disadvantages to each of these methods. In the first instance, the testing is likely to be rather limited and only a few functions will be tested. Of course the main reason for this is clear and understandable. For someone writing a program which implements new ideas, the interest and emphasis is going to be on the original aspects of the work. Coding of test functions, the work needed to augment the basic code to maintain and produce a meaningful set of statistics, and the time needed to perform repetitive testing on a good size group of test functions all make thorough testing a lot of effort, and therefore

* This research was supported by the National Sciences and Engineering Research Council of Canada under grant A8962 and the author wishes to acknowledge this support.

often only a minimal effort is put into these auxiliary tasks. The intent of the package to be described here is to substantially reduce this effort. A general framework will be described into which the test algorithm may be easily inserted. Routines to do many of the standard chores of minimization testing will be automatically supplied with the package. In order to make these routines attractive to use, care has also been taken to ensure that they can be used with minimal effort.

The principal disadvantage of the second approach is that it is not sufficiently accessible. Authors of new algorithms do not usually have convenient access to the testing procedures. The programs I will discuss will provide a portable tool that can be used to test any unconstrained minimization algorithm. I should note that the approach to be described is not limited to minimization. It could be easily applied in other contexts.

Note that, because it is the most widely used language in North America, FORTRAN was chosen as the programming language for the package, which, for want of a better name, I will call TESTPACK. To be portable, all routines are in standard FORTRAN, but, in order to produce programs of tolerable structure and readability, the new ANSI 1977 Standard was chosen instead of the old 1966 standard.

The testing package TESTPACK consists of a number of interconnected parts which provide the various facilities needed by a person wishing to test a minimization code (or codes). In the following sections I will describe each of these parts, demonstrate the input required and illustrate the output provided. This paper is a summary of TESTPACK; a more complete document describing its use is being written.

2. Overview

First I will present a general picture of how the testing package is set up. Then I will discuss each of the parts in more detail. Note that only the programs in the third column are provided by the user of TESTPACK. In fact, of these, only ZZLINK is necessary.

In subsequent sections the purpose of each of these parts will be indicated. Briefly, though, MINTST is the main program. Execution starts by reading in a set of minimization problems to solve. Then the program loops as shown; each function is minimized and statistics are kept. Then a summary is printed and another block of tests is requested. We begin by describing the test functions and problems since they are simplest.

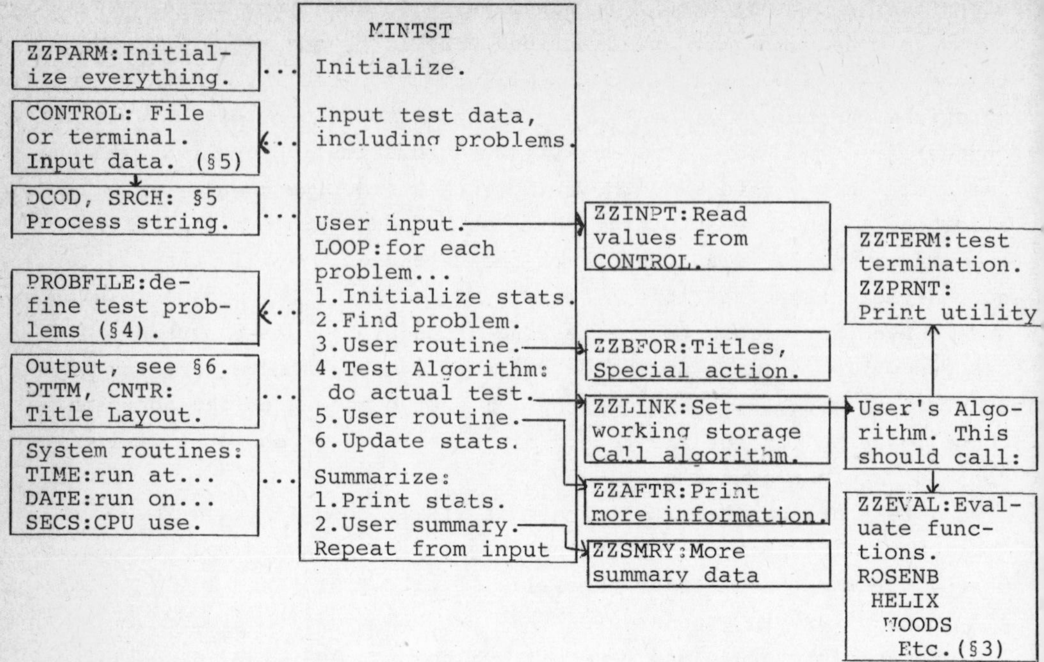

3. Test Functions

Subroutines are provided for evaluating about 40 standard test functions. All of the routines from the Argonne testing package [1] are included. The calling sequence is standard, e.g., SUBROUTINE ROSENB (N,X,F,G,IFL). Note that the function F is evaluated at X if IFL≥0; the gradient G is evaluated if IFL≤0. Clearly IFL=0 gets both. (also see ZZEVAL in §7.) Note that one test function may be used in several test problems (§4).

4. PROBFILE

A collection of standard test problems is provided in a file. These may be referred to by name or number (see§5). Here are 9 typical lines of the file:

```
*ROSENBRK THIS IS THE EVER-FAMOUS ROSENBROCK FUNCTION.
0001 0001 0002 00100
-1.2      1.0
*ROSNBRK2 THIS IS ROSENBROCK'S FUNCTION WITH A DIFFERENT X0.
0002 0001 0002 00100
2.5       -1.6
*HELIX THIS IS THE HELICAL VALLEY FUNCTION, COMMONLY CALLED HELIX.
0012 0013 0003 00100
-1.0      0.0        0.0
```

This specifies 3 problems (of about 70 currently available). The first is named ROSENBRK. The string "THIS IS..." will be in the output for identification (see §6). The data, in order, indicates: this is problem #1, the test function is #1, the dimension is 2, only 100 function evaluations are allowed and the starting point is -1.2, 1.0. The next 3 lines define ROSNBRK2 as problem #2 using the same function #1 of dimension 2. Just the starting point is different. To choose a test problem (see 5), specify its name, e.g., HELIX, or its number, e.g., 12. The problem is found in PROBFILE (§2) and the data are read.

For testing the same group of problems repetitively, a GROUP may be defined. To test problems 1,2, HELIX, 7, 14 and WOODS, add the line "*GROUP, THESE, 1, 2, HELIX, 7, 14, WOODS" to PROBFILE and enter "GROUP= THESE" at input time (see §5).

5. Input

Tests may be run from batch or from a terminal. Minimal input is required; ease of use is foremost. The input routines are quite powerful (at least I think so!); if the input makes sense to you, it is likely acceptable to the system. I will illustrate with a sample terminal session, for it is with this mode in mind that the routines were designed. However the input commands work equally well in batch mode. (The leading "?" is a prompt by CDC CYBER NOS.) The summary appears unless it is turned off.

```
(1)TITLE LINE (OR "END OF RUN")?
?ILLUSTRATE THE TESTING PACKAGE.
(2)BASIC INFORMATION?
?SUBROUTINE=SHANNO, ACCURACY=1.E-3
(3)SPECIAL PARAMETERS?
?METHOD=CG
(4)WHICH PROBLEMS?
?HELIX, 1, WOODS, 17
(5)READY?
?YES
```

```
                ILLUSTRATE THE TESTING PACKAGE.
      TEST BEING EXECUTED AT 9:18 P.M., DECEMBER 16, 1980

STANDARD CONTROL PARAMETERS:

   TERMINATION NORM = 2 (EUCLIDEAN)
   TERMINATION TYPE = 1 (GRADIENT)
   ACCURACY SPECIFIED       .100E-02
```

```
SUMMARY OF PROBLEMS DONE...
PR#  FN#  NAME      DIM  ITS  FNCS  GRDS  FVALUE    GVALUE    MSECS FSECS ER
12   2    HELIX      3   19    43    43   .11E-07  .56E-03    .09   .01  0
 1   1    ROSENBRK   2   26    59    59   .30E-10  .22E-03    .09   .02  0
18   4    WOODS      4   46   102   102   .32E-10  .22E-03    .23   .05  0
17   3    PWING      4   23    47    47   .31E-05  .53E-03    .11   .02  0
                   ITERS FUNCS GRADS RATIO                    MSECS FSECS
      TOTALS        114   251   251  2.20                      .52   .09
NONE OF THE PROBLEMS WERE FLAGGED WITH ERRORS.
THE METHODS WAS NUMBER 0 WITH SHANNO'S ROUTINE.
```

There are a number of relevant comments; a number of these are demonstrated in the sample output in §6 and the summary above. The "title line" will appear in the output for identification. Level (2) sets basic parameters such as the termination accuracy (e.g., 10^{-3}; see ZZTERM in §7).Here subroutine SHANNO (see [2]) is chosen to be used for testing; it is one of 4 available in this sample version. At (3), parameters peculiar to the user's algorithm are entered. The routine SHANNO runs in 2 modes. Here I have chosen Shanno's "memoryless quasi-Newton" algorithm. Step (4) indicates that problems numbered 1 and 17 and named "HELIX" and "WOODS" are to be used for this test. The problems are automatically found in PROBFILE: the order is irrelevant. At (5), you are asked to check that all is in order before beginning the test.

There are a number of features of the input process that this example does not illustrate. Keeping in mind that only one test run may be madé, or that many may be done in sequence, some handy features are: (i) On a low speed terminal, the summary appearing above (i.e., on the terminal) can be eliminated by typing "SUMM=FALSE" at any time. The summary still appears at the end of the main output file (§6). (ii) Keywords may be abbreviated; delimiters may be left out. For "ACCURACY= 1.E-6" enter "AC1.E-6"; for "SUMMARY=FALSE", type "SUM=F", or for "PRINT=10" type "P10" (or something similar in each case). Even keywords can often be omitted; "GRAD" will be understood as "TYPE=GRADIENT". Note that spaces are ignored. (iii) There are many options in the program which can be changed. For example, if you don't want to be cautious, level (5) may be skipped by typing "CHECK=FALSE" (or "CH=F"). Such changes stay in effect. (iv) Error correction is provided. Suppose at (4) you noted that you forgot to reset a value SCALE at (3). Just type "GOTO=3" (or "G3") after entering the problem list. The system will respond again at (3). So type "SCALE=FALSE, GOTO=5" (or "SC=F,5") to reset scale and skip immediately to execution. (v) If the system is busy and response is slow, enter several lines at once. Data for steps (2), (3) and (4) may be entered together as "SUB=SHAN, ACC=1,E-3!

METH=CG!HELIX,1,WOODS,17,GOTO6"; "!" is a line separator. (vi) The list of problems does not have to be retyped at (4) if it does not change. Type "REPEAT", or skip (4) with "...G5" at level (2) or (3). (vii) If several runs have been made with varying parameters, and you wish to restore them to their original values, type "DEFAULT". Of course, all defaults are in effect initially (ZZPARM, §2).

6. Output

One portion of the output produced by the sample test run of §5 is presented on the last page. (It has been slightly photoreduced in order to fit on a single page.) Only the output from one of the test functions is given. The others are similar. The summary, which was also deleted, would be the same as it appears in §5. There are three main parts to the output for each function: (1) the header; (2) the points generated, i.e., every K^{th} according to "PRINT=K" in §5 (default K=10); and (3) the solution. If any error occurs, the statement "THE SOLUTION..." is replaced by an appropriate message.

Note that all of the output is provided by TESTPACK, except that the two lines beginning "COMPUTATION DONE..." on the past page were printed in the user routine ZZBFOR mentioned in §2 (also see §8) and the last line of the summary appearing in §5 was printed in the user routine ZZAFTR. All information needed to identify the output and the values of all parameters used in the run are automatically listed.

In the timing information, it should be noted that FSEC is the time spent in evaluating the function, whereas MSEC is the time spent in the minimization algorithm. It is also worth noting that any time spent in printing has been excluded; that is important since time spent in FORTRAN's format decoding routines can overwhelm the time spent in actually minimizing the function.

7. Routines Supplied

Certain routines are supplied (§2, 4th column) which should be used by the test algorithm. This will require minor changes in the test program if it already exists. One purpose of these routines is to eliminate the need for the programmer of the test algorithm to have to worry about inessential details, e.g., counting function evaluations. Here is a brief description of the function of these routines.

ZZEVAL(FUNCT,N,X,F,G,IFL): This calls FUNCT(N,X,F,G,IFL) (see §2). The time taken for the call is recorded, the number of function and gradient calls are separately counted and termination is forced when the function count reaches a preset maximum. A nonlinear scaling may be applied to the function; evaluation of the function may also be traced. This routine must be used for function evaluations in order to obtain statistics on function calls and time.

ZZPRNT(N,S,F,G,FORCE): This prints the function value F and the function call, gradient call and iteration counts at specified iterations, say every K^{th} (see PRINT in §§5,6). Optionally, the X and G vectors may also be printed. ZZPRNT determines when output is required according to the iteration count and K, except in the case where FORCE is TRUE, which forces printing. This is appropriate, for example, when the solution is found. There is no printing if K=0.

ZZTERM(N,X,G,EPS,LESS): This routine is provided as a convenience so that the termination test in a program may be easily changed. That may be desirable in order to be consistent with results appearing in the literature. ZZTERM returns TRUE in LESS when the required accuracy EPS has been reached. The test used is determined by parameters appearing in COMMON. These may be changed at level (2) of the input (§5). For example, the default test is $||g||_2$ < EPS; typing NORM-ABSSUM (or N=L1) will change it to $||g||_1$ < EPS. In §5, the required accuracy EPS was set to 10^{-3} by the statement "ACCURACY=1.E-3".

8. Using TESTPACK

In order to use TESTPACK to test a specific minimization algorithm, it is necessary of course to link them together. The main routine MINTST will call ZZLINK (§2). It is up to the user to provide the necessary statements to call his algorithm. In the present version, the call to test Shanno's routine CONMIN appears as:

```
IF( NMETH .EQ. 0 ) THEN
  MXSTOR = 5*(N+2)
ELSE
  MXSTOR = N*(N+7)/2
ENDIF
IF ( MXSTOR .GT. LR ) THEN
   ERFLAG = -1
ELSE
   CALL ZZSECS(TT)
   TMIN = TMIN - TT
   CALL CONMIN (FUNC,N,X,F,G,ACC,ERFLAG, RWORK,LR,EPS,NMETH)
   CALL ZZSECS(TT)
   TMIN = TMIN + TT
ENDIF
```

Note that I have included a precautionary check that the working storage provided, LR, is sufficient for that which is required (which varies according to the method defined by NMETH). A call to ZZSECS just before and after the call to CONMIN ensures that the time taken in the minimization is recorded.

In addition to ZZLINK, four other user defined routines appear in §2. These are all optional and can consist of a simple RETURN. They are provided so that MINTST can be tailored to a particular algorithm without the need to modify any of the existing code in TESTPACK. Simple examples of how they can be used is given with the following descriptions.

ZZINPT is called in order to read the data provided at level (3) of the input routine (§5). For example, the statement "METHOD=CG" is particular to Shanno's routine CONMIN, and therefore it is up to the user to write the code to take the appropriate action. The decoding routines will still decode the string; it is just up to the user to use ZZINPT to set the variable NMETH for CONMIN appropriately.

ZZBFOR is called just before each function is minimized. It is probably unnecessary in most cases and can consist of just a RETURN, but an example of its use is in the sample output in §6. My version tests 4 algorithms; in ZZBFOR I print 2 statements ("COMPUTATION DONE...") with each block of output which identify the subroutine used.

ZZAFTR is called just after each function minimization is done. It is similar to ZZBFOR, and it is not used in the current version.

ZZSMRY is called after the summary has been printed. Here (see §5) it is used so that the subroutine used is identified in the summary. It can also be used to output any special statistics which have been accumulated for the subroutine being tested.

In fact, I think it is quite easy to implement an algorithm for testing. For example, I have modified VA08 from the Harwell Library and Shanno's routine CONMIN with no trouble in order to run them with MINTST.

9. Conclusion

I have presented some ideas on how a set of programs can be written which are portable and yet versatile enough to be used to test any minimization algorithm. Facilities are included which allow tailoring to a particular algoirthm. Input of control information is such that it makes use of the testing package very easy. Programming required by someone using this to test an algorithm has been kept to a minimum. Careful attention has been paid to miscellaneous details. For example, all external names begin with "ZZ" to avoid conflict with user defined externals and the standards of coding are high. With all of these things in mind, it is hoped that others will find these programs useful.

At present the programs are running but they are not complete. Improvements are being made and suggestions would be welcome.

10. Acknowledgments

I would like to thank Mr. Patrick Leung for his able assistance in coding portions of TESTPACK, especially the routines for interpreting the input string. Also I would like to express my appreciation to Mrs. Sasha LeNir for her patience as a guinea pig while the programs were being debugged. Thanks are also due to Professors Dave Shanno and Phillippe Toint for their suggestions and encouragement regarding the programs.

11. References

[1] More, Jorge J., Garbow, Burton S. and Hillstrom, Kenneth E., "Testing Unconstrained Minimization Software," Report TM-324, AMD, Argonne National Laboratory (1978).

[2] Shanno, D.F. and Phua, K.H., "A Variable Method Subroutine for Unconstrained Nonlinear Minimization," ACM Transactions on Mathematical Software (in press), (1978).

```
<THE OUTPUT FOR HELIX HAS BEEN DELETED.>

-------------------------------------------------------------------------------
|                                                                             |
|                      ILLUSTRATE THE TESTING PACKAGE.                        |
|                                                                             |
|            TEST BEING EXECUTED AT  9:18 P.M., DECEMBER 16, 1980             |
|                                                                             |
-------------------------------------------------------------------------------
|                                                                             |
|                THIS IS THE EVER FAMOUS ROSENBROCK FUNCTION.                 |
|                                                                             |
-------------------------------------------------------------------------------
|                                                                             |
| STANDARD CONTROL PARAMETERS:                                                |
|     TERMINATION NORM =  2 (EUCLIDEAN)                                       |
|     TERMINATION TYPE =  1 (GRADIENT )                                       |
|     ACCURACY SPECIFIED        .100E-02                                      |
|                                                                             |
-------------------------------------------------------------------------------

ADDITIONAL USER DEFINED CONTROLS:

COMPUTATION DONE WITH SHANNO'S ROUTINE CONMIN.
NMETH =  0
-------------------------------------------------------------------------------

STARTING PROBLEM NUMBER      1.

.....................HAVING REACHED THE POINT NUMBER    0 .....................
THE FUNCTION VALUE IS    .242000000E+02 (    1 FUNCTION EVALUATIONS).
                                        (    1 GRADIENT EVALUATIONS).
THE VARIABLES HAVE THE CURRENT VALUES GIVEN BY
 -.12000000E+01  .10000000E+01
THE GRADIENT AT THIS POINT IS
 -.21560000E+03 -.88000000E+02

.....................HAVING REACHED THE POINT NUMBER   10 .....................
THE FUNCTION VALUE IS    .104531091E+01 (   25 FUNCTION EVALUATIONS).
                                        (   25 GRADIENT EVALUATIONS).
THE VARIABLES HAVE THE CURRENT VALUES GIVEN BY
  .13476345E-01 -.26666477E-01
THE GRADIENT AT THIS POINT IS
 -.18283217E+01 -.53696178E+01

.....................HAVING REACHED THE POINT NUMBER   20 .....................
THE FUNCTION VALUE IS    .927609806E-02 (   46 FUNCTION EVALUATIONS).
                                        (   46 GRADIENT EVALUATIONS).
THE VARIABLES HAVE THE CURRENT VALUES GIVEN BY
  .91213161E+00  .82804041E+00
THE GRADIENT AT THIS POINT IS
  .12631173E+01 -.78873160E+00
-------------------------------------------------------------------------------
|                                                                             |
|                         THE SOLUTION HAS BEEN FOUND                         |
|                                                                             |
-------------------------------------------------------------------------------

.....................HAVING REACHED THE POINT NUMBER   26 .....................
THE FUNCTION VALUE IS    .297665609E-10 (   59 FUNCTION EVALUATIONS).
                                        (   59 GRADIENT EVALUATIONS).
THE VARIABLES HAVE THE CURRENT VALUES GIVEN BY
  .10000026E+01  .10000047E+01
THE GRADIENT AT THIS POINT IS
  .19693603E-03 -.95861421E-04
-------------------------------------------------------------------------------
|                                                                             |
|   SUMMARY OF PROBLEM:                                                       |
|   PR# FN#    NAME    DIM ITS FNCS GRDS    FVALUE    GVALUE MSECS FSECS ER   |
|     1   1 ROSENBRK    2  26   59   59  .30E-10  .22E-03   .09   .01  0    |
|                                                                             |
-------------------------------------------------------------------------------

<THE OUTPUT FOR WOODS HAS BEEN DELETED.>
<THE OUTPUT FOR PWSING HAS BEEN DELETED.>
<THE SUMMARY HAS BEEN DELETED.>
```

TRANSPORTABLE TEST PROCEDURES FOR ELEMENTARY FUNCTION SOFTWARE

W. J. Cody*

Applied Mathematics Division
Argonne National Laboratory
Argonne, Illinois 60439

Abstract

This paper explores the principles and numerics behind the use of identities to test the accuracy of elementary function software. The success of this approach depends critically on proper matching of identity and test interval with the purpose of the test, and on careful implementation based on an understanding of computer arithmetic systems and inherent accuracy limitations for computational software.

1. Introduction

In some respects the testing of software for the evaluation of elementary functions is much simpler than the testing of software for other computational purposes. Certainly the methods for evaluating functions are better understood than methods for, say, minimizing a function of several variables. The evaluation of a function is generally a simple one-to-one mapping from a one or two dimensional domain to a one or two dimensional range, and the concept of accuracy in that mapping is well understood. Function minimization, on the other hand, is a complicated mapping from one high dimensional space to another with a poorly understood concept of accuracy or measure of success for the process. Because function evaluation is relatively easy we expect the software to return precise results. This implies that the software is more sensitive to the computational environment than software for function minimization, and that testing procedures for function software must be more carefully designed and implemented than those for function minimization software. This paper presents one successful approach to testing elementary function software.

The next section summarizes necessary background material under the assumption that the reader is familiar with the basic terminology

---

* This work was supported by the Applied Mathematical Sciences Research Program (KC-04-02) of the Office of Energy Research of the U.S. Department of Energy under Contract W-31-109-Eng-33.

normally associated with computational software (see [4,6]). Thus no attempt is made to define robustness, for example. Section 3 treats testing objectives in general and discusses the details of one widely used but machine sensitive technique for determining the accuracy of function software. Section 4 discusses a less sensitive but highly portable set of Fortran programs, the ELEFUNT package, for performance evaluation of software for the elementary functions.

The material presented here is extracted from a longer discussion presented in Sorrento, Italy in November, 1980 [5], and is based on work described in greater detail in [2,6]. We thank W. Kahan, the late H. Kuki, our colleagues at Argonne National Laboratory, and an anonymous multitude of others who have contributed to our ideas without necessarily endorsing them.

2. Preliminaries

In general the computation of an elementary function involves three steps. First, the given argument is usually reduced to a related argument in some restricted domain together with the parameters describing the argument reduction. For example, in computing the sine function for radian arguments a given argument x might be decomposed into

$$x = |x| \text{ sign } (x)$$

and

$$|x| = N\pi + g,$$

where $|g| \leq \pi/2$ and N is an integer. Here g is called the reduced argument. In the second step, a related function value is calculated for the reduced argument. This might involve the evaluation of a polynomial or rational approximation of some sort, or it might involve a more complex evaluation such as in the CORDIC scheme [9]. Finally, the desired function value must be reconstructed from the results of the first two steps. Thus

$$\text{sine}(x) = (-1)^N \text{ sign}(x) \text{ sine}(g).$$

The accuracy of the elementary function program depends on the accuracy of the reduced argument as well as the accuracy of the approximation used in the second step.

Because of its apparent simplicity, argument reduction is often slighted during implementation with most of the effort going into the approximation. The result is that many elementary function programs cannot determine accurate function values for larger arguments. To illustrate the importance of argument reduction, Table I compares the computation of sine(22) using the built-in function on an HP 65 hand calculator, and the built-in function preceded by careful argument reduction, with a standard value taken from [1]. (All results are given to the working precision of the calculator.) This computation is a severe test of the argument reduction step because the argument is close to an integral multiple of π.

Accuracy of sine(22) on HP 65

Program	sine(22)
Standard Value [1]	-8.8513 09290 E-03
HP 65 (hardware)	-8.8513 06326 E-03
HP 65 (careful arg. red.)	-8.8513 09290 E-03

Table I

To understand the problems in argument reduction we must first understand where errors originate. There are two types of error associated with computational software in general. The first is due to error in the data. Let $y = f(x)$ be a differentiable function of x. Then

$$dy/y = x\, f'(x)/f(x)\, dx/x$$

is an analytic relation between the relative error dx/x in the argument and the relative error dy/y in the function value. The transmitted error dy/y depends solely on the inherited error dx/x and the analytic properties of the function. It does not involve computations internal to the software, and therefore is beyond the control of the software. All error directly attributable to the internal computations, such as rounding, truncation of analytic expansions, and inexact representation of constants is grouped together and called generated error. This is the error that is the sole responsibility of the computational software and is the error that a testing procedure should strive to measure.

The determination of the reduced argument

$$g = |x| - N\pi$$

involves the difference of two nearly equal quantities. Because software cannot detect inherited error, $|x|$ must be assumed to be exact. But $N\pi$ cannot be represented exactly in any finite precision and must involve some rounding error. As $|x|$ becomes larger more and more leading digits of $|x|$ are lost in the subtraction, thus promoting the rounding error in $N\pi$ in importance. The relative error in the reduced argument g is therefore roughly proportional to $|x|$. This error can be controlled if the product $N\pi$ and the difference are both carried out in higher precision arithmetic. The details of how this is done when higher precision arithmetic is not available in the hardware are discussed in [6] along with other subtleties we ignore here.

The argument reduction step is often the best place for two related programs to be merged. The sine program can serve double duty, for example, by being used for the cosine as well. Because

$$\text{cosine}(x) = \text{sine}(x+\pi/2)$$

and

$$\text{cosine}(-x) = \text{cosine}(x),$$

cosine(x) can be calculated by evaluating sine $(|x|+\pi/2)$. But this computation must be done carefully. Simple addition of $\pi/2$ to $|x|$ introduces error, contradicting the assumption that the argument for the sine computation is exact and defeating subsequent careful argument reduction. Full precision is maintained in g if, instead, the value of N is adjusted during argument reduction. Again the details will be found in [6].

3. Testing Procedures

There are two fundamentally different reasons for testing computational software, hence two fundamentally different approaches to the task. Creators of new algorithms are intent on showing that their creations are superior in some sense to other algorithms, and they approach performance testing as a contest. Tests are specifically designed to display whatever superiority the new algorithm may have. There is ordinarily no attempt to uncover or explore weaknesses in the algorithm or its implementation.

Someone selecting programs for inclusion in a library is interested in overall performance, however. If some duplication of purpose is acceptable to the library, the concern may be more with eliminating programs that are unacceptable and in matching programs with problem characteristics than with determining the 'best' program. Tests for this purpose ought to aggressively exercise a program in ways that will detect weaknesses as well as demonstrate strengths, that will explore robustness as well as problem solving ability. Inevitably the results of such testing will be used to compare programs, but the original intent in this approach is that a program will stand or fall on its own merits.

Tests in the sense of contests are too problem dependent to be applied in general. We therefore limit our discussion here to the second type of testing as it applies to function software. If there is a theme in what follows, it is that testing is an important numerical problem. As much attention should be given to designing and implementing objective test programs as is given to designing and implementing software for solving other numerical problems. Each test performed ought to have a specific purpose just as each step in the solution of any other numerical problem should be purposeful. Blind testing is no more useful than attempting to solve a polynomial equation by trial and error.

A complete battery of tests should include special tests to verify robustness, check error returns, and verify simple analytic properties of the function, as well as the usual timing and accuracy tests. Special tests are mostly a matter of ingenuity. For example, because robust programs have been defined as resilient under misuse, tests for robustness should involve deliberate misuse. This includes specifying arguments violating stated constraints for the program, and arguments leading to possible underflow or overflow. Symmetries and other simple properties of a function should be verified computationally. More complicated properties, such as monotonicity and periodicity, especially with an irrational period, are too difficult to verify computationally and are probably best ignored during testing.

The details of timing a program depend critically on the computational environment and all we can do here is to offer general suggestions and guidelines. One widely used technique is to call the program with fixed arguments inside a loop, calling the clock routine immediately before and after the loop. If the program is short enough that the overhead for the loop is significant, the elapsed time can be adjusted by also timing an empty loop. Care must be taken to make each timing loop long enough to compensate for the inherent coarseness of the clock, and to vary the fixed arguments used to insure that each major path through the program is independently timed. Finally, timing may be load dependent, especially under large and complicated operating systems. Under such conditions it is best to repeat the timing runs several times under varying system loads and to average the reported times after discarding any obvious outliers. When several timing runs are averaged, the standard sample deviation adds useful information to the reported average time.

That leaves only the sticky problem of determining the accuracy. We are fortunate, in dealing with function programs, in having a well defined concept of accuracy. A function maps one or more arguments into a unique real or complex value. The accuracy of the computer program is determined by measuring the difference between the correct function value and the computed function value.

Earlier we identified two types of error associated with numerical software: transmitted error and generated error. Ideally, accuracy tests should measure only the error directly attributable to the software, i.e., the generated error. The first requirement, then, is that the test arguments be exact. If they are not exact, transmitted error will creep into the measurement and distort the findings. Thus, data contamination by I/0 conversion processes eliminates comparison against published tables as a viable means of accuracy testing. Sparsity of tabular data is a second reason for rejecting this approach.

Comparison is still possible, but the comparison should be against standard function values generated within the computer under careful control. The technique that we have found most useful and that we see being used almost routinely now by others involves direct comparison against higher precision computations using random arguments. Briefly, the procedure is as follows.

First, the desired test interval is subdivided into a number N of subintervals, and a random argument is generated within each subinterval using the same precision arithmetic as is used in the function being tested. These arguments are then regarded as exact, eliminating inherited error from the test process. Each argument in turn is extended to higher precision by appending trailing zero digits to its machine representation, thus creating a higher precision argument with exactly the same numerical value as the original argument. The higher precision argument is then passed to a higher precision function program and the resulting higher precision function value is taken as the standard result for that argument. Next, the original working precision argument is passed to the program under test and the resulting function value is compared against the standard value. This comparison can be carried out in several ways. The working precision result can be extended to higher precision and the error measured there, or the higher precision result can be rounded back to working precision and the error measured there. The second approach is ordinarily used when standard arguments and function values are precomputed and saved on magnetic tape or other storage devices. Stored values must not pass through I/0 conversion, however. They must be stored and retrieved in machine representation, i.e., in binary or hexadecimal form, to preserve their purity.

Statistics gathered from such tests normally include the maximum relative error, MRE, as well as the root mean square relative error, RMS, where

$$\mathrm{MRE} = \max |F_i - f_i|/f_i,$$

$$\mathrm{RMS} = \mathrm{sqrt}\ \{(1/n)\ \Sigma\ |F_i - f_i|/f_i\},$$

the max and sum are taken over i, n is the number of random arguments used, F_i is the calculated function value and f_i is the higher precision function value for x_i. If the test arguments have been sorted algebraically, the error can be monitored to detect trends pointing to correctable problems in the function being tested.

A different set of statistics is often collected at the same time as the MRE and RMS. It is possible through programming subterfuge to consider the floating-point function values as large integers and to gather statistics on the error measured in units of the last place, or ULPs, by forming integer differences. A tabulation of the frequency of ULP errors can be informative. Table II contains test results of the types just described for a good implementation of the sine function on a VAX 11/780.

Test Results for SIN on VAX 11/780

Argument Range	Frequency of Error in ULPs						
	0	1	2	3	> 3	MRE	RMS
$(0,\pi/2)$	1936	64	0	0	0	6.61 E-08	2.27 E-08
$(2\pi,8\pi)$	1928	70	1	0	1	2.53 E-07	2.45 E-08

Table II

We have glossed over several problems here that should be explored further. First, we have assumed that the higher precision routine is accurate. This implies that it too must undergo testing. Fortunately, this requirement does not lead to an infinite loop, because the accuracy required of the higher precision routine is not that great. For our purposes the higher precision routine need only be shown to be accurate to slightly more than the test precision - accuracy to a few ULPs in the higher precision is not needed. This moderate accuracy can be verified, for example, by simply comparing two higher precision programs implementing different algorithms.

There is also a problem when higher precision arithmetic is not available in the system. In that case either programmed arithmetic must be used, or the standard arguments and function values must be generated in a different system and transmitted to the working system via magnetic tape. This latter approach requires careful control to insure that the arguments and corresponding function values are exactly representable in the target machine even when that representation is not native to the machine generating the values. This is the method that was successfully used in the FUNPACK project [3].

Many of the procedures just described for accuracy testing can be programmed once and for all in subroutines that are easily transported from one environment to another. We now have a collection of Fortran subroutines, for example, that take as arguments the name of the single precision function program to be tested, the name of the double precision function program to be used as a standard, the endpoints of the test intervals, and the number of random arguments to be used. Printed output from these subroutines includes the MRE and RMS errors, and a frequency table of the errors measured in ULPs. The differences between the various subroutines lie in the number and type of arguments expected by the function programs, i.e., one or two, real or complex. It is therefore only necessary to select the proper test subroutine, to provide a driver specifying the functions to be tested and the test intervals, and to provide the master function program when preparing a new set of accuracy tests. Because floating-point numbers are represented differently in different systems, the details of the computation of errors in ULPs must sometimes be modified when transporting the tests. The least significant bit of the significand of a single precision number occurs in the middle of its machine representation on a VAX, for example, instead of at the end as on most other machines, and the subroutine used to generate the data in Table II is different in detail from the corresponding subroutine used on IBM or CDC equipment. Other than that detail, the test programs are reasonably transportable, at least to systems supporting both single and double precision arithmetic. They are not useful, however, for testing the accuracy of double precision programs because few machines support arithmetic of more than double precision.

4. ELEFUNT

The discussion in the last section omitted the possibility of determining accuracy by examining the degree to which functions satisfy mathematical identities. This approach to accuracy testing is not new [7,8], but it has not been particularly successful in the past. The main problem is that the error we want to measure, the generated error from the function subroutine, may be masked by error introduced by the identities. Nevertheless, recent advances have shown that carefully selected identities, implemented with the same care taken in implementing elementary functions, can return meaningful error statistics.

The motivation for the work leading to this discovery was the development of a collection of self-contained transportable test programs for the elementary functions, the ELEFUNT package, intended to complement the comprehensive collection of algorithms and implementation notes for the computation of these functions [6]. The methods described for accuracy testing in the last section were not suitable for this package because they assumed the availability of higher precision arithmetic and of a suitable master function routine. The only alternative was to resort to the use of identities. We illustrate the approach taken by examining the program to test the sine and cosine functions.

The first task is to select an identity that is not likely to be used internally in the program being tested. Our purpose is to measure the accuracy of the calculation of a particular function and not the accuracy of the implementation of an identity. Identities based on the triple angle formulas are likely candidates for the sine and cosine functions. Thus, we propose to measure

$$E = \{\text{sine}(x) - \text{sine}(x/3)[3 - 4\text{sine}^2(x/3)]\} / \text{sine}(x).$$

It is important that the identity not introduce unnecessary error into the computation. The subtraction inside the square brackets is a potential source of significance loss that can be eliminated by proper argument selection. When x is drawn from the interval $[3m\pi, (3m+1/2)\pi]$ for integer m, x/3 lies in the interval $[m\pi, (m+1/6)\pi]$, $4\text{sine}^2(x/3) < 1$, and there is no cancellation of leading significant digits. We assume these limitations on the test arguments.

We stated earlier that testing should be purposeful. There are two distinct computational steps in the evaluation of the sine function that could contribute to the generated error in a computer routine: the argument reduction step and the evaluation of the function for the reduced argument. The accuracy of these steps should be checked separately if at all possible. Restricting the argument to the interval selected by setting m = 0 assures that there will be little, if any, argument reduction involved, and the accuracy statistics gathered will reflect the accuracy of the function evaluation given a reduced argument. Arguments selected from the interval with m = 2 will assure that argument reduction is involved. Significant differences between the results of the test for m = 2 and the results of the previous test must then be attributable to the argument reduction scheme.

Results of these tests applied to a double precision sine function on the IBM 195 are given in Table III under the heading 'Simple Tests'. Master test results for the same function, obtained by comparison against higher precision computations, are also listed in the table. Each test used 2000 random arguments from each interval. Errors are tabulated as the reported loss of base-β digits, where β is the radix for the arithmetic system in the machine. Tests labeled '1' correspond to the case m = 0, and those labeled '2' correspond to the case m = 2.

Accuracy Tests for Sine/Cosine

Test	Machine	β	Library or Program	Reported Loss of Base β Digits in MRE	RMS
Master Values					
1	IBM 195	16	Argonne	1.20	0.48
2	IBM 195	16	Argonne	1.20	0.56
Simple Tests					
1	IBM 195	16	Argonne	1.28	0.78
2	IBM 195	16	Argonne	3.50	2.30
Tests with Purified Arguments					
1	IBM 195	16	Argonne	1.18	0.69
	PDP/11	2	DOS 8.02	1.99	0.10
	Varian 72	2	Fort E3	1.87	0.00
2	IBM 195	16	Argonne	1.16	0.70
	PDP/11	2	DOS 8.02	1.74	0.09
	Varian 72	2	Fort E3	13.54	8.55
3	IBM 195	16	Argonne	1.16	0.69
	PDP/11	2	DOS 8.02	12.63	8.55
	Varian 72	2	Fort E3	12.69	7.31

Table III

These results show that the identity tests we have described are not reliable, especially for the second interval. The test procedure has introduced large errors into the computation completely masking the error generated in the sine routine. To see this, assume for the moment that $|x| \leq \pi/2$. The random argument x is exactly representable in the machine, but there may be a small error ε in the evaluation of the argument x/3. It is easily shown that for $|\varepsilon| << 1$,

$$\text{sine}(x/3+\varepsilon) = \text{sine}(x/3) + \varepsilon$$

to terms of first order. Further assume that relative errors of D and d are made in the evaluation of sine(x) and of sine(x/3), respectively. Substituting these expressions in the expression for E gives

$$E = \frac{\text{sine}(x)(1+D) - \text{sine}(x/3+\varepsilon)(1+d)[3-4\text{sine}^2(x/3+\varepsilon)(1+d)^2]}{\text{sine}(x)\ (1+D)}$$

Using the original identity and the above relation for sine(x/3+ε), and keeping only terms linear in D, d and ε, this simplifies to

$$E = D - d[1-8\frac{\text{sine}^3(x/3)}{\text{sine}(x)}] + \varepsilon[\frac{1}{\text{sine}(x/3)} - 8\frac{\text{sine}^2(x/3)}{\text{sine}(x)}].$$

Because $\text{sine}^3(x/3)/\text{sine}(x)$ is bounded above by 1/8 for the interval under consideration, the coefficient of d is crudely bounded between 0 and 1. However, the coefficient of ε is unbounded. This is the source of error we see in the second test interval. Fortunately, this error can be removed by 'purifying' the argument, i.e., by perturbing the test argument x slightly to x' so that both x' and x'/3 are exact machine numbers and $\varepsilon = 0$. The following Fortran statements accomplish this on most computers.

```
      Y = X/3.0E0
      Y = (Y+X)-X
      X = 3.0E0*Y.
```

The exceptions are those machines in which the active arithmetic registers are wider than the storage registers. On those machines storage of results must be forced at the completion of each arithmetic operation in the purification process. When purified arguments are used, the measured error is approximately

$$E = D - cd,$$

where $0 \le c \le 1$.

A similar test procedure is derivable from the triple angle formula for the cosine function:

$$\text{cosine}(x) = \text{cosine}(x/3)[4\text{cosine}^2(x/3)-3].$$

Analysis similar to that for the sine function shows that x should be drawn from the interval $[(3m+1)\pi,(3m+3/2)\pi]$. The tests we recommend use m = 2. If argument purification is used, the resulting error expression is again

$$E = D - cd,$$

where E, D and d are defined analogously to the previous case, and where $0 \le c \le 2$.

Table III contains results of these tests, as implemented in the ELEFUNT package, on several different programs on several different computer systems. Tests labeled 3 are the cosine tests. Comparison of the IBM results with the master values shows how sharp these tests are. Large errors reported for other machines indicate poor argument reduction or improper meshing of the cosine and sine computations.

Identity tests of the kind just described are seldom as discriminating as direct comparisons with higher precision calculations, but they can be surprisingly sharp and provide extremely useful diagnostic techniques when properly designed and implemented. The test programs in ELEFUNT supplement accuracy tests based on identities with various other tests probling to see whether the function routines preserve important analytic properties of the function, whether the routines are robust, and how error conditions are handled. Thus, ELEFUNT exemplifies the physical examination approach to software testing that we advocate.

5. References

[1] M. Abramowitz and I.A. Stegun, Handbook of Mathematical Functions with Formulas, Graphs, and Mathematical Tables, Nat. Bur. Standards Appl. Math. Series, 55, Washington, D.C., 1964.

[2] W.J. Cody, "Performance testing of function subroutines," AFIPS Conf. Proc., Vol. 34, 1969 SJCC, AFIPS Press, Montvale, N.J., 1969, pp. 759-763.

[3] W.J. Cody, "The FUNPACK package of special function subroutines," TOMS 1, 1975, pp. 13-25.

[4] W.J. Cody, "Basic concepts for computational software," Applied Mathematics Division Technical Memorandum 360, Argonne National Laboratory, November, 1980 (to appear in Proceedings of International Seminar on Problems and Methodologies in Software Production, Sorrento, Italy, November 3-8, 1980).

[5] W.J. Cody, "Implementation and Testing of Function Software," Applied Mathematics Division Technical Memorandum 363, Argonne National Laboratory, November, 1980 (to appear in Proceedings of International Seminar on Problems and Methodologies in Software Production, Sorrento, Italy, November 3-8, 1980).

[6] W.J. Cody and W. Waite, Software Manual for the Elementary Functions, Prentice Hall, Englewood Cliffs, N.J., 1980.

[7] C. Hammer, "Statistical validation of mathematical computer routines," AFIPS Conf. Proc., Vol. 30, 1967 SJCC, Thompson Book Co., Washington, D.C., 1967, pp. 331-333.

[8] A.C.R. Newbery and A.P. Leigh, "Consistency tests for elementary functions," AFIPS Conf. Proc., Vol. 39, 1971 FJCC, AFIPS Press, Montvale, N.J., 1971, pp. 419-422.

[9] S. Walther, "A unified algorithm for the elementary functions," AFIPS Conf. Proc., Vol. 38, 1971 SJCC, AFIPS Press, Montvale, N.J., 1971, pp. 379-385.

TESTING AND EVALUATION OF STATISTICAL SOFTWARE

James E. Gentle

IMSL, Inc.
Houston, Texas

Abstract

Data sets analyzed by statisticians are likely to be ill-conditioned for the kinds of computations ordinarily performed on them. For this reason, most of the activity in testing and validation of statistical software has centered on numerical error analysis. The most generally effective testing and validation method has been test data generation. Once the parameters of numerical condition are identified, methods for systematically generating test data sets can be developed. In the case of least squares computations, for example, collinearity and stiffness seem to be the major components of condition. Test data sets with prespecified collinearity and stiffness are discussed.

The problem of software validation goes beyond the tests of the kernels that perform numerical computations. A program may involve the use of several computational modules, and errors in the program often occur in the setup stages in moving from one computational module to another. A stochastic model for errors remaining in a program after a sequence of tests is presented and discussed.

There are many statistical computations for which perturbation methods can be used easily to assess correctness. Perturbation methods can be employed by the user so as to avoid the bigger question of testing the software; the test is for the performance of the software on the specific problem of interest.

I. Introduction

The past several years have seen the development of a number of software systems for statistical data analysis. While the quality of this software in terms of capability, efficiency and reliability has continued to improve over time, we still often hear horror stories of programs gone awry. And for each known instance of program failure, it is likely there are several unknown failures. Why are there so many unreliable programs still in use? What are the causes of poor software? I will mention five reasons for poor and unreliable software.

1. SOFTWARE PRODUCTION HAS GENERALLY BEEN A COTTAGE INDUSTRY. Many of the programs in widespread use were developed by some graduate student or research assistant, possibly for a project that the person was closely involved with, or possibly at the behest of a researcher

in another department, and for a project that had no intrinsic interest for the person writing the program. Programs developed in this manner get passed around the computation center, then to other computer centers as personnel movement occurs, and frequently find their way into commerically distributed software packages. Software, in such movement, acquires a certain legitimacy and aura of respectability, as each user assumes that the growing pedigree confers quality. The cottage nature of the software industry extends to the commercial houses, many of which have only one or two programmer/analysts. (The third and subsequent persons hired are generally in marketing.) The result of the nature of software development groups is that adequate resources can rarely be devoted to testing and validation.

2. THE MAGNITUDE OF THE NECESSARY DEVELOPMENT EFFORT IS OFTEN OVERLOOKED. The development of high quality software just seems to take longer than most people realize. One reason for this is that a running prototype of almost any program <u>can</u> be developed in a relatively short time. The innumerable details requiring careful attention in the final product are often not appreciated by the manager preparing a Gantt chart or setting the completion deadlines.

3. SOFTWARE PRODUCED AD HOC OFTEN FALLS INTO GENERAL PURPOSE USAGE. Many (most?) software development projects are initiated in response to the needs of other projects. The software often outlives its usefulness, but continues to live because it is used. Why? Because "the computer center has it and we know how to use it."

4. THE ATTITUDE IN THE COMMERCIAL SOFTWARE MARKET IS CAVEAT USITOREM. In the early days software was free or at least cheap. The consumer could not expect much for nothing. Even after software became expensive the production remained in a fragmented cottage industry. The supplier may not be in the same business when the customer realizes the product is not performing as advertised.

5. NO ENTIRELY ADEQUATE TESTING AND VALIDATION METHODOLOGY IS IN COMMON USE. There remain very real difficulties in software testing. Some persons working in the area of testing and validation may feel that adequate methodology is available for bug-free development and thorough testing of software. The only sense in which this is true is tautologically: incorrect programs result from mistakes in program development and failure to determine a program is in error results from mistakes in program testing, in that the tests did not uncover the error; hence, bug-free development occurs if proper methodology is employed.

The problem of testing and validation of software can be viewed in three separate aspects: tests of computational kernels (algorithms), validation of programs, and evaluation of software systems. In each phase certain steps toward automation may be made. A final aspect of validation may be carried out by the user. This consists of consistency checks and perturbation methods to insure that the results for a given problem are correct without regard to whether the program is correct in general.

II. Testing Computational Kernals

In mathematical and statistical applications, certain computational problems occur repeatedly: solution of a full-rank set of equations, best solution of an overdetermined system, optimization of a function, it is necessary to understand what factors determine whether a given problem specification is hard, that is, ill-conditioned. By "condition" of data we mean the extent of expected computational difficulties. Knowing the factors of condition and how to measure them aids not only in the construction of test data, but also yields more direct benefits. The condition may indicate the need to switch to a more accurate, but slower, algorithm; or it may provide the user with information on the confidence to be placed in computed results.

For most statistical computations there are two relevant aspects of condition: "stiffness" and "collinearity". Stiffness is a univariate problem of low coefficient of variation. Consider the following data and associated summary statistics.

$$\begin{array}{r} 9000 \\ 9001 \\ \underline{9002} \end{array}$$

$$\Sigma x_i = 27003 \qquad \Sigma x_i^2 = 243{,}054{,}005$$

$$(\Sigma x_i)^2/n = \underline{243{,}054{,}003}$$

$$\Sigma x_i^2 - (\Sigma x_i)^2/n = \quad 2$$

With seven digits of precision $\Sigma x_i^2 = (\Sigma x_i)^2/n$. So we would have to be careful if we wanted to compute some quantity like $s^2=(\Sigma x_i^2-(\Sigma x_i)^2/n)/(n-1)$. Of course if we remove the mean this example causes no problem

$$\begin{array}{r} -1 \\ 0 \\ 1 \end{array} \qquad \text{easily yields } s^2 = 1$$

(As an aside, we may mention that this problem, as much as any other, led statisticians to the realization that there is some benefit in the analysis of algorithms. Each method suggested above has a clear advantage--efficiency in one case and accuracy in the other--and a clear disadvantage--lack of either efficiency or accuracy. Hence statisticians immediately recognized the superiority of the recursive algorithm for computation of adjusted sums of squares. As a further aside to those familiar with this recursive method, we may mention that in multivariate problems the recursive method loses some of its appeal vis-a-vis the two-pass method if it is desired to use higher precision in the accumulations.)

The stiffness property geometrically is the angle that a data vector makes with the unit vector. The collinearity property is the angles made by data vectors with linear combinations of other data vectors. Consider the data

x_1	x_2
1	0
90	90
900	900
9000	9000

$$\Sigma x_{1i}^2 = 81818101$$
$$\Sigma x_{1i}x_{2i} = 81818100$$
$$\Sigma x_{2i}^2 = 81818100$$

In the seven digits the matrix

$$\begin{bmatrix} \Sigma x_{1i}^2 & \Sigma x_{1i}x_{2i} \\ \Sigma x_{1i}x_{2i} & \Sigma x_{2i}^2 \end{bmatrix}$$

is singular. Again, taking out the means helps. (But not as much: replacing 9000 by 90000 in the data set will again give a singular matrix in seven digits even if the means are removed.)

Stiffness may be measured easily by the number of leading digits which do not vary in a data set. (See Greefield, and Siday, 1980, for further discussion of this property.) In the first example above the stiffness is 3. Collinearity may be measured by the condition number of the data matrix, i.e. by the ratio of the largest singular value to the smallest singular value of the data matrix. In the second example above the condition number is approximately 2×10^4. When transformations are applied to the data these measures may change. It is by observing the change in these measures within the steps of an algorithm that may lead to the identification of poor algorithms. For example, $\sqrt{x_i} - \sqrt{x_j}$ may not be a simple computation if x is stiff, since the stiffness increases for a square root operation; hence the calculation may be performed by evaluating $(\sqrt{x_i} + \sqrt{x_j})^{-1}$. Subtraction of the mean from the untransformed data is a technique that frequently improves the quality of the computations because it decreases the stiffness. A more striking example of the effect of intermediate computations on a measure of condition is the squaring of a matrix condition number when the matrix is multiplied by its transpose. Since the condition number of $X'X$ is the square of the condition number of X, the better algorithms for regression analysis deal with the X matrix rather than the $X'X$ matrix. (This is in the ordinary notation for a regression model: $\underline{y} = X\underline{\beta} + \varepsilon$.) Since linear programming codes for L_1 regression do not use the $X'X$ matrix in any event, statisticians generally find less numerical problems in L_1 regression than in least squares regression on data sets which are ill-conditioned by traditional standards.

After identification of the factors that make a problem hard and definition of quantitative measures of these factors, we wish to provide test data sets that range over relevant values of the quantitative measures. The problem, of course, is in knowing what the true solution is, so that the computed solution can be evaluated for correctness and for accuracy. In some cases a true solution is chosen and the problem is generated from it, and in other cases the true solution is obtained by painstakingly carrying a very large number of digits of precision throughout the computations. Choosing a solution first and then generating a problem is a technique that must be handled with care. As an example, consider testing a linear regression code using data generated from a polynomial model. This technique is used often, since the terms arising from various powers will have high collinearity. But suppose the regression coefficients are all chosen to be, say 1, and that the values of the independent variable are allowed to range from 1 to 20. For a test in single precision on a computer with 32-bit words, by the time the degree of the polynomial is allowed to exceed six, enough error has been introduced into the <u>input data</u>, i.e. the test data, to significantly affect the <u>true solution</u>; so what is thought to be the true solution is no longer it. This sort of problem may be expected to occur often in the construction of test problems, unless care is exercised, since by their nature test problems are ill-conditioned and, in ill-conditioned problems, small perturbations of the input effect large perturbations of the output.

In one method of test problem construction, specific data sets are given, as in Gregory and Karney (1969), Hastings (1972), and Kennedy and Gentle (1980), Chapter 10. In other cases, methods for construction of random data sets are given, as in Velleman and Allen (1976), Hoffman and Shier (1977) and Kennedy and Gentle (1980), Chapter 8.

In a least squares regression analysis on the model $\underline{y} = X\underline{\beta} + \underline{\varepsilon}$, whether or not the model fits the data exactly is of little consequence for the performance of the algorithm. (This does make a big difference in L_1 regression analysis, however.) For test data sets for least squares regression problems, therefore, it is a simple matter of choosing X and β and computing y from them. The following method allows random choice of X within a class of data sets having approximately preassigned stiffness and collinearity. In what follows we assume X is n by p and that the model contains an intercept term.

a) Generate randomly p-1 columns in the n by p matrix Z having all 1's in the first column.

b) Decompose Z:

$$Z = Q_1 \begin{bmatrix} R \\ 0 \end{bmatrix}$$ where Q_1 is n by n and orthogonal and R is p by p upper triangular

c) Let $U = ZR^{-1}$

d) Generate randomly (p-1) by (p-1) matrix W

e) Decompose W:

$$W = Q_2 T$$ where Q_2 is orthogonal

f) Let

$$V = \begin{bmatrix} 1 & 0 \ldots 0 \\ 0 & \\ \cdot & \\ \cdot & Q_2 \\ \cdot & \\ 0 & \end{bmatrix}$$

g) Select the column means of X, call them m_i and let

$$E = \begin{bmatrix} 0 & m_2 & m_3 \ldots & m_p \\ \cdot & \cdot & \cdot & \cdot \\ \cdot & \cdot & \cdot & \cdot \\ \cdot & \cdot & \cdot & \cdot \\ 0 & m_2 & m_3 & m_p \end{bmatrix}$$

h) Select a range for the condition number of X by choosing a_1 and a_p in the lower and upper limits

$$\frac{a_1 - e}{a_p + e} \quad \text{and} \quad \frac{a_1 + e}{a_p - e} \quad ,$$

respectively where $e = \sqrt{n \sum_{i=2}^{p} m_i^2}$ and $a_1 > a_p > e$.

i) Let D be the diagonal matrix with elements $a_1 > a_2 > \ldots > a_p$, where the other a_i are chosen arbitrarily.

j) Let $X = UDV + E$.

This procedure has been used advantageously in tests of regression algorithms for incorporation in the IMSL Library.

In addition to data sets with known solutions, groups of related data sets with known relationships among their solutions can be usefully employed in testing computational kernels. In this technique one data set is obtained from another by a transformation that has a known effect on the solution. If the solutions obtained are not consistent then the computing algorithm failed. We will say more about these perturbation methods in Section 4.

III. Validation of Software

Problems of validation of software go beyond tests of computational kernels--that is, tests of numerical algorithms. A program may involve the use of several computational modules; and errors in the program often as not are due to setup and housekeeping activities preparing input to, or processing output from, the computational kernels. The tutorial collection of papers edited by Miller and Howden (1978) provides a good survey of techniques of software validation.

Recent work by Duran and Wiorkowski (1979) and by Kubat and Koch (1980) develop stochastic models of program correctness along two distinct lines. Duran and Wiorkowski build models around the proportion of data sets for which the program will fail, and the models have application only to programs thought to be correct, whereas Kubat and Koch consider models based on error-free running times, and the models also have application in program development. In this section we describe an alternate model for the proportion of data sets for which the program will execute without error. The model is developed within a program development environment.

Various testing procedures have been discussed in the literature (see Kubat and Koch, 1980, for a description of four procedures). Consider the following procedure: The program is run on test data sets until the program fails or until it has performed successfully on N test sets. When and if the program fails, an attempt is made to correct the program error; and then more test sets are submitted to it. The test continues until the program performs successfully on N successive test cases.

Suppose in this testing procedure that k errors are found. Effectively, then, there are k+1 versions of the program, each version resulting from a correction of the previous version. Let p_i be the (inknown) proportion of data sets from the relevant universe for which the ith version of the program would fail, and let n_i be the number of test cases for which the ith version operated successfully before failing on the (n_i+1)th test case. The model, therefore, is a random sequence of k+1 geometric distributions with parameters $p_1, p_2, \ldots, p_{k+1}$. The problem is to estimate p_{k+1}, that is, the proportion of possible input choices for which the final model will fail.

The probability function describing the model is

$$P(k=0) = 1 - \sum_{j=0}^{N-1} p_1(1-p_1)^j$$

$$P(n_1,n_2,\ldots n_k,k) = \left[1 - \sum_{j=0}^{N-1} p_{k+1}(1-p_{k+1})^j\right] \prod_{i=1}^{k} p_i(1-p_i)^{n_i}$$

for k=1, 2, 3, ...

and n_i=0, 1, ..., N-1

This model, like many describing failure rates, does not yield useful maximum likelihood estimates. (The MLE for p_{k+1} is 0.) The traditional procedure in such cases is either to derive interval estimates or to determine a Baysian solution. Either procedure for this model yields an optimization problem that must be solved iteratively. If the assumption is made that $p_1 \geq p_2 \geq \ldots \geq p_k \geq p_{k+1}$, maximum likelihood conditioned on k yields a flat maximum for p_{k+1}, maximum likelihood conditioned on k yields a flat maximum for p_{k+1} in the interval $[0, \hat{p}_k]$, where $\hat{p}_k$ is the value of p_k that maximizes the geometric objective function

$$\prod_{i=1}^{k} p_i(1-p_i)^{n_i}$$

subject to the inequality restrictions on the p_i's. More work remains to be done on this model and will be reported on elsewhere.

IV. Ad Hoc Testing by the User

The real concern of the user of computer software is whether or not the answers to the user's problems are correct. The larger question of general correctness of the software is interesting, but may possibly be avoided if the user can get assurances of the correctness of the answers to the given problem. A useful way of obtaining such assurances is by perturbation methods and internal consistancy checks. Perturbation methods involve modifications of the input data or the intermediate quantities in the computations. The program is then run on both the original problem and the perturbed problem. If the perturbations ar slight, the extent of the perturbations in the output indicates condition of the data. If the indication is that the data are well conditioned, some confidence may be gained in the computed results for the original problem. Specific methods of slight perturbation are 1) changing the precision of the input and/or the intermediate computations, 2) modifying the sequence of the input data, and 3) changing starting points in algorithms requiring such input.

More useful perturbation methods involve the formation of different problems whose solutions have known relationships to the solution of the given problem. If the computed solutions are consistent (and if the perturbations have actually modified the intermediate computations of the program), then the user can have more confidence in the successful execution of the program on his problem. An example of such a perturbation method in regression analysis of the model

$$\underline{y} = X\underline{\beta} +$$

is, after obtaining an estimate of $\underline{\beta}$, call it $\underline{b}$, and the residuals, $\underline{e}$, multiplication of the $j\underline{th}$ column of X by d_j and addition of this quantity to $\underline{y}$ to form the regression problem

$$\underline{Z} = X\underline{\alpha} + \underline{\varepsilon}.$$

The least squares estimator of $\underline{\alpha}$, call it $\underline{a}$, should satisfy the relationship $a_j = b_j + d_j$, and the residuals from this data set should be exactly the same as those from the original data set. If in applying this method for a few values of j and d_j all the consistency checks are passed, the user can have a high degree of confidence in the computed solution to the original problem. (If not, the user has the same dilemma as any person testing software who finds that it fails.)

These perturbation methods can be so helpful to the end user of software that software developers and numerical analysts should attempt to define such methods for all common problems in numerical computations.

References

[1] Duran, J.W., and Wiorkowski, J.J. (1979). "Quantifying software validity by sampling," UTD Programs in Mathematical Sciences, Technical Report #50.

[2] Gilsinn, J., Hoffman, K., Jackson, R.H.F., Leyendecker, E., Saunders, P., and Shier, D. (1977). "Methodology and analysis for comparing discrete linear L_1 approximation codes," Communications in Statistics, Part B, 6, 399-414.

[3] Greenfield, T., and Siday, S. (1980). "Statistical computing for business and industry," The Statistician, 29, 33-55.

[4] Gregory, R.T., and Karney, D.L. (1969). A Collection of Matrices for Testing Computational Algorithms, Wiley-Interscience, New York.

[5] Hastings, W.K. (1972). "Test data for statistical algorithms: Least squares and ANOVA," JASA, 67, 874-879.

[6] Hoffman, K.L., and Shier, D.R. (1977). "A test problem generator for discrete linear L_1 approximation problems," NBS Working Paper.

[7] Kennedy, W.J., and Gentle, J.E. (1977). "Examining rounding error in LAV regression computations," Communications in Statistics, Part B, 6, 415-420.

[8] Kennedy, W.J., and Gentle, J.E. (1980). Statistical Computing. Marcel Dekker, New York.

[9] Kennedy, W.J., Gentle, J.E., and Sposito, V.A. (1977). "A computer oriented method for generating test problems for L_1 regression," Communications in Statistics, Part B, 6, 21-27.

[10] Kubat, P., and Koch, H.S. (1980). "On the estimation of the number of errors and reliability of software systems," University of Rochester, Graduate School of Management, Working Paper No. 8012.

[11] Miller, E., and Howden, W.E., Editors (1978). Tutorial: Software Testing and Validation Techniques, IEEE Computer Society, Long Beach.

[12] Osterweil, L.J., and Fosdick, L.D. (1976). "Program testing techniques using simulated execution,' Simuletter, 7 (4), 171-177.

[13] Velleman, P.F., and Allen, I.E. (1976). "The performance of package regression routines under stress: a preliminary trial of a regression evaluation method," Proceedings of the Statistical Computing Section, American Statistical Association, Washington, 297-304.

[14] Wampler, R.H. (1978). "Test problems and test procedures for least squares algorithms," Proceedings of Computer Science and Statistics: Eleventh Annual Symposium on the Interface. North Carolina State University, Raleigh, 84-90.

TOOLPACK - AN INTEGRATED SYSTEM OF TOOLS FOR MATHEMATICAL SOFTWARE DEVELOPMENT

Leon Osterweil
Department of Computer Science
University of Colorado, Boulder

ABSTRACT

This paper describes the approach being taken in configuring a set of tool capabilities whose goal is the support of mathematical software development. The TOOLPACK tool set is being designed to support such activities as editing, testing, analysis, formatting, transformation, documentation and porting of code. Tools for realizing most of these functional capabilities already exist, yet TOOLPACK aims to do far more than simply bring them together as a collection of side-by-side individual tools. TOOLPACK seeks to merge these capabilities into a system which is smoothly integrated both internally and from a user's external point of view. The internal integration approach involves the decomposition of all tools into a more or less standard set of modular "tool fragments". The external integration approach involves the creation of a command language and a set of conceptual entities which is close to the conceptual set used by mathematical software writers in the process of creating their software. This paper describes both of these integration approaches as well as the rather considerable and novel software support needed to make them work.

1. Introduction and Overview

This paper describes a project known as TOOLPACK which is being undertaken jointly by eight organizations: Argonne National Laboratory, Bell Telephone Labs, International Mathematical and Statistical Libraries, Inc., Jet Propulsion Laboratory, Numerical Algorithms Group, Ltd., Purdue University, University of California at Santa Barbara, and University of Colorado at Boulder.

The purpose of TOOLPACK is to produce a software system providing strong, comprehensive tool support to programmers who are writing, testing, transporting or analyzing mathematical software. TOOLPACK is currently in the design phase, but already certain important assumptions and design features have been established. These will be discussed here.

This work was supported by NSF Grant #MCS8000017 and DOE Contract #DE-AC02-39ER10718.

The guiding assumptions about TOOLPACK are as follows:

- o TOOLPACK is designed to provide cost effective support for the production by up to 3 programmers of programs whose length is up to 5000 lines of Fortran 77 source text. TOOLPACK may be less effective in supporting larger projects.
- o TOOLPACK is to be designed to provide cost effective support for the analysis and transporting of programs whose length is up to 10,000 lines of Fortran 77 source text. TOOLPACK may be less effective in supporting larger projects.
- o TOOLPACK will support users working in either batch or interactive mode, but may offer stronger more flexible support to interactive users.
- o TOOLPACK itself will be highly portable, making only weak assumptions about its own operating environment. It will be designed, however, to make effective use of large amounts of primary and secondary memory, whenever they can be made available.

In order to achieve these goals, a set of underlying tool capabilities has been tentatively agreed upon. These capabilities include intelligent source text editing, source text formatting, source text structuring, various forms of static analysis, dynamic testing assistance, and various forms of program transformation. Tools to provide these capabilities have already been built (e.g., see [DORR76], [OSTE76], [STUC75], [FEIB80], [BOYL76], [MYER81]). It has been observed, however, that they have not received the acceptance that had been hoped for. We attribute this to a number of factors, most particularly inefficiencies in the tools. lack of portability of the tools, difficulties in understanding and effectively utilizing the tools, and problems in interfacing the tools to each other, as is often desirable.

Hence, a primary motivating goal of the TOOLPACK design is to provide the user strong and comprehensive support in as direct and painless a fashion as is feasible. In particular, the design attempts to relieve the user of having to understand the natures and idiosyncrasies of individual TOOLPACK tools. It also relieves the user of the burden of having to combine or coordinate these tools. Instead the design encourages the user to express his/her needs in terms of the requirements of his/her own software job. The TOOLPACK support system is designed to then ascertain which tools are necessary, properly configure those tools, and present the results of using the tools to the user in a convenient form.

The design encourages the user to think of TOOLPACK as an energetic, reasonably bright, assistant, capable of answering questions, performing menial but onerous tasks and storing and retrieving important bodies of data.

In order to reach this view, the user should think of TOOLPACK as a vehicle for establishing and maintaining a data base of all information important to the user, and using that data base to both furnish input to needed tools and capture the output of those tools. Clearly such a data base is potentially quite large and is to contain a diversity of stored entities. Source code modules would certainly reside in the data base, but so would such more arcane entities as token lists, and flowgraph annotations. In order to keep TOOLPACK'S user image as straightforward as possible the design proposes that virtually all data base management be done internally to the TOOLPACK system, out of the sight and sphere of responsibility of the user. The user, in addition, is to be afforded direct access only to a relatively small number of data base entities -- only those such as source code modules and test data sets which are of direct concern to him/her. The user may create, delete, alter and rename these entities. More important, however, the user is to manipulate these entities with a set of commands which selectively and automatically configure and actuate the TOOLPACK tool ensemble. The commands are designed to be easy to understand and use. They borrow heavily on the terminology used by a programmer in creating and testing code, and conceal the sometimes considerable tool mechanisms needed to effect the results desired by the user.

In order to encourage and facilitate the preceding view of TOOLPACK, the system will support the naming, storage, retrieval, editing and manipulation of the following classes of entities, which should be considered to be the basic objects of TOOLPACK:

1. Program Units: A TOOLPACK program unit (PU) is the same as a Fortran program unit, except that TOOLPACK will require a number of representations of the program unit other than the source code. The identities and significance of these other representations are to be concealed from the user. They will be managed exclusively by the TOOLPACK system.
2. Execution Units: Any set of TOOLPACK program units which the user chooses to designate, can be grouped into a TOOLPACK execution unit (EU). Ordinarily it is expected that an execution unit will be a body of code which is to be tested as part of the incremental construction process.

An execution unit may also include an optionally specifiable transformation specification in order to enable users to painlessly apply canonical transformations to their code.

3. Test Data Collections: A TOOLPACK test data collection (TDC) is a collection of test data sets to be used in exercising one or more TOOLPACK execution units.
4. Options Packets: A TOOLPACK options packet (OP) is a set of directives specifying which of the many anticipated options are to be in force during any given application of a specific tool.

The reason for defining these four entities as being basic to TOOLPACK is that they seem to correspond closely to the concepts programmers use in thinking about and carrying out their program production and analysis work. This can perhaps best be seen by introducing the TOOLPACK command set and indicating roughly how it is to be used to manipulate these named data entities.

The exact syntax of the TOOLPACK command language has not yet been established. It has been tentatively agreed, however, that the command primitives will closely reflect functional capabilities needed by users, and that they will be readily composable into sequences to which names can be attached. As an illustration, a sampling of some of the proposed functional primitives is now presented, using an arbitrarily chosen syntax.

The proposed TOOLPACK command set seems to divide logically into four parts: data base management commands, edit (synthesis) commands, tool application (analysis) commands, and perusal commands.

1. Data Base Manipulation Commands:
   NEW [PU, EU, TCD, OP] entityname
   OLD entityname
   DELETE entityname
   REPLACE [newname]
   RENAME oldname, newname
2. Edit (Synthesis) Commands:
   EDIT entityname
3. Tool Invocation (analysis) Commands:
   FORMAT entityname [, newname]
   STRUCTURE entityname [, newname]
   ANALYZE entityname [, thoroughness-level],[, reporting option]
   This command results in the static analysis of the entity named "entityname". If "entityname" specifies a program unit, then single unit analysis will be performed. If "entityname" specifies an execution unit, then each program unit will be analyzed indivi-

dually and integration analysis will also then be performed.

A thoroughness level may be specified by the user. This specification will cause analysis to go as far as the lexical level, the syntactic level, the static semantic level or the data flow level. If this specification is omitted, the TOOLPACK system will select a default option (probably full data flow analysis).

The results of this analysis will be placed into an entity-attribute-relational data base which will then be available for perusal by a browsing subsystem to be described subsequently. The user may, however, specify that certain subsets of the analytic results are to be printed immediately in addition to being stored in the data base. If omitted, the default specification will probably be to print only fatal errors.

It should be clear that invocation of the ANALYZE command will effect the marshalling and configuration of a considerable assortment of tools and tool fragments. In addition, the stronger forms of analysis will necessitate the use of a number of intermediate images of the source text (e.g., parse tree, flowgraph, callgraph). An important design criterion was that these maneuverings and the materialization of these intermediate images by concealed from the user and made the responsibility of the TOOLPACK system.

EXECUTE TEST EUname, TCDname [, OPname]

This command results in the execution of a collection of test data sets by a specified execution unit. The test data sets comprising the test data collection "TCDname" are fed into the execution module derived from the execution unit "EUname" one at a time, with the results of each execution being used to build an execution history data base.

The user may optionally specify a test options packet (OPname) whose purpose is to select and specify which of the numerous execution monitoring options are to be employed during the test runs. The power and flexibility of the proposed dynamic test monitoring system is considerable [FEIB80]. This is deemed to be necessary, but is also considered to be a serious problem, in that a casual or novice user may be intimidated by the variety of available choices. Hence it is proposed that a set of standard test OP's be prepared by the builders of the dynamic test monitoring system and stored permanently in the TOOLPACK data base. Users could select from among these, tailor them to individual needs, or create their own test OP's from scratch. One of the standard test

OP's would be configured to be the default test OP, enabling the user to do useful dynamic testing without needing to specify any test OP.

4. Perusal Commands

TOOLPACK will contain tools to facilitate the examination of the various objects in the TOOLPACK data base. At present it is unclear whether one tool, such as a text editor, will be responsible for examination of all objects, or whether specially adapted intelligent editors will be built for various entities such as source text, and diagnostic sub-data bases.

Although it is possible that there will be different browsers for browsing the static analysis and dynamic testing data bases, it is expected that they will both be invoked by the same command:

PERUSE [databasename]

The databasename is one which will be automatically created by the TOOLPACK system by a straightforward naming algorithm. For example, the data base produced by test run # n of TDC t applying test OP p to EU e would perhaps be named e/p/t/n/DB. After each test run the user would be supplied this name and the size of the data base itself and offered the opportunity to SAVE the data base. SAVE'd data bases would then be available for subsequent PERUSE'ing.

II. Overview of Implementation Approach

The preceding section was devoted to a presentation of a user's view of the TOOLPACK system. The purpose of this section is to suggest an interior, or implementor's, view of the TOOLPACK system.

Clearly the primary feature of the proposed TOOLPACK system is the central data base of information about the subject program. The user is encouraged to think and plan his/her work in terms of it, and the functional tools all draw their input from it and place their output into it. The data base should be considered to be a single large (perhaps up to several megabytes on large-configuration machines) array made available to the TOOLPACK system as a permanently catalogued random access file by whatever file system and virtual storage mechanism the host machine may furnish. (It should be noted that there is a portable system, written in Fortran, which accomplishes this [HANS80a] [HANS80b]. It is likely that this system will be incorporated into TOOLPACK to provide these capabilities.) This file is to be initialized with the start of a project and remain and grow throughout the lifetime of the project. There is no reason why 2-3 users may not all access this file, although we will make the implicit assumption that the file is accessed by one user at a time in a nondestructive way.

The TOOLPACK system itself will manage this array by allocating contiguous blocks for the storage of named and synthesized TOOLPACK entities and TOOLPACK system tables and directories (see Figure 1). These blocks are accessible to TOOLPACK users and tools alike only by name, never by direct pointers. The blocks will point and link to each other only from their headers so as to enable the TOOLPACK system to effectively manage storage through such techniques as free lists and garbage collection. The bodies of the blocks themselves will contain no direct data base pointers either to other blocks or to places within the same block. All pointers contained in blocks will point to relative locations within the containing block.

A critical characteristic of the TOOLPACK system data base is that it will be maintained as a virtual data base. Both the end user and the tool ensemble will be safe in assuming that any needed named entities and derived images will always be available. It will be the responsibility of the TOOLPACK information management system, however, to either retrieve these items directly or have them created or regenerated (in case storage exigencies precipitated their deletion by TOOLPACK's information management system). This process is to be completely transparent to the end user and tool ensemble. It seems to be the most straightforward way in which to assure that TOOLPACK will run on machine configurations offering varying amounts of storage and not penalize those users blessed with larger configurations.

Perhaps the workings of this virtual data base scheme can best be understood through an example. Suppose one of the functional tools (e.g., the static analyzer) needed access to the parse tree of a particular version of particular PU let us call it SUBR/VER. The tool would request (and subsequently receive) the parse tree through a subroutine call such as

```
CALL DBFTCH ('PARSE', 'SUBR', 'VER', ARRAY, LEN)
```

where ARRAY is the name of the array within the tool which is to receive the parse tree, and LEN is a specification of the length of ARRAY, included in this invocation to guard against inadvertent array overflow.

Subroutine DBFTCH would then use 'SUBR' and 'VER' as index keys to arrive at the block containing the directory of images of SUBR/VER in the data base. The pointer to the parse is to be stored at a fixed offset location in this block. Should the parse tree pointer be non-null, DBFTCH would follow the pointer to the data base block containing the parse. A length specifier stored there would be used to determine whether the invoking tool's array was large enough to hold the parse. If so DBFTCH would need only to read the block into the tool's array.

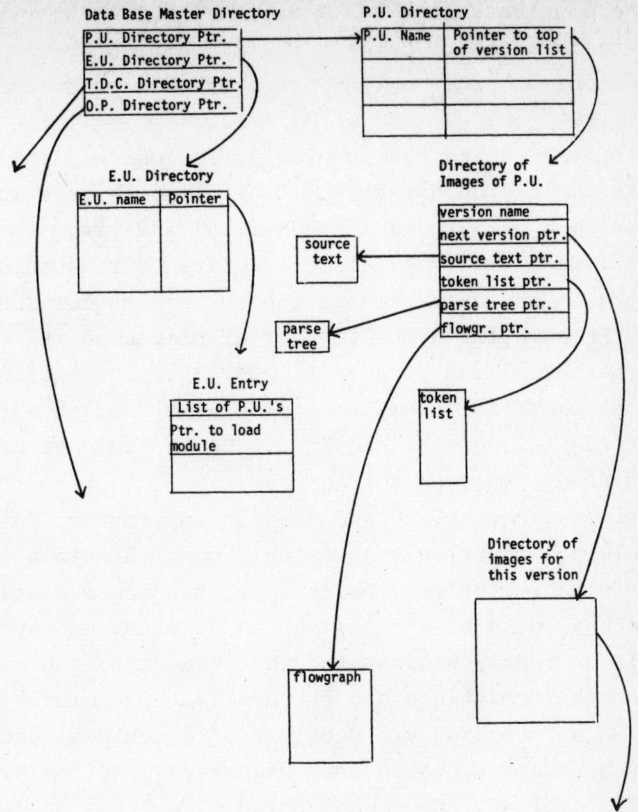

Figure 1
Schematic Layout of TOOLPACK Data Base

If the parse pointer were null, DBFTCH would need to see that a parse was created. Guidance for this process would come from an internal table specifying how the various TOOLPACK derived images are to be derived from each other. This table would be essentially a directed acyclic graph (DAG) with the nodes representing the various data base entities, and the edges representing processing capabilities. In particular, an edge would represent the processing capability needed to produce the entity at its head from the entity at its tail. It is worth noting that the production of some entities (e.g., an annotated flowgraph) might require that more than one process acting on more than one data base entity.

In any case DBFTCH would, from this table, produce an ordered list of the processes and entities which would be needed to produce the requested entity by traversing the dependency DAG. DBFTCH would then proceed down this list looking to see which entities are already stored in the data base. Using this information DBFTCH would then invoke in the correct order only those processes needed to produce the desired entity.

Returning to our example, DBFTCH would look up 'PARSE' in the dependency DAG, which would then show that a parse tree is derived from a token list by a parser and a token list is derived from a source string by a lexical analyzer (lexer). Thus DBFTCH would next check for the existence in the data base of the token list for SUBR/VER. If it is present DBFTCH will invoke the parser producing the required parse tree. If the token list is absent, DBFTCH will first invoke the lexer to produce a token list from the source text, and then invoke the parser. If the source text should not be in the data base, no error message would be passed on to the user.

The most straightforward and elegant implementation of this strategy is one involving recursion. Acceptably comprehensible and efficient solutions are, nevertheless, possible within the capabilities of Fortran.

This virtual data base scheme could be stretched even farther. Although it is currently anticipated that the data base will hold in explicit form any formattings and structuring of a given piece of source text, this is not necessary. Such versions could be recreated by the information management system only when needed by following a procedure such as just outlined. Even whole static analysis or dynamic testing data bases could be regenerated in this way. This gives the information management system the flexibility to purge large files to regenerate storage while still retaining the ability to recreate these files when necessary.

This feature should prove particularly useful in hosting the TOOLPACK system on smaller storage machines. Here it may be necessary to permanently store only source text. Under these circumstances all derived images and intermediate entities will be routinely purged, requiring that they be recreated whenever needed. This will result in extra computation time to meet the user's requests, but seems a very reasonable trade for the significant reduction in required storage.

III. Future Plans

It is anticipated that a system such as TOOLPACK will be capable of strongly impacting the way in which mathematical software is created and maintained. We are of the opinion that the changes effected by the availability and exploitation of high quality, well integrated, broadly ranging tools may for the first time also effect serious, thoughtful evaluation of what is really needed in the way of such tools. Accordingly, the TOOLPACK group is planning for not one, but a series of at least two releases of TOOLPACK. Our intent is to release a prototype version for evaluation and comment during the winter of 1982-1983. After this evaluation, a second release, presumably strongly influenced by public reaction and serious expressions of perceived needs, would follow within a year. Beyond that, many of us believe that subsequent releases of TOOLPACK and successor systems will be necessary periodically as we come to understand our true needs in such critical but relatively new areas as user interfaces, browsing capabilities, and effective use of data bases.

References

[BOYL76] J.M. Boyle and M. Matz, "Automating Multiple Program Realization," MRI Conf. Procd. XXIV, Symp. on Computer Software, Polytechnic Press, pp. 421-456 (1976).

[DORR76] J. Dorrenbacher, et al, "POLISH-A Fortran Program to Edit Fortran Programs," Dept. of Comp. Sci., University of Colorado Technical Report, #CU-CS-050-76 (Rev.) (May 1976).

[FEIB80] J. Feiber, R.N. Taylor and L.J. Osterweil, "NEWTON A Dynamic Testing System for Fortran 77 Programs; Preliminary Report," Univ. of Colorado, Dept. of Comp. Sci. Tech. note (November 1980)

[HANS80a] D.R. Hanson, "The Portable I/O System PIOS," University of Arizona, Dept. of Comp. Sci., Tech. Report #80-6a (April 1980, revised December 1980).

[HANS80b] D.R. Hanson, "A Portable File Directory System," Software-Practice and Experience, 10, pp. 623-634 (August 1980).

[MYER81] E.W. Myers, Jr. and L.J. Osterweil, "BIGMAC II: A Fortran Language Augmentation Tool,' proc. 5th Int'l Conf. on Software Engineering (to appear, March 1981).

[OSTE76] L.J. Osterweil and L.D. Fosdick, "DAVE - A Validation, Error Detection and Documentation System for Fortran Programs," Software Practice and Experience, 6, pp. 473-406 (September 1976).

[STUC75] L.G. Stucki and G.L. Foshee, "New Assertion Concepts in Self-Metric Software," Proc. 1975 Int'l Conf. on Reliable Software, IEEE Cat #75-CHO940-7CSR pp. 59-71.

OVERVIEW OF TESTING OF NUMERICAL SOFTWARE

Lloyd D. Fosdick
Department of Computer Science
University of Colorado, Boulder, CO 80309

ABSTRACT

The purposes of program testing are to expose errors; to insure that standards for portability, robustness, etc. are met; and to evaluate performance. A selection of work in these three applications of program testing is presented here.

A variety of tools and techniques for exposing errors in programs have been proposed. They range from measurements of testing thoroughness, to models for predicting errors, to attempts to prove a limited form of correctness. For various reasons, including cost, availability of tools, and interpretation of results, testing of numerical software for the purpose of exposing errors has made only limited use of these tools and techniques.

The problem of insuring that standards are met is easier to deal with, and one tool in particular, the PFORT verifier, has been widely used to insure portability. Tools for transforming programs to meet certain requirements have also been used in the production of numerical software as for example, in the production of LINPACK.

Program testing for performance evaluation has relied primarily on the use of selected test problems. Collections of test problems for several subject areas -- ordinary differential equations, linear equations, and optimization -- exist, and have received extensive use. Performance profiles have been used to present results of performance testing in a succinct and meaningful way. Modeling has been used to evaluate the performance of an algorithmic idea without incurring the expense of actually executing all of the supporting software.

1. Introduction

Computer programs are tested in order to detect errors, to insure that standards are met, and to evaluate performance. The methods used include controlled execution, direct inspection, formal analysis, and simulated execution. The focus of this overview is on testing to detect errors using controlled execution, and the discussion will assume that "testing" refers to this particular application.

Techniques for the analysis of programs are identified as dynamic or static according to whether they do or do not require execution of the program: from our viewpoint testing is a dynamic analysis technique. As a method for error detection it complements static analysis techniques which are capable of detecting errors such as improper matching of formal and actual procedure arguments, inconsistent use of COMMON storage, and referencing uninitialized variables but; it usually fails to detect computationally dependent errors such as execution of SQRT(X) when X has been assigned a negative value, or executing the array element reference A(I) when the value of I is out of bounds. Testing also exposes design errors not caught in direct inspection of the program: these include incorrect values for mathematical constants, sign errors, and incorrect convergence tests. In a recent study [HOW80a] of errors in a commercial numerical software library it was found that static analysis was the most effective method for detecting about half of the errors, and dynamic analysis was the most effective method for detecting the other half.

2. Why Test Software?

A program is simply a string of symbols, thus it is like a mathematical expression. In fact Davis [DAV72] identifies a program with the proof of a theorem: "The axioms are analogous to the input. The theorem is analogous to the output while the proof is the program." Since the analysis of mathematical expressions by formal methods is familiar and well understood, it is natural to ask -- why test programs, why not simply analyze them (i.e., the "proofs") by formal mathematical methods? The answer is that programs are different in two important ways: they are longer and they are not well-defined. The typical mathematical expression is short and therefore can be easily held and examined by the minds eye as a single entity, not so with programs which may be many thousands of symbols in length. The modern view of program design and construction recognizes this fact and attempts to deal with it by encouraging the designer and the programmer to think in terms of easily conceptualized modules. Nevertheless, suppose the program is viewed as a very long string of symbols -- why should that stop us, let a computer analyze the string. The response to this lies in the other difference, namely that programs are not well-defined. For a complete formal analysis of a program, a proof of correctness, we would need at least a complete specification of the environment of the program (machine, operating system, compiler, etc.) and a guarantee that the specification is correct (i.e. consistent with the actual environment). Without this the program is not well-defined: its definition is buried in its environment. For the present such a specification of the environment is completely beyond us. I suspect it always will be. Besides this, the enormous difficulty of proving, manually or mechanically, even the simplest programs in an ideal environment suggests that testing will always be necessary in establishing the reliability of computer programs. The article by Davis [DAV72] mentioned above and a recent article by DeMillo, Lipton, and Perlis [DLP79] on this subject are particularly interesting.

3. The Special Problem of Numerical Software

Testing uses an oracle. This oracle gives the correct answers for test problems, that is the answers we should get if there are no mistakes in the program. But, alas, the oracle for numerical software must contend with the inherent errors arising from rounding of numbers, discretization of continuous problems, and truncation of infinite sequences. So it gives only fuzzy answers. It does not say "the answer is 7," it says, "the answer is 7+e and I don't know what e is but I do know its magnitude is less than 1/10**6." Moreover the oracle is sometimes tricked by the presence of arithmetic anomalies [COD80], such as 1*x is not equal to x. Thus errors in test results which are caused by mistakes easily can pass unnoticed.

4. Costs

Software costs are a major part of the cost of a computing system. Moreover, 40% to 50% of software development costs are attributed to testing in large systems such as OS 360. There is good evidence showing that these percentages also represent the cost of testing numerical software if high quality is the goal.

One example concerns a collection of subroutines, 4500 lines of Fortran, intended for inclusion in a commercial library of numerical software. The test programs, which consisted of drivers and test data generators, consisted of 7900 lines of Fortran. Thus the size of the software created to test the product was 175% of the size of the product. A similar example concerns LINPACK [DON79]. It consists of 31,343 lines of Fortran. The test routines distributed on the LINPACK tape consist of 17096 lines of Fortran.

A direct cost figure can be drawn from the development of EISPACK [COW77]. It is estimated that the total development cost for the two releases of EISPACK was $1,000,000. The cost of field testing, which does not include testing done during program construction, is estimated at $300,000.

The implications of using unreliable numerical software cannot be ignored, therefore the high cost of testing it is a necessary expense. This provides a strong economic incentive for seeking more efficient methods for testing. Perhaps the biggest advance that can be made now to gain efficiency is to provide good tools to support existing methods and to make them widely available. This is a principle aim of the TOOLPACK project [COW79]. More on the issue of costs to build mathematical software can be found in an article by Cowell and Fosdick [COW77].

5. Methodology

The methods used to test programs are classified according to the criteria used for selecting test data. Three classes of methods are reviewed here: structural testing, functional testing, and mutation testing. With structural testing methods the test data are selected so that some criterion of coverage is satisfied; for example, every statement is executed at least once. With functional testing the view of the program and its components is that they are functions, transforming values from one domain to another. Here data are selected so as to "cover" the input and output domains in an effective way. With mutation testing the data are selected so that errors (mutations), deliberately inserted in the program, for example replacing a "+" by a "-", will cause incorrect results when the program is executed with the selected data.

It is universally agreed that whatever method is used the principal benefit is derived from the thought which must be given to the test data selection problem. If done carefully and conscientiously, it stimulates the tester to make a careful study of what the program is supposed to do, directing the tester's thoughts in an organized and purposeful way. The result is that many errors are discovered even before the test problems are run.

It is customary to insert probes in the program to help localize test results. Usually these probes are in the form of predicates. As with test data, experience has shown that careful attention to the construction of these predicates often reveals errors before the tests are run.

5.1 Structural testing. In structural testing the structures usually considered are statements, branches, or paths. Data are selected so that testing will result in the execution of all instances of the structure in the program. Execution of all statements is considered to be a minimal test. Probes in the program are required in order to record coverage. Tools which automatically insert these probes in the software are available. One example is BRNANL [FOS74] which can be used to measure statement coverage in Fortran programs. Since the number of paths is often indeterminate, it is necessary to impose some restrictions; for example, multiple traversals of a loop are not distinguished, procedures are treated separately, etc. A structure called a linear code sequence and jump has been proposed by Hennell [HEN80]. It is a restricted set of paths. Huang [HUA75] has discussed branch coverage in particular.

Data selected at random, even a rather large set, is not likely to yield 100% coverage. Therefore it is necessary to control the selection of test data. Tools to help do this have been described in the literature [MIL78], but they are not widely available. The problem of test data selection can be formulated as a problem of solving a system of inequalities. This can be seen in the following way. Trace any path through the program. In order for an execution of the program to follow this path certain predicates, namely those arising at each branch point, must be satisfied. These predicates are of the form

<expression> <relational operator> 0;

for example,

$$A - B < 0$$

The variables arising in the "expression" can always be expressed in terms of variables representing the input data. This is straightforward but tedious. Prototype tools for selecting test data by this approach have been built [CLA76]. Schnabel [SCH80] has developed an algorithm which seems to be well suited for solving such systems of inequalities.

It is difficult to attach any quantitative measure of reliability to structural testing methods. While most would agree that a program that had been tested with data selected so that every branch was executed at least once is more reliable than one that had been tested with data that did not meet this condition, just how much more reliable is it? No solution to this problem has appeared in the literature.

5.2 Functional testing. Here the program and its components are viewed as functions. The details of the realization of these functions as programs or procedures, or segments of these units, are ignored. For this reason it is sometimes called a "black box" method. In practice one actually regards the program as a composition of a number of functions whose inputs and outputs can be separately tested. Thus, the underlying program cannot be totally ignored. Howden has described this approach to testing in a recent paper [HOW80b].

Functional testing is actually no more than a formalization of the approach most numerical software developers take in testing their software. Not surprisingly, Howden found [HOW80a] that "Functional testing was significantly more effective in discovering errors ... than structural testing." However it must be pointed out that he advocates the use of both methods.

There are two particularly attractive aspects of functional testing. The first is that it is consistent with the current trend in programming methodology -- that is, a stepwise progression from high level functional specification of the program, through successive refinement steps, reaching finally the code itself. With this approach one can separate the issue of correctness of the functional specification from that of the correctness of the function transformations used in making the refinements. The advantage in making this separation is that the function transformations are quite general. Hence, once the correctness of the transformations is assured, its correctness need not be considered again for each application. The second attractive aspect of functional testing is that it permits a far more natural point of view of the program for the numerical analyst. Thus it is easier for him to guide the testing of the program.

5.3 Mutation testing. This approach to testing has been described by DeMillo, Lipton, and Sayward [DLS78]. It is somewhat akin to the technique of error seeding discussed by Gilb [GIL77], but differs from it in the manner in which the inserted errors are used. In the DLS scheme they are used to guide the selection of test data in the following way. We begin with the program to be tested and from it we generate an ensemble of mutants. Each mutant is derived from the initial program by making some slight change to it. For example the statement

A = B + C

is changed to

A = B - C.

Suppose we now select, in some unspecified way, input data. We execute each member of the ensemble of mutants on this data. Those mutants which yield incorrect results with this data are said to be "killed." Now new input data are selected and we execute the mutants that have not been killed with this data. This process is continued until all mutants are killed -- or until it can be shown that any mutants which remain alive are equivalent to the original program. The resulting set of input data constitutes the test data for the program. So far no independent experience with this technique has been reported in the literature. The method has the potential for being very expensive to use and it remains to be seen whether or not the benefits that can be derived justify the cost.

6. Effectiveness

Howden [HOW80a] compared the effectiveness of a number of methods, including structural and functional testing, for detecting errors. He studied 98 known errors in edition five of the IMSL library. These errors were documented by IMSL and corrected in edition six of the library. Howden found that functional testing was "the best or most appropriate method" for the detection of 32% of the errors. He found that structural testing was the best method for 11% of the errors. While some of the errors that could have been found by structural testing could also have been found by functional testing, he noted that structural testing was necessary for the detection of some of them.

Even though careful testing is performed it is a common experience that when the program is moved to a new system difficulties are encountered. To illustrate the nature of these difficulties I will describe the experiences of implementing the 4500 lines of Fortran for a commercial library which I referred to in section 4. These experiences occurred during implementation of the program on seventeen distinct computing systems. The small number of problems described below is indicative of the care that went into testing the original software.

A. When the implementation was made on an IBM 370 it was necessary to change variables from type REAL to type DOUBLE PRECISION. This was anticipated of course and a programming standard that had been imposed was that all variables be declared (though Fortran does not require this). This makes it easy to redeclare the variables as DOUBLE PRECISION when moving the program to a system which has short words, as was the case here. Unfortunately, a few variables had not been included in the declarations. This led to a storage alignment fault.

B. Although a tolerance factor had been carefully selected to be suitable for all systems on which it was expected the program would be used, it turned out that this tolerance factor had to be adjusted in order to make all of the test problems run successfully.

C. Use of the expression X*DSIGN(1.ODO,Y) caused problems on some machines because multiplication of X by +/-1 changed the magnitude of X. The problems caused by this arithmetic anomaly included slow performance and illegal memory addresses.

D. Destructive underflows were encountered on some machines when executing reasonably well-scaled problems. This led to performance failures, including division by zero.

7. Conclusion

This brief review has focused on the use of testing to detect errors in numerical software. The principal methodologies used are structural and functional testing. Another methodology receiving attention lately is mutation testing. No one method can be called best: a careful regime of testing uses a variety of complementary methods.

Testing depends heavily on heuristics and special case considerations. Much work needs to be done in order to establish a sound theoretical base to supporting testing.

8. References

[CLA76] Clarke, L.A., "A System to Generate Test Data and Symbolically Execute Programs," IEEE Trans. SE 2,3 (May 1976), 215-222.

[COD80] Cody, W.J., "Towards Sensible Floating-Point Arithmetic," Argonne National Laboratory, AMD (1980).

[COW77] Cowell, W.R., and L.D. Fosdick, "Mathematical Software Production," pages 195-224 in Mathematical Software III, J.R. Rice, ed., Academic Press (1977).

[COW79] Cowell, W.R., and W.C. Miller, "The TOOLPACK Prospectus," Argonne National Laboratory, AMD Tech. Memo. 341 (September 1979).

[DAV72] Davis, P.J., "Fidelity in Mathematical Discourse: Is One and One Really Two?", Amer. Math. Monthly, 79, 3 (March 1972), 252-263

[DLP79] DeMillo, R.A., R.J. Lipton, and A.J. Perlis, "Social Processes and Proofs of Theorems and Programs," ACM Comm. 22, 5 (May 1979), 271-280.

[DLS78] DeMillo, R.A., R.J. Lipton, and F.G. Sayward, "Hints on Test Data Selection: Help for the Practicing Programmer," IEEE Computer 11, 4 (April 1978), 34-41.

[DON79] Dongarra, J., C.B. Moler, J.R. Bunch, and G.W. Steward, "LINPACK User's Guide," SIAM (1979).

[FOS74] Fosdick, L.D.,"BRNANL, A Fortran Program to Identify Basic Blocks in Fortran Programs," University of Colorado, Dept. of Computer Science, Technical Report 40 (1974).

[GIL77] Gilb, T., Software Metrics, Winthrop (1977).

[HEN80] Hennell, M.A., and Prudom, A., "A Static Analysis of the NAG Library," IEEE Trans. SE., 6, 4 (July 1980), 329-333.

[HOW80a] Howden, W.E., "Applicability of Software Validation Techniques to Scientific Programs," ACM TOPLAS, 2, 3 (July 1980), 307-320.

[HOW80b] Howden, W.E., "Functional Program Testing," IEEE Trans. SE, 6, 2 (March 1980), 162-169.

[HUA75] Huang, J.C., "An Approach to Program Testing," ACM Computing Surveys, 1, 3, (September 1975), 113-128.

[MIL78] Miller, E., W.E. Howden, "Software Testing and Validation Techniques," IEEE Catalog No. EHO 138-8 (1978).

[SCH80] Schnabel, R., "Determining Feasibility of a Set of Nonlinear Inequality Constraints," Univ. of Colorado, Dept. of Computer Science, Technical Report 172 (February 1980).

THE APPLICATION OF HALSTEAD'S SOFTWARE SCIENCE DIFFICULTY MEASURE TO A SET OF PROGRAMMING PROJECTS

Charles P. Smith
International Business Machines Corporation
General Products Division
Santa Teresa Laboratory
San Jose, California

ABSTRACT

The difficulty measure, as defined by Halstead [1], is shown to have useful applications in the measurement of certain properties in code implementations. This difficulty measure reflects the effort required to code an algorithm, to test it, to inspect and review it, and to understand it later when the code needs alterations.

This paper explains how the difficulty metric reveals insights to the structure of a program module and also to some possible code impurities within the module. It also shows that assembler language programs are significantly more difficult than higher-level PL/S language programs. The author proposes that a maximum level (or threshold) of difficulty should be established to help manage the complexity of programs.

1. Introduction

Maurice H. Halstead [1,2,3] has suggested a comprehensive set of software metrics that IBM's Santa Teresa Laboratory is using to examine the characteristics of existing program products [4.5]. These software metrics do not hinge upon traditional lines-of-code counts but rather upon the counting of operations and operands. This report explains how the Software Science difficulty metric, as defined by Professor Halstead, relates to the world of IBM data gathered from large programming products.

2. Fundamental Metrics

The four basic metrics of software science are:

η_1 = The number of unique or different operators

η_2 = The number of unique or different operands

N_1 = The total usage (number of occurrences) of operators

N_2 = The total usage (number of occurrences) of operands

Operands are defined as the constants or variables that an implementation of code employs. Operators then are the op-codes, delimiters, arithmetic symbols, punctuation, etc., which act upon the operands.

In order to automate the process of automatically counting large volumes of code, a very specific set of counting rules had to evolve from these two definitions such that a program counting tool could precisely parse tokens that represent software science definitions of operators and operands. These definitions, and also the specialized procedures for tuning the rules, form an independent study that warrants special attention [6].

For this study, and at this point in time, we have counted 1637 modules (see Table 1) which have been coded in either of two programming languages-assembler and PL/S. A "module" is defined as an independently compilable, externally callable piece of code.

	No. of Modules	Implementation Language
Project A	211	Assembler
Project B	514	Assembler
Project C	176	Assembler
Project D	63	PL/S
Project E	82	PL/S
Project F	54	PL/S
Project G	354	PL/S
Project H	90	PL/S
Project J	93	Assembler
TOTAL	1,637	

TABLE 1: MODULES ANALYZED IN THIS TECHNICAL REPORT

The projects which constitute this study cover a wide spectrum of diverse applications and systems programs. Projects A, B, and C are data base products. Projects D, E, and F are data management access method products. Project G is an interactive system used for problem solving. Projects H and J are text processing products. Projects A, B, C, and J were coded in IBM 360-370 assembler language. Projects D, E, F, G, and H were coded in PL/S, which is a systems-oriented subset of PL/1.

3. Definition of the Difficulty Measure

Difficulty is defined by software science to be the product of two factors:

$$D = \frac{\eta_1}{2} \quad \frac{N_2}{\eta_2}$$

The first term $\eta_1/2$ represents the total number of unique operators in a program divided by the least possible number of unique operators, which is the constant 2. The second term expresses the ratio of total occurrences of all operands over the number of unique operands in a program. This ratio reflects the average number of times the operands are used and shows the effect of repetition upon difficulty. Difficulty therefore is a unitless number that ranges from 1 to k where k is determined by the number of unique operators and total operands a programmer chooses to use in his code.

4. Observed Difficulty Values

Figures 1 and 2 display the difficulty values sorted from high to low for the modules from two "families" of products which we have counted. Figure 1 contains the difficulty curves for the data base family of products (A, B and C), and Figure 2 shows the difficulty curves for the three data management access method products (D, E, and F). Basic Assembler Language (BAL) project A had one module with an exorbitantly high difficulty value of 11074; this data point was not plotted because it distorted (compacted) the three curves in Figure 1. In the data base family of assembler language products (Figure 1), project "A" has a higher difficulty value even though it is relatively small in size (211 total modules). In the data management access method family of PL/S products (Figure 2), project "D" likewise stands out as having a higher difficulty value even though it is not the largest product. This is true for their average total difficulty values and also for module-by-module comparisons. This observation bears a strong correlation to the fact that products "A" and "D" have historically exhibited considerably poorer quality track records than their peer products. A study which relates difficulty to quality is currently underway and will be the subject of a future technique report.

Table II summarizes the difficulty values for each project. The average difficulty column is a straight numerical average while the weighted average column has been weighted by volume (the size metric defined in references 1 and 5) so that the larger modules bear a greater influence upon the difficulty value.

Figure 3 demonstrates this fact by showing how difficulty grows with size for the 993 assembler modules. (Figure 3 excludes the largest module from project A). Figure 4 shows the same difficulty growth for all 643 of the PL/S modules (projects D, E, F, G and H). In these two figures, each box represents a collection of modules in four dif-

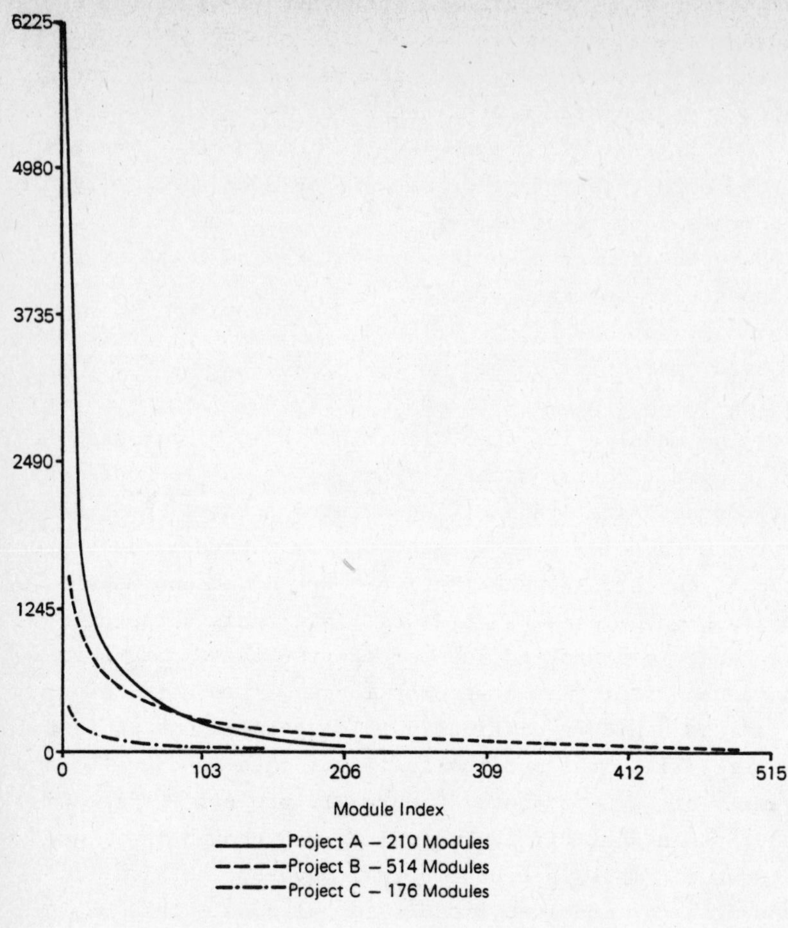

FIGURE 1. Difficulty Ranges-Assembler Language Modules Data Base/Date Communications Family of Products

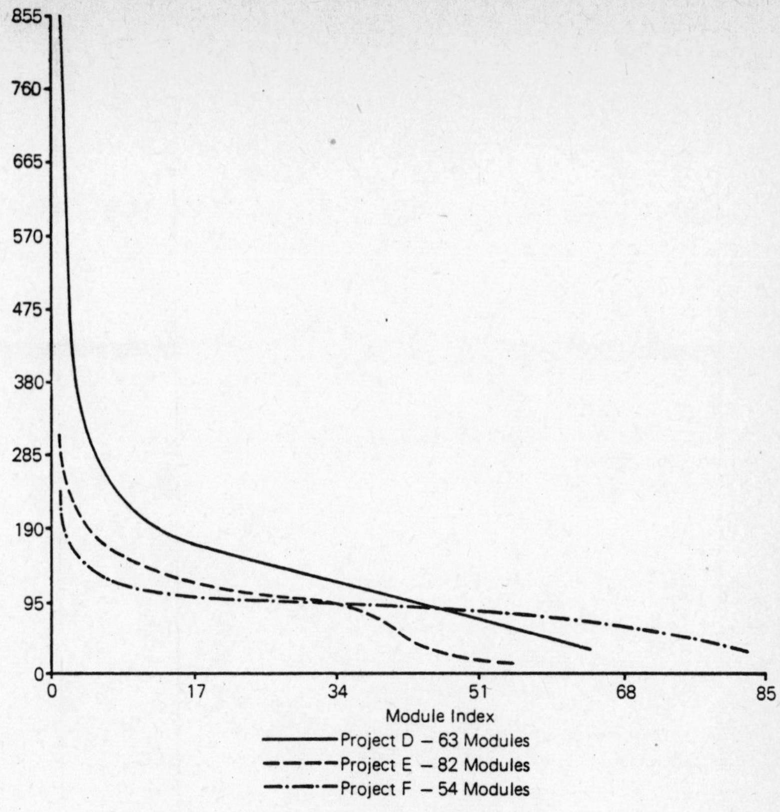

FIGURE 2. Difficulty Ranges-PL/S Modules Data Management Access Method Family of Products

	Low Value	Average (Mean)	Weighted Average	High Value	Standard Deviation
Project A	5	561	2,019	11,075	1,048.3
Project B	8	208	741	3,956	263.1
Project C	7	91	199	499	86.5
Project D	30	151	237	851	121.5
Project E	24	88	103	241	34.1
Project F	18	107	146	310	60.7
Project G	5	88	150	544	69.4
Project H	16	115	204	501	97.6
Project J	3	166	420	920	184.0

TABLE II. Ranges of Difficulty Values

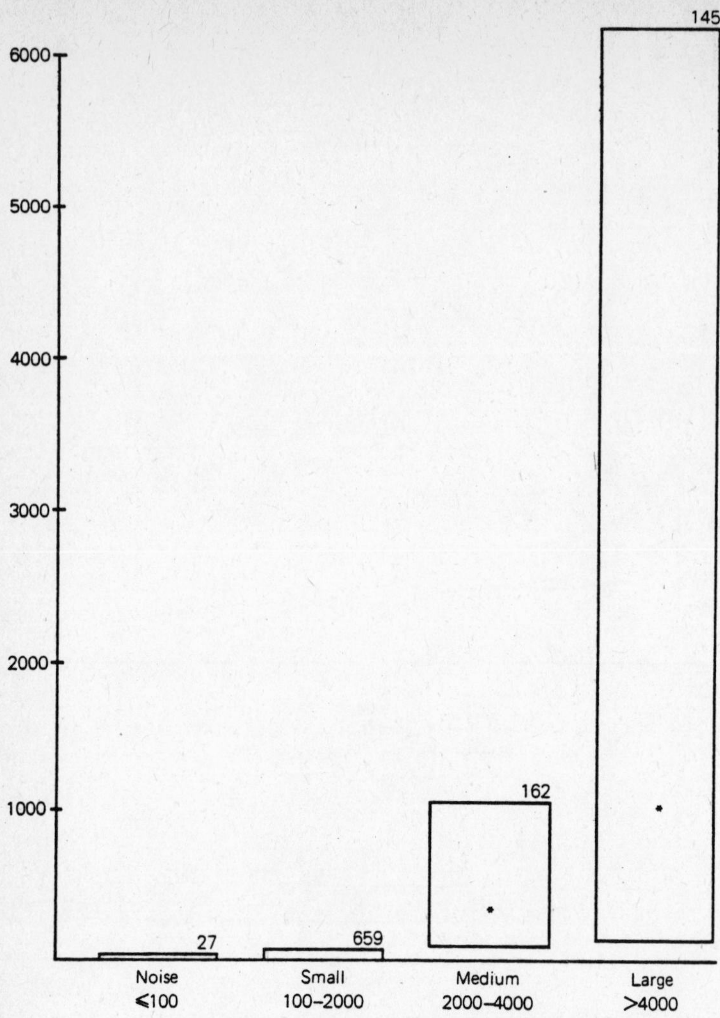

FIGURE 3. Difficulty vs. Size-Assembler Modules

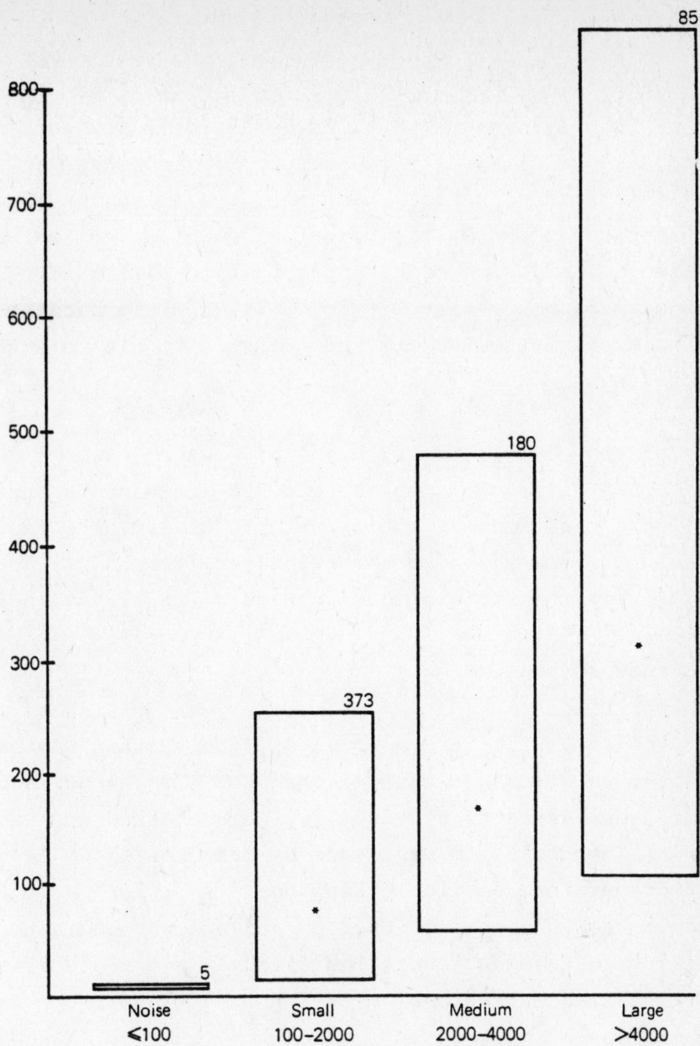

FIGURE 4. Difficulty vs. Size-PL/S Modules

ferent size categories. The term "size" represents the length or total count of all operators and all operands added together ($N_1 + N_2$). The number of modules resting above each box reflects the number of modules contained in that size category, i.e., the sample size. On both of these bar charts, as in each one of the individual projects we have counted, the average difficulty (denoted by *) increases as the size of the module increases. Also, the range of difficulty values within each size category, shown by the height of each box, grows wider with increased size. Thus, some modules have a high difficulty value no matter what size category they fit into. We therefore consider it important to understand the factors that cause a module to earn a high difficulty rating.

5. Causes of High Difficulty

Although the difficulty metric tends to increase as the size of the module increases, module size is not the determining factor. Modules that are the same size can have widely different difficulty values depending upon the style of code. High difficulty values are caused by:

- poor structure, and
- code impurities.

Structured code will have a lower difficulty than unstructured programs of the same size because of the lack of branches and GO TO statements. Each branch to a unique label is counted as a new unique operator [6]. Therefore, an abundance of branches or GO TO's will drive η_1 (and therefore difficulty) higher.

Difficulty can also be affected by and understood in terms of code impurities. Six code impurities cited in reference 7 are:

- Self-cancelling operations
- Ambiguous use of operands
- Synonymous use of operands
- Common subexpressions
- Unwarranted assignments
- Unfactored expressions.

Any of these impurities, when present in a code implementation, will actually increase the difficulty factor.

To further understand the difficulty metric, we should examine its composite parts. Figures 5 and 6 display the two factors that constitute difficulty. In these two figures the difficulty values of the modules are plotted after sorting them from the lowest difficulty value to the highest value. The corresponding unique operator usage

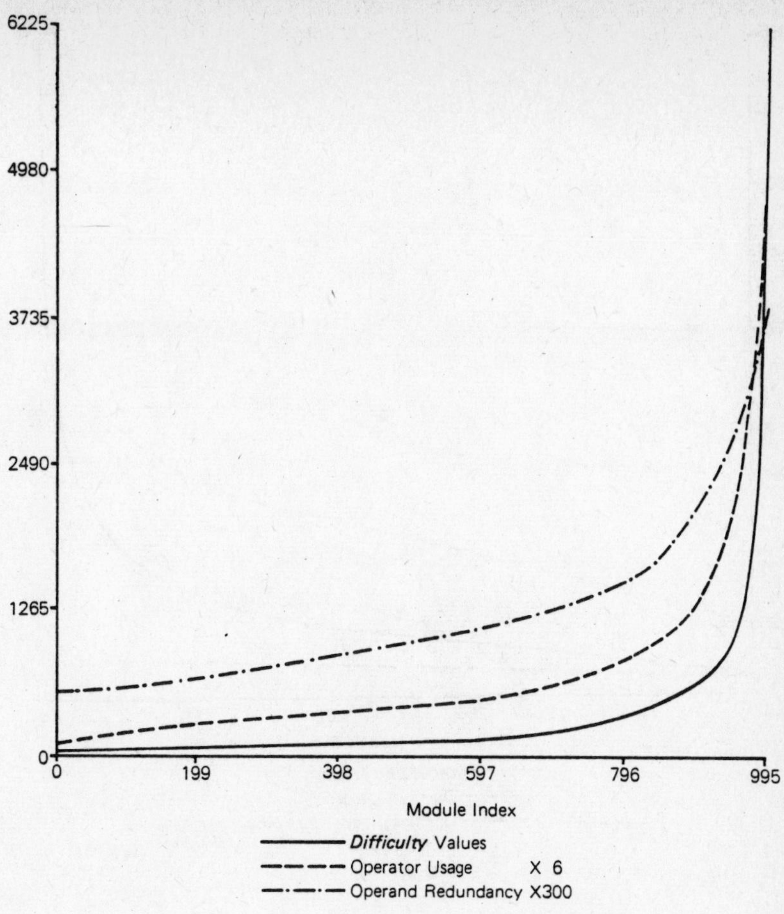

FIGURE 5. Difficulty vs. Operator Usage and Operand Redundancy 993 Assembler Language Modules

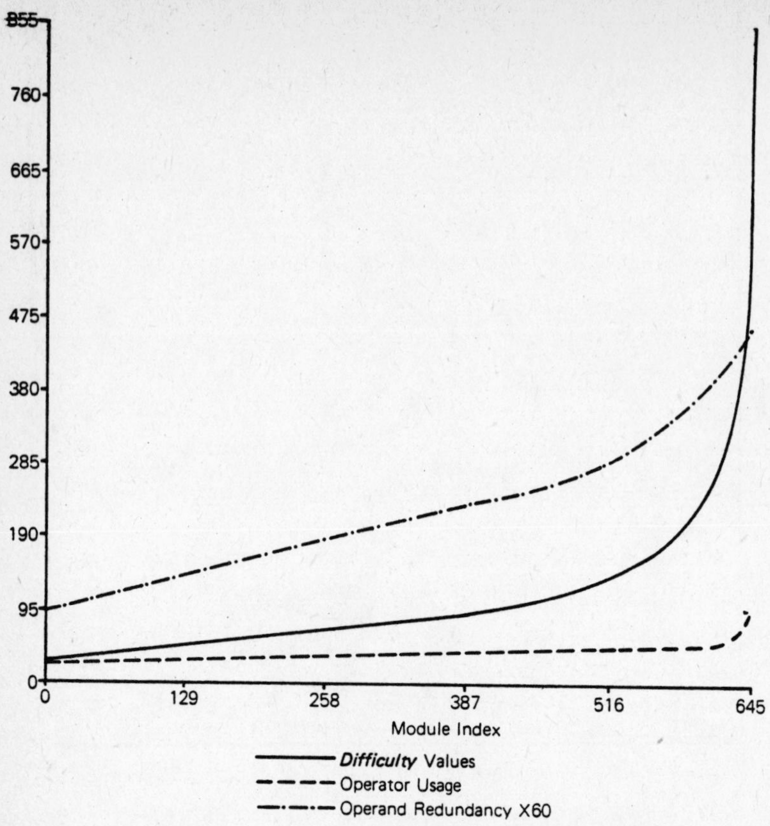

FIGURE 6. Difficulty vs. Operator Usage and Operand Redundancy 643 PL/S Modules

$\left(\frac{\eta_1}{2}\right)$ factor and the average operand redundancy factor $\left(\frac{N_2}{\eta_2}\right)$ are then plotted after adjusting by a multiplier to amplify the curves. This number is usually quite low for a module in either assembler or PL/S language, ranging from 2 to 4 in most instances. The multipliers for this particular curve in each figure had to be significantly large in order to make the operand redundancy factor visible on the same graph with the wide ranging difficulty values. The important question here is, "What affects difficulty?"

6. Language Implications

The assembler language set of modules (Figure 5) reveals that both operator and operand usage steadily increases and that both factors contribute towards increased difficulty. The PL/S set of modules (Figure 6) exhibits a peculiar characteristic: operator usage remains flat [4]. Thus, operand usage drives up difficulty in the PL/S language. This is true for all modules except those few modules with the highest difficulty rating; in these cases high operator usage triggers the highest difficulty values. An examination of the code for these modules has revealed a general lack of structure.

When comparing Figures 3 and 4, we can see that programs written in PL/S have significantly lower difficulty values than programs of the same size written in assembler language. For example, if we examine the set of programs in the medium length range ($2000 < \text{length} < 4000$), we see that average difficulty values are three times higher for assembler language code. Likewise, in the set of large modules, these difficulty values are approximately four times higher for the assembler language modules when compared to the PL/S modules. These comparisons reinforce our intuitive feelings about the two languages. Indeed one can see (in Figures 5 and 6) that the higher difficulty values for assembler language modules stems from the fact that the vocabulary of unique operators (η_1) a programmer must deal with is much larger for assembler than for the higher-level PL/S language. Consequently, for two programs of the same size, an assembler language program will have a higher difficulty value.

7. Utility of the Difficulty Measure

With the above understanding of difficulty, we are in a position to consider how the value of difficulty can be useful to the programmer. An upper boundary of the difficulty metric should be chosen to quantify the particular value of difficulty which a programmer does not want his or her code to exceed. In an initial attempt to do this, one can examine enlarged, more detailed versions of the plots of Figures 1 and 2 as well as the individual module values summarized in Table II to determine some threshold of difficulty where modules become noticeably different. Our research to date [4,5] has resulted in suggested values of 115 to 160 for PL/S code and 300 to 400 for assembler code. These suggested values are not precise; more experimentation and work must be done to advance the utility of this metric. But for now we think that programmers should definitely be concerned about, and want to understand why, their program modules exhibit too high a difficulty value when compared to the average value for their project or for other similar projects. At the Santa Teresa Laboratory, programmers have a tool that allows them to measure the difficulty of their own code and compare their implementation to a norm for their project. If the difficulty value of the program exceeds this norm, then the code is scrutinized very carefully. We currently view this procedure as a way to alleviate some of the drudgery from our development process and as an assist to programmers in the production of high quality code.

8. Conclusions

Difficulty is affected by 1) the structure of the code and 2) impurities in the code. For programs of the same size, assembler language programs are significantly more difficult than higher-level PL/S language programs; therefore, a maximum level (or threshold) of difficulty should be considered to help manage the complexity of programs.

9. References

[1] Halstead, Maurice H., "Elements of Software Science", Elsevier, North-Holland Inc., 1977.

[2] Halstead, Maurice H., "Software Physics: Basic Principals", Technical Report RJ1582, IBM Research, San Jose, California, 1975.

[3] Halstead, M. H., "Management Prediction - Can Software Science Help?" Proceedings IEEE COMPSAC '78 (Computer Software and Applications Conference) Chicago, Illinois, pp. 126-128, November 13-16, 1978.

[4] Fitsos, George P., "Vocabulary Effects in Software Science", IBM Technical Report TR03.082, January 1980, IBM Santa Teresa Laboratory, San Jose, California.

[5] Smith, Charles P., "A Software Science Analysis of IBM Programming Products", Technical Report TR03.081, January 1980, IBM Santa Teresa Laboratory, San Jose, California.

[6] Fitsos, George P., "Software Science Counting Rules and Tuning Methodology", IBM Technical Report TR03.075, September 1979, IBM

Santa Teresa Laboratory, San Jose, California.

[7] Bulut, Necdet and M. H. Halstead, "Impurities Found in Algorithm Implementations", ACM SIGPLAN Notices, March 1974.

NOTE: Readers interested in pursuing the topic of Software Science will enjoy the following reference:

[8] _IEEE Transactions on Software Engineering_, Commemorative Issue in honor of Dr. Maurice H. Halstead, Volume SE-5, Number 2, March 1979.

MATHEMATICAL PROGRAMMING ALGORITHMS IN APL

by

Harlan Crowder*

Abstract

The APL programming language offers a way of succinctly expressing data processing algorithms. We will introduce and demonstrate those APL concepts which we find useful for designing, implementing, and testing mathematical programming procedures.

1. Notation

This section briefly introduces the APL concepts of constants, variables, and functions. While similar in many respects to conventional algebraic notation, there are important differences. Our treatment of these concepts here is brief and informal. A more formal discussion is given in [3].

Constants are numeric or character data items; they have the following characteristics:

- Numeric data are expressed in conventional decimal notation except that negative numbers are preceded by a raised minus sign as in ¯5.1 and ¯2. The raised minus, like the decimal point, is part of the constant. Exponential notation is also used, as in 1.755*E*2 for 175.5 and ¯314159*E*¯5 for ¯3.14159.

- Character data are written as literal strings enclosed in single quotes, as in '*THIS IS A STRING*'.

Variables are used to symbolically reference data items; they have the following characteristics:

- Variable names are alpha-numeric strings, the first character of which must be alphabetic (ie, *A*-*Z*, *A̲*-*Z̲*).

- Values are assigned to variables using the symbol ←. Thus, the expression *A*←5 is read 'Assign *A* the value 5', or 'Let *A* equal 5'.

---

*IBM Thomas J. Watson Research Center, Yorktown Heights, NY 10598

This note is dedicated to my colleagues in the mathematical programming community to whom I continually extol the virtues of APL.

For our purposes, a *function* processes data arguments (constants or variables) and produces data results; functions have the following characteristics:

- Functions have one argument (called *monadic* functions), or two arguments (called *dyadic* functions). The symbol for a monadic function precedes its argument and the symbol for a dyadic function is placed between its two arguments. For example, the monadic function for absolute value is denoted by the monadic use of | and addition is denoted by the dyadic use of +. Thus, |¯4.4 is 4.4 and 3+2 is 5.
- The order of execution in an expression containing several functions is governed by parentheses in the usual manner; in the absence of parentheses, functions are executed in order from right to left. Thus, 2+4x6 and (6x4)+2 are 26, and 2x4+6 is 20.
- *Primitive functions* are defined in the APL language and are represented by special symbols; the next section of this note introduces a subset of the APL primitive functions. In most cases, the APL function symbols represent two functions, one monadic and one dyadic, which are generally related. For example, the (dyadic) division function and (monadic) reciprocal function both use ÷; 5÷2 is 2.5 and ÷5 is 0.2.
- *Defined functions* are programs which are composed of primitive functions and defined functions; they are either monadic or dyadic, depending upon how they are defined. Their syntax is identical to that of primitive functions.

We will use two types of representations for defined functions: *canonical representation*, which is the form of defined functions used on APL computers, and *direct definition*, which is used, when possible, for notational convenience. Given a defined function in direct form, we can obtain an equivalent function in direct form, or we can obtain an equivalent function in canonical form. A more extensive treatment of direct function definition is given in [4]. In the remainder of this section, we show both forms of defined functions for several examples.

The dyadic primitive power function is denoted by *. Thus 3*2 is 9, 2*3 is 8, 16*.5 is 4, and 27*÷3 is 3. The monadic defined function

SQRT computes the square root of its right argument using the primitive power function. For example, on an interactive APL computer, the use of *SQRT* would yield the following sequence:

```
      SQRT 9
3
      SQRT 5.76
2.4
```

Note that the input expression is indented six spaces from the left margin and the response is flush left. This convention will be followed throughout whenever we demonstrate a command-response pair.

The canonical form of *SQRT* is

```
      ∇ Z←SQRT R
[1]     Z←R*0.5
      ∇
```

Note that the parameter *R* in *SQRT* assumes the value of the right argument when the function is invoked.

The direct definition of *SQRT* is

```
SQRT:  ω*.5
```

Note that the symbol ω assumes the value of the function's right argument, and the resulting value of the function is obtained by evaluating the expression following the colon.

The dyadic function *F* is defined informally as follows: the result is the sum of the left argument and the square of the right argument. Thus 2 *F* 3 is 11 and 3 *F* 2 is 7. The canonical representation of *F* is

```
      ∇ Z←L F R
[1]     Z←L+R*2
      ∇
```

and the direct definition is

```
F:  α+ω*2
```

Note that the symbol α assumes the value of the left argument and ω assumes the value of the right argument when *F* is invoked.

From these examples we see that in direct definition, a function defined in terms of ω only is monadic while a function defined in terms of both α and ω is dyadic.

Finally, consider the monadic function *FACT* which computes the factorial of a nonnegative interger *N* recursively as follows: if *N*>0 then the result is *N*×*FACT* *N*-1, and if *N*=0, the result is 1. For example,

```
      FACT 5
120
      FACT 1
1
      FACT 0
1
```

The function *FACT* uses the primitive dyadic function equal which is denoted by = (not to be confused with the assignment function ← discussed above). Equal is a relational function which compares its arguments and results in 1 if they are equal and 0 otherwise. Thus, 3=3 is 1 and 3=5 is 0.

The direct definition of *FACT* is

```
FACT:ω×FACT ω-1:ω=0:1
```

The general form of this type of direct definition is

```
FNAME:  0-EXP :  C EXP :  1-EXP
```

where *FNAME* is the function name, *C-EXP* is a conditional APL expression which results in either 0 or 1, and 0-*EXP* and 1-*EXP* are APL expressions. If the evaluation of *C-EXP* results in 0, the value of *FNAME* is the result of evaluating 0-*EXP*. If *C-EXP* is 1, the value of *FNAME* is 1-*EXP*.

The canonical form of *FACT* is

```
      ∇ Z←FACT N
[1]   →(N=0)/L1
[2]   Z←N×FACT N-1
[3]   →0
[4] L1:
[5]   Z←1
      ∇
```

2. APL Introduction

```
      3+5
8
      3 4 2+5 1 7
8 5 9
      3+5 1 7
8 4 10
      3⌈5 1 7
5 3 7
      1 2 3*2
1 4 9
      2⍟1 2 4 8 16
0 1 2 3 4
      M
 1 2 3
 4 5 6
      M×2
  2  4  6
  8 10 12
      M+M
  2  4  6
  8 10 12
```

Dyadic functions such as +, −, ×, ÷, *, ⌈ (max), ⌊ (min), and ⍟ (log) operate on scalars and extend to arrays in a systematic manner. Two array arguments of a function must have the same *shape* (ie, vectors must have the same number of elements, matrices must have the same number of rows and columns). If one argument of a function is a scalar, it is applied to each element of the other argument.

```
      -3 ¯5 0 ¯2
¯3 5 0 2
      ×3 ¯5 0 ¯2
1 ¯1 0 ¯1
      ⌈3.5 ¯2.1 2
4 ¯2 2
      ⌊3.5 ¯2.1 2
3 ¯3 2
      ○1 2 3
3.1416 6.2832 9.4248
```

Monadic functions such as −, |, × (signum), ⌈ (ceiling, i.e., smallest integer greater or equal to number), ⌊ (floor, i.e., largest integer less than or equal to number) and ○ (pi times) operate on arrays and produce results with the same shape as the argument.

```
      3=3
1
      3=5
0
      3≥1 2 3 4 5
1 1 1 0 0
      3≠1 2 3 4 5
1 1 0 1 1
      3 5 7<4 3 6
1 0 0
      ~1 0 0 1
0 1 1 0
```

Relational functions follow the same rules. The result is 1 for true, 0 for false.

```
      ⍴3 5 7 9
4
      M
 1 2 3
 4 5 6
      ⍴M
2 3
      5⍴2
2 2 2 2 2
      3 2⍴1 3 5 7 9 11
  1  3
  5  7
  9 11
```

Shape, denoted by the monadic form of ⍴, produces the shape of its argument. *Reshape,* the dyadic form of ⍴, reshapes its right argument into an array with shape given by its left argument.

```
      V←5 3 2 1 4
      +/V
15
      ×/V
120
      ⌈/V
5
      ⌊/V
1
```

Summation of a vector V is performed by +/V, which is equivalent to writing the + function between each pair of elements in the array. This operation is *reduction* and it can be applied with dyadic functions other than +.

```
      A←2 3ρ5 1 3 4 6 2
      A
 5 1 3
 4 6 2
      +/[1]A
9 7 5
      +/[2]A
9 12
      +/A
9 12
      ⌈/A
5 6
      ×/[1]A
20 6 6
```

Reduction of a matrix presents two choices: reduce the first coordinate (the rows) or the second coordinate (the columns). The choice is made by indicating which coordinate to reduce. If a choice is not indicated, the last coordinate is reduced.

```
      ⍳5
1 2 3 4 5
      ⍳8
1 2 3 4 5 6 7 8
      V←5 1 4 6
      ρV
4
      ⍳ρV
1 2 3 4
```

Index generator, denoted in general by ⍳N, produces the vector of indices of a vector of length N.

```
      V←11 8 12 10 9
      V[2]
8
      V[3 5]
12 9
      V[2 5 4 1 3]
8 9 10 11 12
      I←2 2ρ1 2 4 5
      V[I]
 11  8
 10  9
```

Vector indexing, denoted in general by V[I], selects the elements of V indicated by the index elements in I. The shape of the result is the same as the shape of I.

```
      A←3 4ρ10+⍳12
      A
 11 12 13 14
 15 16 17 18
 19 20 21 22
      A[1;1]
11
      A[1 2;2 3]
 12 13
 16 17
      A[;2 4]
 12 14
 16 18
 20 22
```

Matrix indexing, denoted in general by M[I;J], selects the rows of M indicated by the index elements in I, and the columns indicated by the index elements in J. If either I or J are elided, all rows or columns are selected respectively.

```
      V←8 4 7 5 6 3
      3↑V
8 4 7
      ¯2↑V
6 3
      2↓V
7 5 6 3
      ¯3↓V
8 4 7
      A
 11 12 13 14
 15 16 17 18
 19 20 21 22
      2 3↑A
 11 12 13
 15 16 17
      1 ¯1↓A
 15 16 17
 19 20 21
```

Take and *drop,* denoted by ↑ and ↓ respectively, are used to select contiguous elements of an array. Note that for matrices, the left argument must be a vector of length 2, indicating both the number of rows and number of columns to be selected.

```
      V
8 4 7 5 6 3
      1 0 0 1 0 1/V
8 5 3
      (6⍴1 0)/V
8 7 6
      A
 11 12 13 14
 15 16 17 18
 19 20 21 22
      1 0 1/[1]A
 11 12 13 14
 19 20 21 22
      1 0 0 1/[2]A
 11 14
 15 18
 19 22
      1 0 0 1/A
 11 14
 15 18
 19 22
```

Compression selects elements from an array based on a boolean vector. Note that for matrices, we must indicate which coordinate to operate over. If no coordinate is indicated, the selection is performed over the last coordinate.

```
      0 1 2 3 ∘.+ 0 1 2 3
0 1 2 3
1 2 3 4
2 3 4 5
3 4 5 6
      V←0 1 2 3 4 5
      V∘.×V
 0  0  0  0  0  0
 0  1  2  3  4  5
 0  2  4  6  8 10
 0  3  6  9 12 15
 0  4  8 12 16 20
 0  5 10 15 20 25
      V∘.⌈V
0 1 2 3 4 5
1 1 2 3 4 5
2 2 2 3 4 5
3 3 3 3 4 5
4 4 4 4 4 5
5 5 5 5 5 5
      V∘.≥V
1 0 0 0 0 0
1 1 0 0 0 0
1 1 1 0 0 0
1 1 1 1 0 0
1 1 1 1 1 0
1 1 1 1 1 1
      'CAT'∘.='CATFAT'
1 0 0 0 0 0
0 1 0 0 1 0
0 0 1 0 0 1
```

Outer product is used to build a table for a particular function. For example, ∘.+ is used to build an addition table. The table is built by applying + to each pair of elements taken one from the left argument and one from the right. The symbols ∘. can be combined with other dyadic functions to build tables.

```
      V←2 4 6
      U←3 2 1
      +/U×V
20
      U+.×V
20
      ⌈/U+V
7
      U⌈.+V
7

      B←2 3⍴3 2 1 4 3 2
      B
 3 2 1
 4 3 2
      1 2+.×B
11 8 5
      B∧.=3 2 1
1 0

      C←3 4⍴⍳12
      C
  1  2  3  4
  5  6  7  8
  9 10 11 12
      B+.×C
 22 28 34 40
 37 46 55 64
      B⌊.+C
 4 5 6 7
 5 6 7 8
```

Inner product is a generalized form of reduction. For two vectors of the same length *U* and *V*, the reduction expression +/*U*×*V* is equivalent and yields the same result as the inner product expression *U*+.×*V*. In general, for arrays *A* and *B*, the expression *A*+.×*B* is valid if the last coordinate of *A* is equal in length to the first coordinate of *B*. (That is, if ¯1↑⍴*A* is 1↑⍴*B*). Inner product can be used with dyadic functions other than + and ×.

Index of, denoted in general by $V\iota A$, gives the index in the vector V of the first occurrence of each element in the array A. If an element in A does not occur in V, the result for that element is the first index outside the range of V.

```
        V←2 4 6 8
        V⍳4
2
        V⍳10
5
        V⍳⍳5
5 1 5 2 5
        A
 11 12 13 14
 15 16 17 18
 19 20 21 22
        V⍳A
 5 5 5 5
 5 5 5 5
 5 5 5 5
```

Catenation, denoted by the dyadic use of comma, is used to catenate arrays together. When more than one direction of catenation is possible, the coordinate to be extended must be specified. If none is specified, the last coordinate is extended.

```
        U←10 20 30
        V←40 50
        U,V
10 20 30 40 50
        A←2 3⍴⍳6
        A
 1 2 3
 4 5 6
        A,A
 1 2 3 1 2 3
 4 5 6 4 5 6
        A,[1]A
 1 2 3
 4 5 6
 1 2 3
 4 5 6
        A,V
  1  2  3 40
  4  5  6 50
        A,[1]U
  1  2  3
  4  5  6
 10 20 30
        A,7
 1 2 3 7
 4 5 6 7
```

Ravel, denoted by the monadic use of comma, is used to obtain a vector of the elements of an array in row-major order.

```
        A
 1 2 3
 4 5 6
        ,A
1 2 3 4 5 6
        ,A,A
1 2 3 1 2 3 4 5 6 4 5 6
        ,A,[1]A
1 2 3 4 5 6 1 2 3 4 5 6
```

Indexed specification occurs when an indexed variable appears to the left of a specification arrow.

```
        V←1 3 5 7 9
        V[2 4]←6 0
        V
1 6 5 0 9

        A←2 3⍴⍳6
        A[1;]←2 4 7
        A
 2 4 7
 4 5 6
```

```
      ∇ Z←WHATSIT N
[1]     →(N≠⌊N)/FR
[2]     →(0=2|N)/EV
[3]     Z←'ODD'
[4]     →0
[5]   EV:
[6]     Z←'EVEN'
[7]     →0
[8]   FR:
[9]     Z←'FRACTION'
      ∇

      WHATSIT 5
ODD
      WHATSIT 18
EVEN
      WHATSIT 3.14
FRACTION
```

Branching in a canonical function definition can take the forms →*L* or →(*EXP*)/*L*, where *L* is a function statement label and *EXP* is an APL expression which evaluates to 0 or 1. The first case is an unconditional branch. In the second case, the branch is taken if *EXP* evaluates to 1. If *EXP* evaluates to 0, the execution sequence continues with the function statement following the branch statement. The branch statement →0 acts as a control flow return, terminating execution of the function.

3. Mathematical Programming Examples

Matrix inversion, solutions of linear equations, least squares fit

```
      A
 3 0 0 1
 0 2 1 0
 2 0 0 2
 1 2 0 4
```

Coefficient matrix.

```
      AI←⌹A
      AI
 5.0E¯1   0.0E0   ¯2.5E¯1  ¯1.4E¯16
 7.5E¯1   0.0E0   ¯1.4E0    5.0E¯1
¯1.5E0    1.0E0    2.7E0   ¯1.0E0
¯5.0E¯1   0.0E0    7.5E¯1   2.7E¯16
```

Matrix inverse of *A*.

```
ZS:ω×1E¯6<|ω
```

Zero suppressor function

```
      ZS AI
  0.5     0     ¯0.25     0
  0.75    0     ¯1.375    0.5
 ¯1.5     1      2.75    ¯1
 ¯0.5     0      0.75     0
```

Cleaner *AI*.

```
      ZS AI+.×A
 1 0 0 0
 0 1 0 0
 0 0 1 0
 0 0 0 1
```

Matrix multiplication of *AI* and *A*.

```
      B
17 10 14 19
```

Right hand side

```
      X←B⌹A
      X
5 3 4 2

      A+.×X
17 10 14 19
```

Linear equation system solution.

<pre>   DAT
  1 3
  2 3
  3 6
  2 9
  3 3
  5 9</pre>
	x-y data points.

<pre> 1 1 1 1 1 1 ⌹DAT 0.17536 0.06951</pre>	Coefficients of equation for linear least squares fit to data.
<pre>LINFIT:((1↑⍴ω)⍴1)⌹ω</pre>	Linear least squares function.
<pre> LINFIT DAT 0.17536 0.06951</pre>	
<pre> DAT[;2]⌹DAT[;1]∘.*0 1 2 3.9 ¯0.05 0.2</pre>	Coefficients of equation for quadratic least squares fit to data.
<pre>QUADFIT:ω[;2]⌹ω[;1]∘.*0 1 2</pre>	Quadratic least squares function.
<pre> QUADFIT DAT 3.9 ¯0.05 0.2</pre>	

Longest subtree in an acyclic directed graph

<pre> I 0 1 0 0 1 0 0 1 0 0 0 0 0 1 0 0 0 0 0 0 0 0 0 0 0</pre>	Incidence matrix of a directed graph with no cycles. In effect, indicates location of all length-1 subtrees.
<pre> I[1;]∧I[;3] 0 1 0 0 0</pre>	Where are the length-2 subtrees between nodes 1 and 3?
<pre> ∨/I[1;]∧I[;3] 1</pre>	Is there *any* length-2 subtree between nodes 1 and 3?
<pre> ∨/I[1;]∧I[;4] 0</pre>	Is there any length-2 subtree between nodes 1 and 4?
<pre> I∨.∧I 0 0 1 0 0 0 0 0 1 0 0 0 0 0 0 0 0 0 0 0 0 0 0 0 0</pre>	In general, what are the length-2 subtrees in the graph? Nodes 1 to 3, and 2 to 4.
<pre> I∨.∧I∨.∧I 0 0 0 1 0 0 0 0 0 0 0 0 0 0 0 0 0 0 0 0 0 0 0 0 0</pre>	What are the length-3 subtrees? Node 1 to 4.

```
      Iv.∧Iv.∧Iv.∧I
 0 0 0 0 0
 0 0 0 0 0
 0 0 0 0 0
 0 0 0 0 0
 0 0 0 0 0
```

What are the length-4 subtrees? None.

```
LTREE:1+α LTREE αv.∧ω:v/,ω:0
```

Function to compute the longest subtree in a directed graph.

```
LONGTREE:ω LTREE ω
```

Cover function requiring one argument.

```
      J
 0 1 1 0 0 0 0 0 0 0
 0 0 0 0 0 0 0 0 0 0
 0 0 0 1 1 0 0 0 0 0
 0 0 0 0 0 0 0 0 0 0
 0 0 0 0 0 1 1 0 0 0
 0 0 0 0 0 0 0 0 0 0
 0 0 0 0 0 0 0 1 1 0
 0 0 0 0 0 0 0 0 0 0
 0 0 0 0 0 0 0 0 0 1
 0 0 0 0 0 0 0 0 0 0
```

A bigger example.

```
      LONGTREE J
5
```

Application of *LONGTREE*.

Shortest path in a directed graph

```
      W
 100   3   7
 100 100   2
 100 100 100
```

A matrix of edge costs. In effect, the cost of one-step transitions between nodes.

```
      100 3 7 + 7 2 100
107 5 107
```

The cost of a two-step transition from node 1 to 3.

```
      ⌊/100 3 7 + 7 2 100
5
```

Cheapest two-step transition from node 1 to 3.

```
      7⌊⌊/100 3 7 + 7 2 100
5
```

Minimum of one-step transition and cheapest two-step transition.

```
      W⌊W⌊.+W
 100   3   5
 100 100   2
 100 100 100
```

Minimum cost of two-step transition for all node pairs.

```
      Q
100   2 100   7   5
100 100   3 100 100
100 100 100   1 100
  2 100 100 100 100
100 100 100   4 100
```

A bigger example.

```
      V←Q⌊Q⌊.+Q
      V
  9   2   5   7   5
100 100   3   4 100
  3 100 100   1 100
  2   4 100   9   7
  6 100 100   4 100
```

Minimum cost of two-step transitions.

```
      V←V⌊Q⌊.+V
      V
  9   2   5   6   5
  6 100   3   4 100
  3   5 100   1   8
  2   4   7   9   7
  6   8 100   4  11
```

Minimum cost of three-step transitions

```
SP:α SP Z:∧/,ω=Z←ω⌊α⌊.+ω:ω
```

Recursive function for shortest path.

```
SHORTPATH:ω SP ω
```

Cover function requiring one argument.

```
      SHORTPATH Q
 8  2  5  6  5
 6  8  3  4 11
 3  5  8  1  8
 2  4  7  8  7
 6  8 11  4 11
```

Application of *SHORTPATH*

```
      ∇ Z←SHTPTH A;B
[1]   ⍝ SHORTEST PATH.
[2]     B←A
[3]   L1:
[4]     →(∧/,B=Z←B⌊A⌊.+B)/0
[5]     B←Z
[6]     →L1
      ∇
```

Iterative function for shortest path.

Simplex method for linear programming

We will consider a linear programming problem of the form $\{\text{minimize } cx \mid Ax \leq b, x \geq 0\}$, with the restriction $b \geq 0$. The algorithm we demonstrate is the elementary tableau simplex method. Hence, while it is useless as a computational tool except for the smallest of problems, it offers a concise outline of the basic procedures embodied in the simplex algorithm. The procedures shown here are extendible to more sophisticated versions of the simplex method by the introduction of appropriate data structures.

```
      A
 3  2
 4 ¯2
 2  5
      B
6 4 10
      C
¯1.5 ¯1
```

The constant data items of a linear programming problem in the form indicated above.

```
      TB←A BUILDTAB B,C
      TB
   3   2  1  0  0   6
   4  ¯2  0  1  0   4
   2   5  0  0  1  10
¯1.5  ¯1  0  0  0   0
```

BUILDTAB constructs a canonical linear programming tableau. Because of the restriction $b \geq 0$, this tableau is primal feasible.

```
      ∇ TAB←A BUILDTAB RC;RH;CO
[1]     RH←(M←1↑⍴A)↑RC
[2]     CO←M↓RC
[3]     TAB←A,(I̲ M),RH
[4]     TAB←TAB,[1] CO,(M+1)⍴0
      ∇

      I̲ 4
 1 0 0 0
 0 1 0 0
 0 0 1 0
 0 0 0 1
```

BUILDTAB uses the function *I̲* which returns an identity matrix.

```
      ∇ Z←I̲ N
[1]     Z←(2⍴N)⍴1,N⍴0
      ∇

      J←PRICE TB
      J
1
```

PRICE picks the pivot column, using the reduced costs (the last row of the tableau, exclusive of the last element). If *PRICE* returns a 0, the tableau is dual feasible, hence optimal.

```
      ∇ J←PRICE TAB;DJ;MDJ
[1]     DJ←¯1↓,TAB[1↑⍴TAB;]
[2]     J←DJ⍳MDJ←⌊/DJ
[3]     J←1↑(MDJ<¯1E¯6)/J
      ∇

      I←J ROW TB
      I
2
```

ROW picks the pivot row, given the tableau and the pivot column index. *X* is the value of the current basic variables and *Y* is the pivot column. Statements 4-7 try to make the best of degenerate pivots. If *PRICE* returns a 0, the problem is unbounded.

```
      ∇ I←J ROW TAB;X;Y;T
[1]     X←¯1↓,TAB[;1↓⍴TAB]
[2]     Y←¯1↓,TAB[;J]
[3]    ⍝ CHECK FOR DEGENERACY...
[4]     I←(0.00001>|X)∧0.00001<|Y
[5]     →(~∨/I)/L1
[6]     I←(I/⍳⍴Y)[Y⍳⌈/Y←|I/Y]
[7]     →0
[8]   L1:
[9]     I←(X>0.00001)∧Y>0.00001
[10]    →(~∨/I)/L2
[11]    I←(I/⍳⍴Y)[T⍳⌊/T←(I/X)÷I/Y]
[12]    →0
[13]  L2:
[14]  ⍝ NO PIVOT...
[15]    I←0
      ∇
```

```
      TB←TB PIV I,J
      TB
  .00   3.50 1.00 ¯.75   .00 3.00
 1.00  ¯.50   .00  .25   .00 1.00
  .00   6.00  .00 ¯.50  1.00 8.00
  .00  ¯1.75  .00  .38   .00 1.50
```

PIV performs a pivot at the row and column picked by *ROW* and *PRICE*. Note that the last element in the last row of the tableau is the objective function value.

```
      ∇ Z←TAB PIV IJ;I;J;D
[1]     I←IJ[1]
[2]     J←IJ[2]
[3]     D←TAB[I;]÷TAB[I;J]
[4]     Z←TAB-TAB[;J]∘.×D
[5]     Z[I;]←D
      ∇
```

```
      TB←A SIMPLEX B,C
OPTIMAL
      TB
  .00 1.00   .29 ¯.21   .00  .86
 1.00  .00   .14  .14   .00 1.43
  .00  .00 ¯1.71  .79  1.00 2.86
  .00  .00   .50  .00   .00 3.00
```

SIMPLEX combines these procedures to implement the tableau simplex method.

```
      ∇TAB←A SIMPLEX RHSCOST;J;I
[1]     TAB←A BUILDTAB RHSCOST
[2]   L1:
[3]     →(0≠J←PRICE TAB)/L2
[4]     'OPTIMAL'
[5]     →0
[6]   L2:
[7]     →(0≠I←J ROW TAB)/L3
[8]     'UNBOUNDED PROBLEM'
[9]     →0
[10]  L3:
[11]    TAB←TAB PIV I,J
[12]    →L1
      ∇
```

References

[1] Gilman, L., and A.J. Rose, APL -- An Interactive Approach, Wiley, 1974.

[2] Polivka, R.P., and S. Pakin, APL: The Language and Its Usage, Prentice - Hall, 1975.

[3] Falkoff, A.D., and K.E. Iverson, APL Language, IBM Corporation, CG26-3847, 1978.

[4] Iverson, K.E., Elementary Analysis, APL Press, Swarthmore, PA, 1976.

SOLUTION STRATEGIES AND ALGORITHM BEHAVIOR IN LARGE-SCALE NETWORK CODES

Richard S. Barr
Edwin L. Cox School of Business
Southern Methodist University
Dallas, Texas 75275

ABSTRACT

This paper surveys results of a recent set of experiments testing different solution strategy parameters and implementation schemes for the U.S. Department of Treasury's in/out-of-core, primal-simplex-based solution system for optimally merging microdata files. Numerous runs were implemented to study the effects of pricing and pivoting rule, data storage technique, compiler, data page size, and problem density on algorithm behavior. Improvements over previous performance levels are noted.

1. Introduction

As detailed in recent reports [1,2,3], the problem of optimally merging microdata files results in the need to solve extremely large uncapacitated transportation problems. Problems with dimensions of over 20,000 constraints and tens of millions of variables are not uncommon and must be optimized on a regular basis. In the mid-1970's, an in/out-of-core, primal-simplex-based solution system was devised to meet this need for the U.S. Department of the Treasury [1].

This study reports the results of recent experimentation with this system to test various solution strategy parameters and implementation schemes. Numerous runs were made to explore the effect on solvability of such considerations as pricing and pivoting rule, data storage technique, compiler, data page size, and problem density. In addition, computer graphics were used to study algorithm behavior during the solution process.

2. The Testing Environment

2.1 Problem Overview. A microdata file is a collection of sample observation records, often based on a national survey. Two such files may be combined, or merged, by mating records from one file with similar records in the other file. Hence, the resultant composite file will consist of enhanced data records with each record containing items from both of the original surveys. The merged file provides an "enriched" data source for use in microanalytic models and policy research.

The optimal set of record pairings may be found by minimizing a linear interrecord dissimilarity function subject to transportation constraints [2]. In their fullest form, these problems are enormous. The merge model includes one structural variable, or network arc, for each pair of records which may be matched plus a network node constraint for each record in each file to avoid over-matching or excluding records. So, recently, when a 75,000-record IRS file was to be merged with a 61,000-record Census file, the full problem involved 136,000 constraints and over 4.5 billion variables.

2.2 Computing Environment. All runs reported herein were made on the U.S. Treasury's Univac 1100/81 mainframe with 400K 36-bit words of primary storage, running under the EXEC 8 operating system. Both the FORTRAN V and ASCII FORTRAN compilers were employed, each permitting buffered input/output from the system disks.

2.3 Matrix Generator. Because of its size, direct solution of a full merge problem is typically well beyond mathematical programming's state-of-the-art, even with the advantageous network structure. For this reason, the matrix generator employs two techniques to reduce problem dimensions. First, the files are partitioned into several sub-files which are merged separately and combined, thus limiting the number of constraints to be considered simultaneously. Secondly, within each sub-problem, nondense problems are generated by including arcs for the "p-best" matches for each record, thus reducing the number of variables but making "infeasible" problems possible.

In designing for problems with up to 50,000 constraints, machine wordsize limits the magnitude of the dual variables and, hence, the problem costs. For this reason, the arc costs are scaled to range from 0 to 63. Finally, in order to accommodate percentage of optimality calculations for intermediate solutions (not used, see [2]), the arcs are sorted in ascending cost order.

2.4 Optimization Software. The solution system was designed with minimal primary storage needs to accommodate large problems. A primal simplex algorithm was deemed best because of the low memory requirements and greater efficiency relative to other methodologies. Furthermore, only a spanning-tree basis need be maintained in-core with the arc cost data paged in piecewise from secondary storage. The system was coded entirely in FORTRAN to simplify maintenance and enhance portability.

The heart of the system is an efficient primal-simplex transportation code which uses an independently-derived variant of the ATI algorithm [4] and maintains the basis in four node-length arrays. By packing, the basis data can be represented in only two node-length arrays. An initial basis is constructed from a single pass of the arc data and any artificials driven out of solution by either the Phase I-II or the Big-M (if the data is not packed) method.

Figure 1 depicts the general system logic and indicates the parameters used to control arc pricing. Note the "double buffering" technique which allows a page of arc data to be read in parallel with the pricing and pivoting operations, thus reducing the real-time, if not CPU-time, requirements.

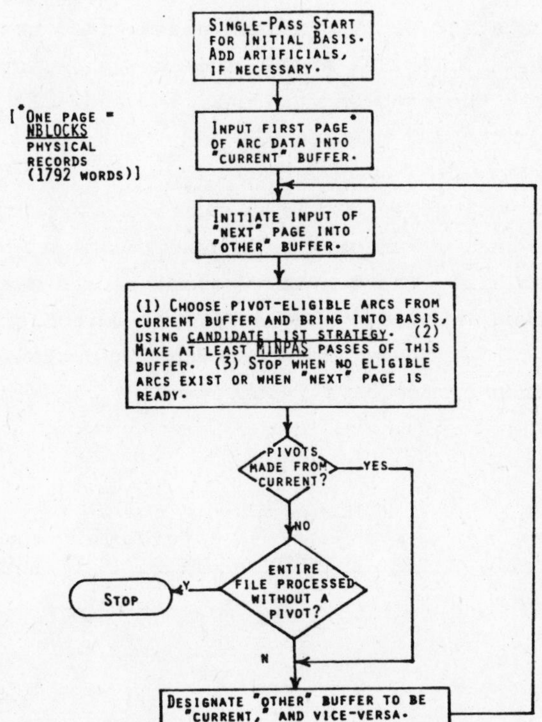

FIGURE 1. Solution System Flow Diagram

In the pricing process, arc data is brought into core one page at a time; NBLOCK defines the page size in multiples of 1792 words. Arcs from the "current" page are selected for pivoting based on the "candidate list" technique of Mulvey which uses two parameters, K and L. With a K/L strategy, (1) L pivot-eligible arcs are selected and placed on the list, (2) arcs on the list are priced and the one with the most negative marginal value is chosen for pivoting into the solution, and (3) step two is repeated K times before a new list is made (see [5] for details). Candidate lists are built and processed until either MINPAS complete passes are made of the page, or until no pivot-eligible arcs remain. This process is applied anew to each page of arc data, and the values of NBLOCK, K, L, and MINPAS define a pricing strategy for the algorithm.

3. Experimentation

3.1 The Base Problem. For test purposes, a relatively small test problem was used, consisting of 3115 origin nodes, 1463 destination nodes (4578 constraints), and 623,000 arcs. The initial solution contained 291 artificial arcs, the problem was feasible, and solution times ranged from five to ten CPU minutes.

3.2 Pricing and Pivoting Tests. To evaluate the effects of the pricing parameters, several strategies were used to solve the base problem, as described in Table 1. The FORTRAN V compiler was used, the basis data stored in packed form, and the Phase I-II method employed. Elapsed central processor (CP) time is shown for problem optimization only but may include a portion of the data input time.

Run No.	MINPAS (Maximum Passes/Page)	NBLOCK (No. of Data Blocks Per Buffer)	K/L (Candidate List Strategy)	Number of Pivots	CP Time (Seconds)
1.	99	4	20/40	32,071	497
2.	99	1	20/40	33,051	453
3.	99	4	1/1	45,010	601
4.	3	4	20/40	28,848	403
5.	3	1	20/40	30,625	433

TABLE 1. Pricing and Pivoting Tests Using the Base Problem

This data indicates that solution times tend to improve when making fewer page passes (MINPAS=3) instead of processing a page of arc data until no eligible arcs remain (MINPAS=99). This is perhaps due to the decreasing rate of solution improvement with continued pricing of a page. Figure 2 illustrates this effect for a set of pivots chosen from early in Phase II of run 1. For each new "current page" inspected, a "pass number" is initially set to zero and incremented by one each time the pricing routine reaches the end of the page. In Figure 2, these pass numbers are plotted against the objective function improvement from pivots made during the page pass. As might be expected, the improvement tends to diminish as the page is priced repeatedly. Hence, a strategy using a smaller MINPAS accepts a good improvement rather than seeking the maximum improvement per page.

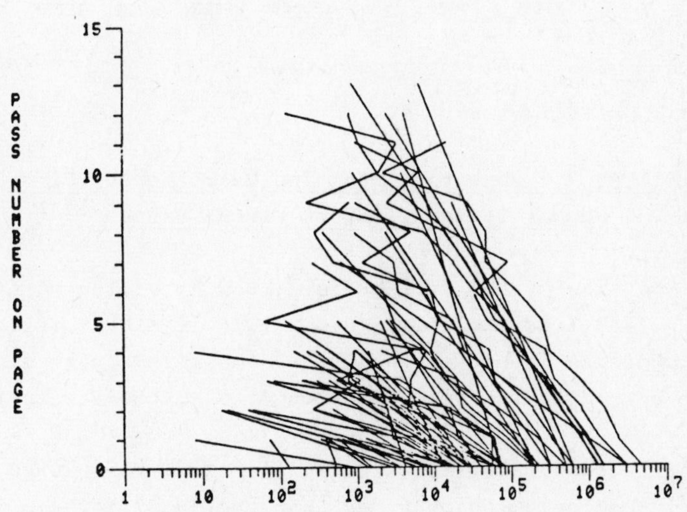

FIGURE 2. The Effect of Multiple Page Passes on Objective Function Value (MINPAS=99)

Figure 3 illustrates the objective function behavior versus pivot number during the solution process with different MINPAS values (runs 1 and 4). The objective function increases are due to the search for a feasible solution in Phase I. The smaller MINPAS strategy achieves feasibility sooner, with the Phase II solution improvement rates approximately the same in both cases.

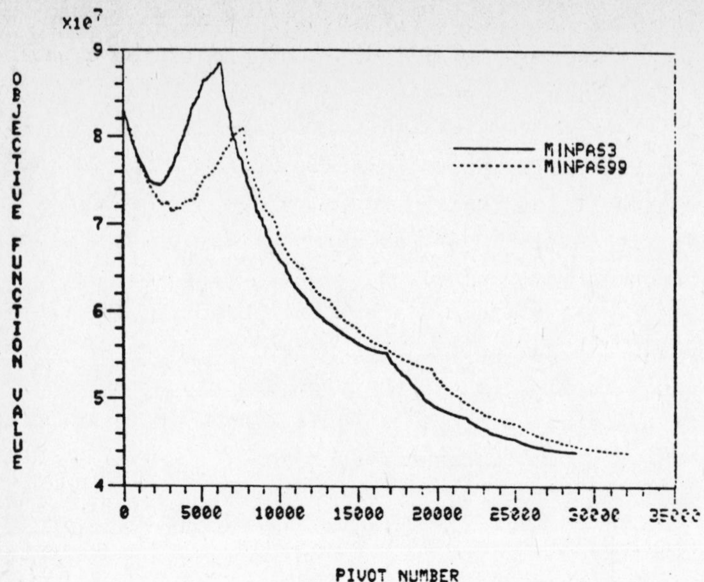

FIGURE 3. Objective Function Value Versus Pivot Number Using Different MINPAS Values

Varying the size of the data buffer (NBLOCK) yields mixed results (see run 1,2,4, and 5 in Table 1). Generally, if MINPAS is exhaustive, the smaller page size is better; otherwise the larger page size is preferable. The best overall time, from run 4, used the larger NBLOCK value.

In comparing candidate list strategies, the larger 20/40 list tended to yield better solution times than the "modified row minimum" 1/1 strategy. Comparing runs 1 and 3 in Table 1, the 1/1 strategy was 20% slower and required 40% more pivots.

3.3 Compiler, Data Storage, and Methodology Tests. Two compilers are available on the Univac 1100: FORTRAN V, a mature product, and a newer ASCII FORTRAN compiler. Both packages generate well-optimized code. Table 2 shows the results of a series of runs using both compilers, with runs 1 and 2 indicating that the ASCII-generated code is slightly faster for the base problem.

The effect of packing the basis data can be seen by contrasting runs 2 and 3a from Table 2. By eliminating the unpacking operations and storing the basis in normal TYPE INTEGER format, solution time is reduced by 20 percent.

Run	Compiler	Basis Data	Solution Method	Pivots	CP Seconds
1.	FORTRAN V	Packed	Phase I-II	28,848	403
2.	ASCII	Packed	Phase I-II	28,848	371
3a.	ASCII	Unpacked	Phase I-II	29,401	297
3b.	ASCII**	Unpacked	Phase I-II	29,401	451
4a.	ASCII	Unpacked	Big-M	34,185	306
4b.	ASCII**	Unpacked	Big-M	34,185	397

*MINPAS=3, NBLOCKS=4, 20/40 candidate list.

**Recompiled for core references beyond 65K.

TABLE 2. Tests of Compiler, Data Storage Technique and Solution Method on Base Problem*

The 1100-series architecture affords greater efficiencies if all program data can be stored in 65K words or less. To reference array locations beyond this limit, a compiler option directs the generation of code which does not use the index registers for indirect addressing. A comparison of runs 3a, 3b, 4a, and 4b indicates that invoking this option escalates solution times by from 30 to 50 percent.

The method by which artificial arcs are handled can have an effect on both the number of pivots and time required to reach optimality. Contrasting runs 3a and 4a of Table 2, there seems to be little difference between the Big-M and Phase I-II methods, but runs 3b and 4b indicate a distinct advantage for the Big-M method. An extremely large value of Big M was used in these runs which, as discovered in the testing described below, was not the best choice since smaller artificial costs yielded a 15 percent improvement in this solution time.

3.4 Problem Density. To investigate the effect of problem density on solvability, the base problem was regenerated with all arcs included -- not just those for the 200 best matches per records. Table 3 indicates that while the totally dense problem has seven times as many arcs, it requires only three times the solution time. Moreover, the denser problem required fewer pivots for optimization, perhaps due to the higher dimensionality, but both runs had few degenerate pivots as is characteristic of transportation problems.

	13.6% Density	100% Density
Origin nodes	3115	3115
Destination nodes	1463	1463
Node constraints	4578	4578
Arcs	623,000	4,361,000
Cost range	0-63	0-63
MINPAS	3	3
Candidate list strategy	20/40	20/40
NBLOCK	3	7
Pivots made	29,401	25,534
Degenerate Pivots	255	273
CP Times (seconds)		
Start	14	98
Phase I	71	55
Phase II	212	782
Total (minutes)	297 (4.9)	935 (15.5)

TABLE 3. Comparative Solution Statistics for the Base Problem and a Totally Dense Problem

3.5 Big-M Values. In the folklore of mathematical programming, smaller Big-M values are preferable to larger ones. To test this hypothesis, the runs described in Table 4 were made. In each case, the cost assigned to basic artificials was set to an initial value; when the pricing routine completed a pass of all pages of the arc data, the cost of any remaining artificials was increased by a multiple of the previous cost. In the second run shown, for example, the first four values of Big M were 62, 80, 104, and 135. Generally the folklore seems to be true. The longest solution time was with the largest Big-M cost and the best strategy began with a small cost that increased gradually, yielding a six percent improvement in solution time over the worst case.

Big-M Value**			
Initial Value	Multiplier at End-of-File	Pivots Made	CP Time (Seconds)
3	0	33,377	257
1	1.3	30,454	255
1.1	1.3	29,741	251
1.5	1.5	31,876	259
2	0	32,222	257
10	0	34,168	266

*MINPAS = 3, NBLOCK = 4, 20/40 candidate list strategy.

**Expressed in multiples of 62, the largest cost.

TABLE 4. Solving the Base Problem with Various Big-M Strategies*

3.6 Summary. Large-scale testing is made difficult by the high cost of a single observation and the resultant limit on the number of strategies to be considered. However, even this modest amount of testing led to the identification of solution strategies which solved a 19,406-constraint, 4.5-million-variable problem from a recent merge by making 298,893 pivots in 185 minutes. This is a considerable improvement over previous experience where, for example, 382 minutes were required to solve a comparable problem on the slightly slower Univac 1110 [1].

4. References

[1] Barr, Richard S., "Design and Evolution of a Mathemtaical Programming System for Ultra-Large-Scale Transportation Problems," research report, Edwin L. Cox School of Business, Southern Methodist University, Dallas, TX (1981).

[2] Barr, Richard S. and J. Scott Turner, "A New, Linear Programming Approach to Microdata File Mergine," in 1978 Compendium of Tax Research, U.S. Government Printing Office, Washington, D.C. (1978), 131-155.

[3] Barr, Richard S. and J. Scott Turner, "Optimal Microdata File Merging through Large-Scale Network Technology," to appear in Mathematical Programming Studies (1981).

[4] Glover, Fred, Darwin Klingman, and Joel Stutz, "Augmented Threaded Index Method for Network Optimization," INFOR 12,3 (1974) 293-298.

[5] Mulvey, John M., "Pivot Strategies for Primal-Simplex Network Codes," JACM 25, 2 (1978) 266-270.

RECURSIVE PIECEWISE-LINEAR APPROXIMATION METHODS FOR NONLINEAR NETWORKS

R. R. Meyer*
Computer Sciences Department
University of Wisconsin
Madison, Wisconsin

Abstract

The use of recursive piecewise-linear approximations in network optimization problems with convex separable objectives allows the utilization of fast solution methods for the corresponding linear network subproblems. Piecewise-linear approximations also enjoy many advantages over other types of approximations, including the ability to utilize simultaneously information from infeasible as well as feasible points (so that results of Lagrangian relaxations may be directly employed in the approximation), guaranteed decrease in the objective without the need for any line search, and easily computed and tight bounds on the optimal value. The speed of this approach will be exemplified by the presentation of computational results for large-scale nonlinear networks arising from econometric and water supply system applications.

Introduction

In a previous paper [Kao and Meyer (1979)] convergence theorems were developed for recursive separable programming algorithms for problems of the form

$$(1.1)\qquad \min_{x} \sum_{i=1}^{n} f_i(x_i) \quad \text{s.t. } x \varepsilon C \cap [\ell, u],$$

where $x = (x_1, \ldots, x_n)^T \varepsilon R^n$, C is a closed convex set, $[\ell,u]$ denotes the hyper-rectangle corresponding to the constraints $\ell \leq x \leq u$, and each f_i is a continuous convex function on the interval $[\ell_i, u_i]$. Here we will consider computational aspects and experience with these and newly developed algorithms for the special case in which $C = \{x | Ax=b\}$, where A is an m×n node-arc incidence matrix, so that the constraints are those of a network flow problem.

It is notationally convenient to denote the feasible set of (1.1)

*Research supported by National Science Foundation Grant MCS-7901066.

by S. To avoid trivial cases we will assume that S is non-empty (if it was empty, this fact would be established on the first iteration) and that the bounds satisfy $\ell < u$. Under the assumptions made, (1.1) has an optimal value, denoted by z**. Finally, we denote $\sum_{i=1}^{n} f_i(x_i)$ by f(x).

Problems of this form arise in numerous applications, including econometric data fitting [Bachem and Korte (1977)], electrical networks [Rockafellar (1976)], water supply systems [Collins, et al (1978)], and logistics [Saaty (1970)]. Computational experience with problems arising from several of these areas will be described.

At each iteration, in addition to the generation of a feasible solution (and a corresponding upper bound for the optimal value) a lower bound on the optimal value may also be computed by using only function value information. (If each f_i is differentiable and if equations of the form $f_i'(x) = \alpha$ may be easily solved, lower bounds based on Lagrangian relaxation may be computed instead.) The algorithm may thus be terminated when the difference between the upper and lower bounds is less than a prescribed tolerance. This termination criterion guarantees that the feasible solution giving rise to the upper bounds differs (in an objective function sense) from the optimal solution by less than the tolerance.

2. The Network Subproblems

The initial method to be described for the given problem involves the construction (at each iteration) of a piecewise-linear "local" approximation of at most two segments for each of the f_i. Except for the initial iteration, (in which a feasible starting solution is not assumed), the piecewise-linear approximations $\tilde{f}_i$ are determined by function values at points $(\tilde{\ell},\tilde{m},\tilde{u})$, where $\tilde{m}$ is a feasible solution (the optimal solution of the preceding iteration), and $\tilde{\ell}$ and $\tilde{u}$ are "temporary" lower and upper bounds with the admissibility properties:

(2.1) $\quad \ell \leq \tilde{\ell} \leq \tilde{m} \leq \tilde{u} \leq u,$

(2.2) $\quad \tilde{m}_i > \tilde{\ell}_i$ if $\tilde{m}_i > \ell_i$, and

(2.3) $\quad \tilde{m}_i < \tilde{u}_i$ if $\tilde{m}_i < u_i$.

From a computational viewpoint, these admissibility properties may be thought of as allowing a decrease in x_i (thought of as starting the iteration with a value $\tilde{m}_i$) if x_i is not at the true lower bound ℓ_i, and allowing an increase if x_i is not at the true upper bound. Methods for generating $\tilde{\ell}$ and $\tilde{u}$ at each interation will be described below. (Note that when $\ell_i = \tilde{\ell}_i = \tilde{m}_i$ or $\tilde{m}_i = \tilde{u}_i = u_i$, f_i is approximated by the secant determined by two function values rather than by a two-segment approximation determined by three function values.)

The key property of $\tilde{f}_i$ that guarantees monotonicity of the algorithm is that $f_i(x_i) \leq \tilde{f}_i(x_i)$ for $x_i \varepsilon [\tilde{\ell}_i, \tilde{u}_i]$. To exploit this property, the constraints $x \varepsilon [\tilde{\ell}, \tilde{u}]$ are imposed in the subproblem. Thus, the triple $(\tilde{\ell}, \tilde{m}, \tilde{u})$ not only determines the approximations $\tilde{f}_i$, but determines additional constraints as well. For this reason the corresponding subproblem denoted by $P(\tilde{\ell}, \tilde{m}, \tilde{u})$ may be written as

$$P(\tilde{\ell}, \tilde{m}, \tilde{u}) \equiv \begin{cases} \min\limits_{x} \sum\limits_{i=1}^{n} \tilde{f}_i(x_i) \\ \\ \text{s.t. } x \varepsilon S \cap [\tilde{\ell}, \tilde{u}]. \end{cases}$$

In order to exploit the very fast codes currently available for linear network optimization [Bradley, Brown and Graves (1977), Glover and Klingman (1975), Grigoriadis (1979)] we take advantage of the fact that problems of the form $P(\tilde{\ell}, \tilde{m}, \tilde{u})$ may be reduced to ordinary linear network problems by taking advantage of the form of the functions $\tilde{f}_i$. More precisely, each single arc on which the cost function is a two-segment, convex, piecewise-linear function $\tilde{f}_i$ is equivalent (by a well-known transformation) to a pair of arcs with linear costs. (Since these two arcs differ only in cost coefficients and flow bounds, it is really not necessary from a data structure viewpoint to represent them as separate arcs in the network; it suffices to keep track of the pair of cost coefficients and flow bounds.) Hence, we will refer to $P(\tilde{\ell}, \tilde{m}, \tilde{u})$ as a <u>linear network flow problem</u> with the understanding that it may be reduced to and solved as such a problem.

For the <u>initial</u> subproblem, in which a feasible starting solution $\tilde{m}$ is <u>not</u> assumed, the problem $P(\ell, (\ell+u)/2, u)$ may be solved. This is the problem corresponding to the two-segment approximations generated by the endpoints and midpoint of each segment $[\ell_i, u_i]$. (Note that the feasible set for this problem is simply the <u>original</u> feasible set, so the feasibility of the initial subproblem is a consequence of the feasibility of the original problem.) If estimates or "target" values (to be described below) are available for any of the x_i, they may be used in the first iteration instead of the midpoint values $(\ell_i+u_i)/2$, which may be thought of as "default" values. It should be emphasized that the values used in constructing the initial approximation need <u>not</u> correspond to a feasible solution. (In later interations $\tilde{m}$ will be feasible, but the values used for $\tilde{\ell}$ and $\tilde{u}$ need not be.)

3. <u>An Overview of the Algorithm</u>

If x^k denotes the optimal solution of the k-th iteration, it may be shown (see [Meyer (1979)] that the iterates have the property that $f(x^k) \geq f(x^{k+1})$, and that convergence of the sequence $\{f(x^k)\}$ to the optimal value of (1.1) is guaranteed when the values of $\tilde{\ell}$ and $\tilde{u}$ are chosen by a procedure called contraction search. The idea underlying contraction search is that a feasible solution $\tilde{m}$ is an optimal solution for the original problem (1.1) if and only if it is optimal for a family of problems of the form $P(\tilde{\ell}^j,\tilde{m},\tilde{u}^j)$, $(j=1,2,\ldots)$, where the initial triple $(\tilde{\ell}^1,\tilde{m},\tilde{u}^1)$ is admissible and the others are defined by $\tilde{\ell}^j \equiv \tilde{m} - \beta(m-\tilde{\ell}^{j-1})$, $\tilde{u}^j \equiv \tilde{m} + \beta(\tilde{u}^{j-1}-\tilde{m})$ for $j \geq 2$, where $\beta \ll 1$ is a given constant (typically $\beta = 1/2$). Thus, unless $\tilde{m}$ is optimal for the original problem, as the box $[\tilde{\ell}^j,\tilde{u}^j]$ contracts toward $\tilde{m}$, a problem of the form $P(\tilde{\ell},\tilde{m},\tilde{u})$ will be generated with the property that $\tilde{m}$ is not optimal for $P(\tilde{\ell},\tilde{m},\tilde{u})$. In that case, if x^* is an optimal solution of $P(\tilde{\ell},\tilde{m},\tilde{u})$, it follows that $f(x^*) \leq \tilde{f}(x^*) < \tilde{f}(\tilde{m}) = f(\tilde{m})$, so that x^* is a feasible solution with a better objective value than $\tilde{m}$. In order to obtain an efficient implementation of contraction search, the initial triple at each interation should have the properties that $\tilde{\ell}^1$ and $\tilde{u}^1$ are neither "too far" from $\tilde{m}$ (if they are "too far" away, $\tilde{m}$ will be optimal for $P(\tilde{\ell}^1,\tilde{m},\tilde{u}^1)$ and a contraction will be required) nor "too close" to $\tilde{m}$ (if they are "too close" only a "small move" away from $\tilde{m}$ will be possible). A procedure that has been found to perform quite well with respect to these criteria involves the use of Lagrangian relaxation with "safeguards".

For every m-vector π we may define a Lagrangian relaxation $P_\pi(x)$ by multiplying the equality constraints of (1.1) by π and subtracting this product from the objective:

$$P_\pi(x) \quad \left\{ \begin{array}{l} \min\limits_{x} f(x) - \pi(Ax-b) \\ \\ \text{s.t. } \ell \leq x \leq u \end{array} \right.$$

We denote by $L(x,\pi)$ the objective function $f(x) - \pi(Ax-b)$ and by $\omega(\pi)$ the optimal value of $P_\pi(x)$, and observe that, for any π, $\omega(\pi) \leq f(x^{**}) - \pi(Ax^{**}-b) = z^{**}$, so that $\omega(\pi)$ is a lower bound for the optimal value z^{**}. From duality theory (see, for example, [Rockafellar (1970)]) it is easily shown that there exists a π^{**} such that $\omega(\pi^{**}) = z^{**}$, and therefore "good" choices of π will provide tight lower bounds on z^{**}.

From a computational viewpoint, the solution of the approximating problem $P(\tilde{\ell},\tilde{m},\tilde{u})$ as an LP provides a set of optimal values π^* for the dual variables corresponding to the constraints $Ax = b$. Moreover, the separability of f implies that an optimal solution of $P_\pi(x)$ may be obtained by separately solving n one-dimensional optimization problems, since $L(x,\pi)$ is also separable and the constraints of $P_\pi(x)$ are simply bounds on the individual variables. In order to give a geometric interpretation to these problems, we define $s_i^* = \pi^* A^i$, where A^i is the ith column of A, and let $P_{\pi^*}(x_i)$ denote the problem

$$\min_{\ell_i \le x_i \le u_i} f_i(x_i) - (s_i^* x_i + k_i^*),$$

where k_i^* is a constant chosen so that $\tilde{f}_i(x_i^*) = s_i^* x_i^* + k_i^*$. (Observe that the value of the constant term in $P_{\pi^*}(x_i)$ has no effect on the set of optimal solutions of this problem.) For notational convenience we denote an <u>optimal solution</u> of $P_{\pi^*}(x_i)$ as $x_i^*(\pi^*)$ and the <u>optimal value</u> of that problem as $\omega_i(\pi^*)$. Note that $-\omega_i(\pi^*) = \max_{\ell_i \le x_i \le u_i} [(s_i^* x_i + k_i^* - f_i(x_i)]$, and that $-\omega_i(\pi^*) \ge 0$ since the functions $s_i^* x_i + k_i^*$ and $\tilde{f}_i(x_i)$ agree at x_i^*. Thus, the value $-\omega_i(\pi^*)$ may be interpreted as the <u>maximum</u> amount (which must be non-negative) by which $s_i^* x_i + k_i^*$ over-estimates $f_i(x_i)$ on the interval $[\ell_i, u_i]$, and $x_i^*(\pi^*)$ is the point at which this maximum error occurs. It is shown in [Meyer (1980)] that the lower bound $\omega(\pi^*)$ may be written as $\tilde{f}(x^*) - \sum_{i=1}^{n} (-\omega_i(\pi_i^*))$, so that it is simply $\tilde{f}(x^*)$ <u>minus</u> the error bounds associated with the approximation of the $f_i(x_i)$ by $s_i^* x_i + k_i^*$. From the standpoint of reducing the error estimate term in the lower bound, $x_i^*(\pi^*)$ is therefore the best point to use in a new approximation. However, keeping in mind the criteria described previously for an efficient contraction search, the use of $x_i^*(\pi^*)$ as an initial lower or upper bound for the next iteration is subject to "safeguard" factors that prevent the interval for the variable x_i from expanding or contracting too greatly from one iteration to the next. Details of these safeguards are given in [Meyer (1980)].

4. <u>Computational Experience</u>

A brief summary of the computational experience for nonlinear networks in [Meyer (1980)] will be given. (That report also gives experience for problems of the form (1.1) in which integrality constraints are present and problems with general linear constraints (as opposed to network constraints).)

[Test Problem 1]

This problem is a quadratic transportation problem arising from an econometric application described in [Bachem and Korte (1977)]. The problem format is

$$\min_x \sum_{i,j} (x_{i,j} - t_{i,j})^2$$

$$\text{s.t.} \quad \sum_{i=1}^{k} x_{i,j} = d_j \quad (j=1,\ldots,k)$$

$$\sum_{j=1}^{k} x_{i,j} = s_i \quad (i=1,\ldots,k)$$

$$x_{i,j} \geq 0,$$

where the $t_{i,j}$ are constants giving "target" flows on the arcs, d_j is the demand at node j, and s_i is the supply available at node i. The constants are k = 10 (yielding 100 arcs, 20 nodes),

$t_{i,j} = 10 + i + j$, $s_i = \sum_{j=1}^{10} (21-i+j)$, $d_j = \sum_{i=1}^{10} (21-j+i)$. The optimal solution with $x^*_{i,j} = 32 - (i+j)$ (optimal value 6600) was obtained in 9 iterations (each of which was the solution of a linear network problem). For this class of problems, it is natural to use the (infeasible) target flows $t_{i,j}$ in place of the mid-range default values in the initial approximation.

[Test Problem 2]

This is a model of the water supply network for Dallas. The data (see [Collins, et al (1978)]) were supplied to us by Jeff Kennington of SMU. The problem has 666 nodes and 906 arcs, and an objective function involving 18 linear functions, 16 terms of the form

$\sqrt{\beta/2}(x\sqrt{\alpha-x^2} + \alpha \sin^{-1} x/\sqrt{\alpha})$, and 872 terms of the form $c|x|^{2.85}$. A feasible solution whose objective value was guaranteed correct (via the lower bound) to 6 figures was obtained in 16 iterations.

5. Directions for Further Research

Although excellent computational experience has been obtained with the algorithms in their present form, there are a number of ideas under study that may further improve efficiency. One obvious strategy is to terminate the solution of the subproblems $P(\tilde{\ell},\tilde{m},\tilde{u})$ prior to optimality, particularly in the initial major iterations in which these problems require numerous pivots. In this case the termination criterion for the early subproblems could be a fixed number of pivots or a tolerance on the reduced cost of candidates to enter the basis or a combination of these two strategies. Such a tolerance would avoid pivots that would have only a marginal effect on the objective function value in favor of pivots (in the next major iteration) with a more significant effect. In later major iterations the termination criterion could be the achievement of a certain percentage reduction of the error bound. Note that the limiting case of this strategy would be the use of only one pivot per iteration (except for the first iteration, in which this approach would be postponed until feasibility had been attained.) While it would probably be inefficient in this limiting case to calculate new objective function values of all variables after each pivot, the algorithm could be further modified by calculating a new value for a variable only if the corresponding flow is driven to a temporary bound $\tilde{\ell}_i \neq \ell_i$ or $\tilde{u}_i \neq u_i$ as a result of the pivot, in which case the new evaluation of f_i would allow the flow to continue its change past the temporary bound $\tilde{\ell}_i$ or $\tilde{u}_i$ (provided that the arc corresponding to the new segment of the cost function priced out with the proper sign). This approach is equivalent to what might be termed an "implicit grid" strategy in which the function values at the implicit grid points are calculated only as needed when a variable reaches the limits of its initial range $[\tilde{\ell}_i,\tilde{u}_i]$. Note that with respect to price out strategy, this approach suggests that the arcs corresponding to each variable be priced out consecutively as the flows reach their bounds.

In a strategy in which the subproblems are not solved to optimality, the use of a line search could be expected to be of greater value than it is in the current strategies. Note that convexity guarantees that any point yielding an improvement in the approximating function furnishes a descent direction.

The idea of using local piecewise-linear approximation may, of course, be easily extended to non-separable objective functions as described in [Kao and Meyer (1979)] and to nonlinear constraints as well. However, the convergence properties of such algorithms in the nonlinear constraint case are still under study.

Acknowledgement

The RNET minimum cost network flow subroutines (see [Grigoriadis and Hsu (1979)]) used to solve the linear network subproblems were provided by M.D. Grigoriadis.

References

[1] Bachem, Achim and Korte, Bernhard, "Quadratic Programming over Transportation Polytopes", Report No. 7767-OR, Institute für Okonometrie und Operations Research, Bonn, 1977.

[2] Beale, E.M.L., Mathematical Programming in Practice, Sir Isaac Pitman & Sons Ltd., London, 1968.

[3] Bradley, Gordon H., Brown, Gerald G., and Graves, Glenn W., "Design and Implementation of Large Scale Primal Transshipment Algorithms", Man Sci., 24, pp. 1-34, 1977.

[4] Collins, M., Cooper, L., Helgason, R., Kennington, J., and LeBlanc, L., "Solving the Pipe Network Analysis Problem Using Optimization Techniques", Man. Sci., 24, pp. 747-760, 1978.

[5] Geoffrion, A.M., "Objective Function Approximations for Mathematical Programming", Mathematical Programming, 13, pp. 23-27, 1977.

[6] Glover, F. and Klingman, D., "Real World Applications of Network Related Problems and Breakthroughs in Solving Them Efficiently", ACM Transactions on Mathematical Software, 1, pp. 47-55, 1975.

[7] Grigoriadis, Michael D. and Hsu, Tau, "RNET The Rutgers Minimum Cost Network Flow Subroutines", SIGMAP Bulletin, pp. 17-18, April, 1979.

[8] Kao, C.Y. and Meyer, R.R., "Secant Approximation Methods for Convex Optimization", University of Wisconsin-Madison Computer Sciences Technical Report #352, 1979, to appear in Math. Prog. Study, 14.

[9] Meyer, R.R., "A Class of Nonlinear Integer Programs Solvable by a Single Linear Program", SICOP, 15, pp. 935-946, 1977.

[10] Meyer, R.R., "Two-Segment Separable Programming, Man. Sci., 25, pp. 385-395, 1979.

[11] Meyer, R.R., "Computational Aspects of Two-Segment Separable Programming", University of Wisconsin-Madison Computer Sciences Department Technical Report #382, 1980.

[12] Meyer, R.R. and Smith, M.L., "Algorithms for a Class of 'Convex' Nonlinear Integer Programs", in Computers and Mathematical Programming, ed. by W.W. White, NBS Special Publication 502, 1978.

[13] Müller-Merbach, H., "Die Methode der 'direkten Koeffizientanpassung' (μ-Form) des Separable Programming", Unternehmensforschung, Band 14, Heft 3, pp. 197-214, 1970.

[14] Rockafellar, R.T., Optimization in Networks, Lecture Notes, University of Washington, 1976.

[15] Saaty, T.L., Optimization in Integers and Related Extremal Problems, McGraw-Hill, New York, 1970.

[16] Thakur, L.S., "Error Analysis for Convex Separable Programs: The Piecewise Linear Approximation and the Bounds on the Optimal Objective Value", SIAM J. App. Math., pp. 704-714, 1978.

COMPUTATIONAL TESTING OF ASSIGNMENT ALGORITHMS

Michael Engquist
University of Texas at Austin
Austin, Texas 78712

ABSTRACT

The methods used in an empirical evaluation of several recently proposed assignment algorithms are discussed. Brief descriptions of the algorithms tested are also included.

1. Introduction

There have been several recent studies in which new algorithms and computer codes for solving the assignment problem have been developed [1], [8], [10].

In this paper we present a computational comparison of these approaches with an implementation of a successive shortest path algorithm (SSP). Since SSP is described in detail in [6], we will give only a very general description here. SSP goes through a series of modified assignment problems in which some destinations are permitted to have demands greater than one, while some demands are set to zero. The algorithm proceeds from the optimal solution of one of these modified problems to the optimal solution of the next via a related shortest path subproblem. The algorithm terminates when the modified problem coincides with the original assignment problem. This algorithm is a refinement of the Dinic-Kronrod algorithm [4], and it is closely related to the algorithms developed independently by Hung and Rom [8] and Gribov [7].

Weintraub and Barahona [10] have based their work on the minimum cost flow algorithm of Edmonds and Karp [5]. Although this approach has some similarities to SSP, it is evidently a different algorithm.

Barr, Glover, and Klingman [1] developed a new version of the primal simplex algorithm called the AB algorithm which examines only certain bases (called the alternating path bases) representing a given extreme point. Even with this improvement to the primal simplex algorithm, over 90% of the pivots are degenerate.

The algorithms tested, other than the AB algorithms, substitute a more complicated procedure for the primal simplex pivot so that nondegenerate flow change and progress toward optimality are guaranteed at each iteration.

2. Computational Considerations

We have used computational testing in two ways. First, in the development of our implementation of SSP, which we call SPAN, a number of preliminary tests were conducted in order to help determine the final configuration of the computer code. Two such tests are briefly discussed in this section. Second, computational tests comparing SPAN with other codes were conducted as described in the next section.

In SPAN, the data for the assignment problem is stored in reverse star form, and a pointer list is employed to indicate the entry position for the block of memory locations assigned to each reverse star. The shortest path subproblems are solved by a label-setting approach based on the Dijkstra algorithm [3]. Our implementation of the Dijkstra algorithm employs a modular sort list and a single radix sort as described in [2].

The initial modified assignment problem for SSP is created by making a tentative (infeasible) assignment. This tentative assignment simply puts a unit flow on a least cost arc from each forward star. A destination node which receives no assignment under this procedure is called _deficient_. At each iteration of SSP one of the remaining deficient nodes is given an assignment.

There may be more than one way to choose a least cost arc from a given forward star. We developed a heuristic for breaking ties in such a way that the number of deficient destinations in the initial assignment is decreased. On the basis of limited testing we concluded that our heuristic was of benefit only for assignment problems in which the cost range is small. For this reason, we did not include the heuristic in SPAN.

At each iteration of SSP, the deficient node which gets the next assignment must be determined. We did some experimentation but were unable to develop a more efficient strategy than simply choosing deficient node j where j is as small as possible. This is the strategy used in SPAN.

3. Comparative Computational Tests

We have tested SPAN against implementations of several other algorithms. These include the codes of Weintraub and Barahona [10], Hung and Rom [8], and Barr, Glover, and Klingman [1]. The first two codes we call DUAL and RELAX, respectively. The third is known as AP-AB. The four codes tested are written in FORTRAN. The problems used in the tests were n×n assignment problems and were randomly generated using NETGEN [9]. In SPAN, there is a single parameter called NBUC which equals the length of the sort list. For all the tests reported in this section, NBUC was set at 200. In DUAL, a parameter called NSQR was set, as suggested in [10], to be about $\sqrt{n}$. We did not set any other parameter values for the codes tested. All computer runs were carried out on the CDC Dual Cyber 170-175 using the FTN compiler during periods of comparable computer use. All solution times reported are exclusive of input and output. Each time reported in Tables 1-4 is the average of times for three runs on a single problem. The actual run times varied from the time reported by as much as 14% for the smallest problems tested, but such variation was generally much less.

Based on total solution times for the problems shown in Tables 1 and 2, SPAN is about 3 times faster than AP-AB and roughly 6 times faster than RELAX. The closest competing code is DUAL; however, because DUAL did not achieve optimality on some problems, we much be somewhat cautious with regard to the solution times reported. It appears that there may be some defect in the code, and this could cause solution times to increase when it is corrected. Nevertheless, we have run SPAN and DUAL on some additional problems with the results shown in Table 3. Based on the total solution time for all the problems on which DUAL achieved optimality, we conclude that SPAN is 25 to 30 percent faster than DUAL.

Although SPAN was much more efficient than RELAX for the tests on sparse problems reported in Tables 1 and 2, it is important to point out that RELAX was developed for totally dense problems. For this reason, we conducted further tests with the results shown in Table 4. These results indicate that RELAX is more efficient than SPAN on totally dense problems. Since assignment problems encountered in practice are invariably sparse, the question arises as to whether some new implementation of the relaxation algorithm might be more efficient than SPAN on sparse problems.

Next, we compare the number and size of arrays required by the various codes. SPAN uses 2 arc length and 10 n-length arrays along with the sort list. DUAL required 6 arc length arrays, 21 n-length arrays, and 3 arrays for which we were unable to determine the length except that it must be more than n. RELAX requires an $n \times n$ matrix and 9 n-length arrays. AP-AB requires 2 arc length and 6 n-length arrays.

For sparse problems AP-AB uses the least array space with SPAN running a close second. Both RELAX and DUAL use a lot of array space for sparse problems; however, RELAX is considerably better off when it comes to dense problems. When the arc density of a problem is about 50%, SPAN and RELAX require roughly the same amount of space. We remark that, as indicated by the results in Table 4, SPAN is more efficient than RELAX on such 50% dense problems.

	NUMBER OF ARCS				
CODE	1500	2250	3000	3750	4500
SPAN	.085	.182	.159	.280	.187
DUAL	.178	did not achieve optimality	.272	.292	.342
RELAX	1.364	1.459	1.117	1.053	1.154
AP-AB	.490	.604	.631	.685	.921

Table 1
Solution Times in Seconds
for 200 x 200 Assignment
Problems with Cost Range 1-100

	NUMBER OF ARCS				
CODE	1500	2250	3000	3750	4500
SPAN	.126	.194	.191	.265	.383
DUAL	did not achieve optimality	.273	did not achieve optimality	.433	.450
RELAX	.882	1.152	1.027	1.148	1.353
AP-AB	.513	.570	.630	.663	.945

Table 2
Solution Times in Seconds
for 200 x 200 Assignment
Problems with Cost Range 1-10000

CODE	NUMBER OF ARCS 3000	3500	4000	4500
SPAN	.278	.279	.355	.314
DUAL	.330	.342	.477	.448

Table 3

Solution Times in Seconds
For 300 x 300 Assignment
Problems with Cost Range 1-100

CODE	NUMBER OF ARCS 2500	5000	7500	10000
SPAN	.083	.165	.216	.305
RELAX	.238	.257	.212	.258

Table 4

Solution Times in Seconds
For 100 x 100 Assignment
Problems with Cost Range 1-100

4. Summary and Conclusions

In this paper, we have presented computational results which show that a successive shortest path algorithm [6] is more efficient for solving large sparse assignment problems than several other recently developed algorithms including the best primal simplex algorithm [1].

Our computational results raise the question of whether some extension of successive shortest path methods will enjoy similar success on transportation and transshipment problems. An extension of this type already exists [7], and it seems that the question will ultimately be settled through further computational testing.

References

1. R. Barr, F. Glover, and D. Klingman, "The Alternating Basis Algorithm for Assignment Problems," Mathematical Programming, Vol. 13, pp. 1-13 (1977).

2. R. Dial, F. Glover, D. Karney and D. Klingman, "A Computational Analysis of Alternative Algorithms and Labeling Techniques for Finding Shortest Path Trees," Networks, Vol. 9, pp. 215-248 (1979).

3. E. Dijkstra, "A Note on Two Problems in Connexion with Graphs," Numerische Mathematik, Vol. 1, pp. 269-271 (1959).

4. E. Dinic and M. Kronrod, "An Algorithm for the Solution of the Assignment Problem," Soviet Math. Doklady, Vol. 10, No. 6 (1969).

5. J. Edmonds and R. Karp, "Theoretical Improvements in Algorithmic Efficiency for Network Flow Problems," Journal of the Association for Computing Machinery, Vol. 19, pp. 248-264 (1972).

6. M. Engquist, "A Successive Shortest Path Algorithm for the Assignment Problem," Research Report CCS 375, Center for Cybernetic Studies, The University of Texas, Austin, TX (1980).

7. A. Gribov, "Recursive Solution for the Transportation Problem of Linear Programming," Vestnik of Leningrad University, Vol. 33, No. 19 (1978) (in Russian).

8. M. Hung and W. Rom, "Solving the Assignment Problem by Relaxation," Operations Research, Vol. 28, No. 4, pp. 969-982 (1980).

9. D. Klingman, A. Napier, and J. Stutz, "NETGEN-A Program for Generating Large-Scale (Un)Capacitated Assignment, Transportation, and Minimum Cost Flow Network Problems," Management Science, Vol. 20, No. 5, pp. 814-821 (1974).

10. A Weintraub and F. Barahona, "A Dual Algorithm for the Assignment Problem," Publication No. 79/02/c, Departamento de Industrias, Universidad de Chile - Sede Occidente, Santiago, Chile (1979).

"ESTABLISHING A GROUP FOR TESTING MATHEMATICAL PROGRAMS"

A PANEL DISCUSSION

CHAIRMAN

John M. Mulvey
Princeton University

PANELISTS

Ron Dembo, Yale University

Richard Jackson, National Bureau of Standards

Darwin Klingman, University of Texas, Austin

Saul Gass, University of Maryland

Roy Marsten, University of Arizona

Ken Ragsdell, Purdue University

Ronald Rardin, Georgia Institute of Technology

Provided here is a summary of the comments made during this panel discussion. While every attempt was made to insure accuracy, J. Mulvey, who functioned as chairman during this discussion, claims full responsibility for any inaccuracies or misrepresentations which may persist. Names and addresses of all contributors to the discussion appear in the Appendix.

J. Mulvey: To conclude the meeting, I have asked the conference chairmen to assess the feasibility of forming a group for the purpose of testing mathematical programming techniques. This group would perform a service similar to the American Society for Testing and Materials (ASTM); it would evaluate, using standardized procedures, MP algorithms and software. Results of these tests would be distributed through a regular publication, such as the COAL newsletter.

An analogy can be made between the current status of MP computational testing and the engineering profession 200 years ago. At that time people who built suspension bridges and other structures had little or no theory to support their intuitions. By observing successful bridge designers and mimicking those designs that withstood the test of time, they were able to continue building. Those bridges that collapsed were considered failures. Gradually, the idea of how a structure should be built became understood, and a set of standards was developed. Today, each component of a structure is analyzed very carefully using a large body of scientific theory relating to strength of materials and mechanics. Yet even today, there are important failures -- the Hartford, Conn. Coliseum. These failures become subjects for intense scrutiny and study; conferences are devoted to them.*

In contrast, researchers who evaluate MPs tend to concentrate on the success stories. Journal publication standards contribute to this tendency. Unless a new algorithm is shown to be substantially better than a previous one, the editors of leading journals will generally reject the article. Unfortunately, authors react to this state of affairs by overstating their conclusions, for instance, and by careful selection of test problems.

Returning to the engineering profession, we find an organization devoted to testing, called the American Society for Testing and Materials. They test new materials, such as concrete, and publish the empirical results. The tedious and expensive task of testing is borne by the entire membership through dues and grants. Perhaps in the future, we would see an analogous organization for testing MP's, -- the International Society for Testing Mathematical Programming Software -- and testing centers which would carry out the computational experiments. What does the panel think of this idea?

---

*See, for example, David Billington, "History and Esthetics in Suspension Bridges," Journal of the Structural Division, August, 1977.

S. Gass: In terms of a society for testing software, I propose that until a procedure is set for developing software standards, the engineering analogy is excellent because it has pre-established procedures, standards, and guidelines. Before we get involved in organizing a group to test software, I think we must first address the more basic problem of what good software really is. I have problems with the name testing center because I'm not sure whether there should be a group concerned with testing algorithms per se. I think that it is appropriate for some agency, possibly a federal government office like the National Bureau of Standards, to be concerned with developing initial software guidelines. While there are some guidelines from computer programming software which must be extended into the algorithmic development area, a group like NBS can formulate important problems and begin making them available for testing. In general, a test center should not be concerned with the codes and algorithms being developed at universities which claim to be better than previous models. Instead, a test center should set the stage in terms of how algorithms are to be tested, how they should be reported, and what type of test problems should be generated.

Any modeling activity has at least one, if not two or three, types of algorithms imbedded in the system, subsequently, researchers who develop algorithms appear to be concerned with the same issues as researchers who develop models. Is this modelling system good? Has it addressed pertinent problems? Are the mathematics correct? Is the documentation complete? Hence, there should be a stronger mesh between activities in the general model evaluation assessment area and those in algorithmic testing.

D. Klingman: I agree that there is a definite need for strict guidelines and a need for an agency to collect test data; however, I don't agree that it would not be constructive to have a testing center to conduct "unbiased" testing and provide public dissemination of this information. Consider the situation of writing a paper and having a referee review it. As a referee, you tend to gain an insight that as the developer you did not possess. A testing center could offer this same kind of constructive feedback to developers and thus improve future development. However, if a center is established, its success or failure will depend primarily on the personnel in charge and their dedication to doing a good job. I think that there are a number of people who want to do testing and contribute ideas to testing, but who are not interdisciplinary enough -- in the sense of being good computer scientists, good math programmers, and good data processing people -- to be able to conduct comprehensive testing. Unless they are associated with organizations that have a pool of people to assist them as well as accessability to computer hardware within an affordable budget cost, people become excluded from this type of research. Even at this conference there is a cadre of people excluded because they are not at the right organization to become involved; they do not belong to large companies that are funded heavily with government research monies or to major state universities which work with state money on low cost computing facilities. In this case, a center would help alleviate the problems of both getting people who are interdisciplinary across a number of areas and also allowing people from situations possessing less extensive facilities to become involved in testing.

R. Meyer: The issue of computational testing becomes an extremely expensive operation if one wants to analyze a reasonably large set of test problems. If a testing center were set up, this burden could be removed from the individual program developer. However, I suggest that if such a policy is adopted, one approach to consider would be to ask that anyone interested in submitting a code for testing must first use his code to solve a small selection of test problems supplied by the test center. If he can't get his code to work on that small selection of problems within a reasonable amount of time, he should not send his code into the center, saving several thousands of dollars in CPU time that would be spent testing it on a larger set.

It is important that a larger group of real-life test problems be developed, since nothing like this currently exists. If these test problems were available they would serve as a valuable screening device and, at the same time, encourage people who have promising codes.

R. Jackson: One potential difficulty with a test center is the loss of cross-fertilization. Many project inputs result from discussions that one has with colleagues. In the past, we saw people developing algorithms, producing codes and publishing their results. Now we're beginning to see researchers who are not only inventing algorithms and producing codes but are also doing testing and providing the necessary feedback for improving codes. If this testing function is taken away from academic institutions and large laboratories and put into a central place, we may end up with a bureaucracy. I'm concerned that we are going to lose cross-fertilization; the knowledgeable people who can perform testing and exercising of these codes and algorithms are not going to be involved at the testing site.

D. Klingman: I don't feel that establishing a center would inhibit or restrict testing at universities or cross-fertilization of ideas. Instead, a testing center would serve to provide standards for collecting and disseminating a large number of test problems and, perhaps, act as a final depository for codes. It could assist those people whose only viable means of doing testing would be to come to the center and utilize its facilities directly.

R. Jackson: I worry that by removing the burden of demonstrating feasibility and the concern with computational accuracies from algorithm developers we might take a great leap backwards.

D. Klingman: In terms of helping people who do not want to become fully interdisciplinary across all testing aspects, the center can provide important assistance because its purpose is not to remove them from the process. Instead, it will draw these people into appreciating and understanding the value of testing and the merits of different data structures. Even the Federal government could afford to support that kind of testing effort.

A. Miele: If a testing center is set up, there is one grave problem i.e., the people who are running the center. The center can be organized with either two categories of people: (1) people who have never developed an algorithm on their own -- in which case, you have a staff of almost incompetent people -- or alternatively, (2) you can staff the center with people who have developed algorithms, but then the problem of bias enters in.

S. Gass: It is a question of whether you are testing one algorithm against another, or whether you are testing an algorithm to see if it is worthwhile from a computational point of view, for example, in terms of robustness. I don't view algorithmic comparisons as a major activity at the test center, because that would create too many difficulties.

A. Miele: As a case in point, the director of the center who does not like a specific approach might find every possible excuse to delay testing a particular algorithm.

L. Fosdick: I haven't heard one point mentioned about networks. You don't have to carry programs around in a suitcase, and you don't physically have to visit Purdue Campus to test codes. There is the ARPANET which some people have access to; there have been discussions about linking together computer science departments of universities. There are a number of possibilities. All that is needed, I would think, would be some place to coordinate these activities. About testing codes, I think there is a definite advantage to running programs on different machines in various places if it is your interest to test software, rather than algorithms. If you're interested in algorithms and models, the issues are different.

R. O'Neill: One area that has been neglected is experimental design. Also, there is no effective mechanism to report results. Hence, it is difficult to make definitive statements and say "we had this test problem and this algorithm performed better." In reality, there is a battery of problems, and in the end, there are no statistically reportable results.

K. Hoffman: I would like to ask the question of whether there are any ideas on how to go about collecting a set of real-world test problems so that a large group of researchers could actually work on these problems, since nothing of this sort currently exists.

E. Eason: I am a consultant operating in the area of engineering design optimization. If COAL is interested in collecting any of the real problems that I can provide, I would be happy to supply them, once I clear it with my clients.

R. Jackson: If you can do that, would you please write down how you were able to convince your clients: if you disguised the data, if you just twisted their arm, or if you promised to make their problems infeasible the next time you run them. Most companies claim that their data is propietary. Your success in this area would be a valuable contribution.

At a series of sessions at ORSA conferences, Dick O'Neill and I brought in major manufacturers, of mixed and pure integer programming packages, for example IBM, and CDC. We asked them to talk about their plans for the future. At the same time, we proposed a systematic evaluation of programs. Unfortunately, these major manufacturers, whom we thought had everything to gain by sharing information, declined to participate. Apparently they felt there would be no benefit.

K. Ragsdell: In the NSF study that began in '73, we had several industries who participated by supplying codes and problems, but, as has been indicated, there was one overriding difficulty. The reason is pure and simple: the problems contain models which are typically proprietary and often represent many man-years of development. Companies view this work as a competitive edge and are unwilling to release it. Consider two examples from our study -- Honeywell and Whirlpool. I spent summers with each of these companies helping to develop problems and working with their people. In each case, after the work was completed, a hired researcher spent six months purifying the model so that it no longer represented the real model. It still had the same contents from my viewpoint (i.e., the same number of variables, the same degree of non-linearity, and the same level of difficulty) but the specifics were altered before we were allowed any further use. In finding a mechanism for obtaining practical problems, I think we must be prepared to protect the proprietary rights of the people who supply the problems and recognize that they do have a legitimate complaint.

J. Mulvey: Let us return to the topics of testing centers and future plans for COAL. Does anyone in the audience have additional comments regarding these issues?

R. Meyer: A number of people have commented that the test group should avoid making value judgements on codes. My own opinion is that it is important to have a reasonable selection of test problems for running codes and putting out results. As far as making judgements is concerned, the group should avoid comparing codes by means of end results, and instead, let other people make these judgements.

R. Dembo: My concern is that there is a definite need for standard procedures and guidelines comparing algorithms and presenting results.

R. Jackson: I agree that guidelines need to be produced, and we see them coming. We're beginning to hear of substantial algorithmic code comparisons which are being conducted. We are beginning to see experimental design concepts and statistical approaches to testing. At the Bureau of Standards, we have always kept in mind a methodology.

Rather than putting together a formal statement about future COAL plans, we ought to encourage ourselves to continue research in this area. We've been talking about the establishment of testing centers, but already these are ad hoc testing centers. Ken Ragsdell at Purdue has gotten substantial support to conduct testing. The Bureau of Standards has been able to locate Federal funds to carry out testing and to be concerned with the development of methodologies. Sandia Laboratory has been able to set aside rather large amounts of money. Perhaps, we've already got testing centers.

The important thing for us to be evaluating is not simply where the centers ought to be or who the people are that ought to run them, but how can we continue to get funding for this activity. This is the issue that the Committee on Algorithms should focus on.

J. Mulvey: There is evidently a variety of opinions concerning the establishment of a testing center. Those who are interested in continuing this discussion should, perhaps, get together and form a sub-committee.

Finally, I'd like to commend everyone for perservering. Thank you all for coming to this first conference devoted to evaluating mathematical programming techniques.

Appendix

A description of the conference program and a list of participants are contained in the Appendix. The conference brought together MP algorithmic researchers, computer scientists, MP users and consultants, and software developers from industry to examine how mathematical programming techniques ought to be evaluated.

It was generally agreed that strides had been made in experimental mathematical programming. Much additional research effort will be required, however, as novel computer systems and MP algorithms are implemented.

Several related reports are included in the Appendix. First, K. Schittkowski gives his views about comparative experiments in "A Model for the Performance Evaluation in Comparative Studies." Next, A. Miele and S. Gonzalez provide additional remarks about comparative studies. These papers are followed by K. Ragsdell's report "The Evaluation of Optimization Software for Engineering Design."

Description of Conference Program

List of Speakers (January 5, 1981)

Opening remarks - John M. Mulvey (Princeton University)

Keynote address - Professor Darwin Klingman (University of Texas at Austin)
"Computational Advances in Networks and Their Transfer to Other Mathematical Programming Areas"

DESIGN AND USE OF PROBLEM GENERATORS AND HAND SELECTED TEST CASES

*Ronald Rardin (Georgia Institute of Technology)

Ronald Rardin (Georgia Institute of Technology) and Benjamin W. Lin (Rutgers University)
"Standard Test Problems for Computational Experiments - Issues and Techniques"

Joyce Elam (University of Texas at Austin) and Darwin Klingman (University of Texas at Austin)
"NETGEN-II: A System for Generating Structured Network-Based Mathematical Programming Test Problems"

Jerrold H. May (University of Pittsburgh) and Robert L. Smith (University of Michigan)
"The Definition and Generation of Geometrically Random Linear Constraint Sets"

Gideon Lidor (The City College of New York)
"Construction of Nonlinear Programming Test Problems with Known Solution Characteristics"

---

*Session Chairman

STATE OF NONLINEAR OPTIMIZATION CODES AND EMPIRICAL TESTS

*Leon Lasdon (University of Texas at Austin) and *Ken Ragsdell (Purdue University)

E.D. Eason (Failure Analysis Associates)
"Evidence of Fundamental Difficulties in Nonlinear Optimization Code Comparisons"

Eric Sandgren (IBM Corporation)
"A Statistical Review of the Sandgren-Ragsdell Comparative Study"

Klaus Schittkowski (Institut für Angewandte Math. and Stat.)
"Nonlinear Programming Methods with Least Squares Subproblems"

APPROACHES TO SOFTWARE TESTING FROM OTHER DISCIPLINES

*Richard Jackson (National Bureau of Standards)

K.L. Hoffman and R.F. Jackson (National Bureau of Standards)
"In Pursuit of a Methodology for Testing Mathematical Programming Software"

James E. Gentle (IMSL, Inc.)
"Testing and Evaluation of Statistical Software"

W.J. Cody (Argonne National Laboratory)
"Transportable Test Procedures for the Elementary Function Software"

INTEGER PROGRAMMING AND COMBINATORIAL PROBLEMS

*Roy Marsten (University of Arizona)

Fred Shepardson (Stanford University) and Michael G. Chang
"Evaluating a New Integer Programming Algorithm: A Case Study"

Robert L. Sielken, Jr. (Texas A and M) and William J. Riley
"Which Options Provide the Quickest Solution"

William Stewart (College of William and Mary)
"A Computational Comparison of Five Heuristic Algorithms for the Euclidean Traveling Salesman Problem"

COMPARATIVE COMPUTATIONAL STUDIES IN MATHEMATICAL PROGRAMMING

*Ron Dembo (Yale University)
Participants:
A. Miele (Rice University)
S. Gonzalez (Department of Simulations, Palmira, Moreles)
J. Bus (Stichting Math Center)
R. Schnabel (University of Colorado)
K. Ragsdell (Purdue University)
K. Schittkowski (Institut fur Angewondte Mathematik and Statistik)
D. Himmelblau (University of Texas)

Tuesday (January 6, 1981)

REDUNDANCY

Mark Karwan, Vahid Lotfi, Jan Telgen and Stanley Zionts (State University of New York at Buffalo)
"A Study of Redundancy in Mathematical Programming"

A. Boneh (Universitv of California at Santa Barbara)
"Method for Identifying Redundancy in Nonlinear Sets of Inequalities"

*Session Chairman

COMPARATIVE STUDIES

Jacques Bus (Stichting Mathematisch Centrum)
"A Methodological Approach to Testing NLP Software"

Kathy Hiebert (Sandia Laboratory)
"Testing Mathematical Software Which Solves Nonlinear Systems of Equations"

APPROACHES TO SOFTWARE TESTING FROM OTHER DISCIPLINES

*Richard Jackson (National Bureau of Standards)

Leon Osterweil (University of Colorado)
"TOOLPACK - An Integrated System of Tools for Mathematical Software Development"

Lloyd D. Fosdick (University of Colorado)
"Overview of Testing of Numerical Software"

Chuck Smith (IBM Corporation)
"The Application of Software Science to Testing"

DESIGN OF EXPERIMENTS

*Richard O'Neill (U.S. Department of Energy)

Richard O'Neill (U.S. DOE)
"A Comparison of Real-World Linear Programs and Their Randomly-Generated Analogs"

Lunch - COAL BUSINESS MEETING (Jackson)

Future direction and goals of COAL

SPECIAL TOPICS IN TESTING OF MP ALGORITHMS (MICRO-PROCESSORS, APL, OTHERS)

*Harlan Crowder (IBM Corporation)

Uwe Suhl (Free University, Berlin)
"Implementing an Algorithm: Performance Considerations and a Case Study"

A. Buckley (Concordia University)
"A Portable Package for Testing Minimization Algorithms"

Harlan Crowder (IBM Research)
"Design and Validation of Mathematical Algorithms Using APL"

ADVANCES IN NETWORK RELATED ALGORITHMS

*Darwin Klingman (University of Texas at Austin)

Michael Engquist (Eastern Washington University)
"Computational Testing of Assignment Algorithms"

R.R. Meyer (University of Wisconsin)
"Recursive Piecewise-Linear Approximations Methods for Nonlinear Networks"

R. Barr (Southern Methodist University)
"Solution Strategies and Algorithm Behavior in Large Scale Network Problems"

PANEL DISCUSSION ON ESTABLISHING A GROUP FOR TESTING MP TECHNIQUES

*John Mulvey (Princeton University)

Participants: (Dembo, Jackson, Klingman, Gass, Marsten, Mulvey, Ragsdell, Rardin)

---

*Session Chairman

COAL Conference Participants

Professor Richard Barr
Southern Methodist University
Box 333
Dallas, Texas 75275

Professor Arnon Boneh
University of California at
Santa Barbara
Computer Science Department
Santa Barbara, California 93106

Professor A. Buckley
Dept. of Mathematics
Concordia University
7141 Sherbrooke Street, W.
Montreal, CANADA H4B 1R6

Dr. Jacques Bus
Mathematical Centre
P.O. Box 4079
1009 AB
Amsterdam
THE NETHERLANDS

Dr. Gilberto Calvillo
Banco De Mexico S.A.
Condesa 5, 4° Piso
Mexico, 1, D.F.
MEXICO

Professor Michael G. Chang
Dept. of Decision Sciences
Wharton School
University of Pennsylvania
Philadelphia, PA 19106

Dr. William J. Cody
Applied Mathematics Division
Argonne National Laboratory
Argonne, Illinois 60439

Dr. Harlan P. Crowder
IBM
Thomas J. Watson Center
P.O. Box 218
Yorktown Heights
New York, NY 10598

Professor Ron S. Dembo
School of Organization and
Management
Yale University
Box 1A
New Haven, Connecticut 06520

Dr. Ernie Eason
Failure Analysis Associates
2225 E. Bayshore Road
Palo Alto, CA 94303

Professor Joyce Elam
Analysis, Research and Computation
P.O. Box 4067
Austin, TX 78765

Professor Michael Engquist
University of Texas
BEB 600
Austin, TX 78712

Professor Lloyd D. Fosdick
Dept. of Computer Science
Campus Box 430
University of Colorado
Boulder, Colorado 80309

Professor Saul I. Gass
University of Maryland
8809 Maxwell Drive
Potomac, Maryland 20854

Dr. James E. Gentle
IMSL, Inc.
7500 Bellaire
Houston, TX 77036

Dr. Salvador Gonzalez
Institute of Electrical Investiga-
tions
Division of Power Systems
Dept. of Simulation
Palmira, Morelos, MEXICO

Dr. Richard V. Helgason
OREM Dept. - Caruth Hall
Southern Methodist University
Dallas, Texas 75275

Dr. Kathie Hiebert
Sandia National Labs
Org 5642
Albuquerque, New Mexico 87111

Dr. David M. Himmelblau
Dept. of Chemical Engineering
University of Texas
Austin, Texas 78712

Dr. Karla L. Hoffman
Center for Applied Mathematics
National Bureau of Standards
Washington, D.C. 20234

Dr. Richard H.F. Jackson
Center for Applied Mathematics
National Bureau of Standards
Washington, D.C. 20234

Professor Darwin Klingman
BEB 600
University of Texas
Austin, Texas 78712

Mr. Mike Kupferschmid
Math Science Department
RPI
Troy, NY 12180

Professor Gideon Lidor
Dept. of Computer Sciences
The City College
New York, NY 10031

Professor Roy Marsten
Dept. of MIS
University of Arizona
Tucson, Arizona

Professor Jerrold H. May
Graduate School of Business
University of Pittsburgh
2110 C.L.
Pittsburgh, PA 15260

Professor Robert R. Meyer
Dept. of Computer Science
University of Wisconsin
1210 W. Dayton
Madison, Wisconsin 53706

Professor Angelo Miele
Aerospace Group
Rice University
Houston, TX 77001

Professor John M. Mulvey
Engineering Management Systems
School of Engineering/
Applied Science
Princeton University
Princeton, NJ 08544

Professor Robert M. Nauss
School of Business Administration
University of Missouri
8001 Natural Bridge Road
St. Louis, Missouri 63121

Dr. Richard P. O'Neill
Oil and Gas Analysis Division
EI-522, Room 4520
MS 4530
12th & Pennsylvania Avenue, N.W.
Washington, D.C. 20461

Professor Leon Osterweil
Dept. of Computer Science
Campus Box 430
University of Colorado
Boulder, Colorado 80309

Professor Uwe Pape
Technische Universitat Berlin
Fachbereich 20
Lehreinheit Angewandte
Electronische Datenverar-
beitung
Berlin 10, WEST GERMANY

Professor K.M. Ragsdell
School of Mechanical Eng'g
Purdue University
West Lafayette, Indiana 47907

Professor Ronald L. Rardin
School of Industrial and
Systems Eng'g
Georgia Institute of Technology
Atlanta, GA 30332

Dr. Eric Sandgren
Dept. of Mechanical & Aerospace
Engineering
University of Missouri
Engineering Building 2008
Columbia, Missouri 65211

Dr. Patsy Saunders
Building 101, Room A427
National Bureau of Standards
Washington, D.C. 20234

Dr. Klaus Schittkowski
Dept. of Operations Research
Stanford University
Stanford, CA 94305

Professor Robert L. Sielken
Institute of Statistics
University of Texas
College Station, TX 77843

Dr. Robert Smith
Dept. of Industrial and
Operations Engineering
University of Michigan
Ann Arbor, Michigan 48103

Professor Bill Stewart
School of Business
College of William and Mary
Williamsburg, Virginia 23185

Professor Uwe Suhl
IBM
Thomas J. Watson Center
Yorktown Heights
New York, NY 10598

Dr. Jan Telgen
Dept. of Applied Mathematics
Laan Van Eikenstein 9 (ZL-G-170)
3705 AR ZEIST
THE NETHERLANDS

Dr. Almos P. Zob
P.O. Box 794
Boeing Computer Service Co.
Renton, Washington 98057

A MODEL FOR THE PERFORMANCE EVALUATION IN COMPARATIVE STUDIES

by

Klaus Schittkowski
Stanford University

Abstract

The paper describes a model for the performance evaluation in an arbitrary comparative study of mathematical algorithms. Distinctions will be made between "static" testing, i.e. evaluation of the design of the corresponding programs and "dynamic testing", i.e. evaluation of the performance of these programs when implemented and run on a digital computer. For both levels, relevant criteria for choosing among codes are described. Advantages and disadvantages of real life, hand selected, and randomly generated test problems, respectively, are outlined. It is furthermore shown how a decision maker can combine a variety of performance measures into a final score for each major criterion and program.

1. Introduction

One should start with a precise definition of a 'comparative study'. What we will mean throughout this paper by this expression is a numerical evaluation of the performance of algorithms which can only be obtained in an experimental way. For example, it is supposed that bounds for the number of arithmetic operations cannot be determined by theoretical considerations. It is assumed that all algorithms are designed for the same underlying problem type, e.g. an optimal control problem, a linear or nonlinear programming problem, or the solution of differential equations. The intention of a comparative study should be to describe the organization of the programs and their numerical performance. The results give a decision maker a basis for deciding which program could solve his individual practical problems in the most effective way. This decision will, naturally, be dependent upon his priorities. The mathematical programmer should be able to interpret the results of any computational study in a way that allows him to draw conclusions about the mathematical methods.

Algorithmic performance is evaluated on two levels. Before starting the numerical tests, it is possible to describe the design and the organization of a program. Technical details of this kind are summarized in Section 1. A list of possible performance criteria which can only be evaluated by numerical experiments is given in Section 2.

Since the quality of numerical experiments is based on a suitable choice of test examples, the advantages and disadvantages of different types of test problems are outlined in Section 3. The last section describes ways one could combine the various performance measures into a final score or scores for each program tested.

2. Organizational Details of Competing Programs

The organizational details of competing programs is relevant to report in a comparative study, only if the algorithms have been implemented by different authors. If, on the other hand, all algorithms are programmed by one author and are designed in the same way, the subsequent information is only of secondary interest. In contrast to the experimental performance evaluation, the organizational details of a program can be described before testing it numerically, and I propose that any comparative study should present at least the following information about the codes:

a) Availability: First of all, a decision maker wants to know whether a program is available, or, more precisely, the conditions of availability (e.g. costs, maintenance) and the source (e.g. library, institution).

b) Implementation: Details of the practical implementation of a code are specifically important when a program has to be incorporated into an existing system or a program library. The description of programs should contain therefore at least the following details: programming language, arithmetic (single, double precision), length of the program, dimensioning of arrays (variable, fixed dimensions), design of the driving routine (main, subprogram), and storage requirements.

c) Documentation: The first contact of a user with a computer program consists of investigating the documentation. A clear and complete description of the usage facilitates the solution of a problem. A comparative study should contain information about the quality of the documentation. To get such a measure, one could use priority theory as applied in [3].

d) Provision of problem data, parameters, and functions: One has to expect that the programs under consideration use different strategies to provide the code with problem data (dimension, number of constraints), parameters (stopping tolerances), and problem functions (separately, in blocks). A comparative study should explain the organization of these input data.

e) Control of the solution process: By an external control of the solution method (different update rules, stopping criteria, interpolating schemes), the user has a chance to solve even those problems for which the first solution attempt failed with the standard version.

f) Special problem types: The solution of a problem is often facilitated when the user can take advantage of special structures in the problem, for example, constraints in form of upper and lower bounds or a quadratic objective function.

g) Additional capabilities: Some programs have additional options which facilitate the solution of particularly difficult problems making it easier for unsophisticated users to become comfortable with the package.

3. Performance Criteria

The main feature of a comparative study is the experimental evaluation of the performance of different algorithms and presentation of the results. Important criteria are listed below which have been developed as part of a specialized comparative study of nonlinear programming codes, cf. [3]; these could be used for other types of mathematical models as well. They should be considered as possibilities for a test designer to choose among. This choice depends on the individual objectives of the comparative study.

a) Efficiency: This criterion attempts to measure whether one algorithm solves a problem 'better' than another one. For the definition of 'better', one could choose some of the following subcriteria:

(i) Execution time.

(ii) Iteration number.

(iii) Number of function evaluations (separately for objective function and constraints, if available).

(iv) Number of gradient evaluations (separately for objective function and constraints, if available).

The presentation of iteration count is meaningful if the algorithms possess a similar mathematical structure. For example, this subcriterion should not be used in the comparative study of a variety of nonlinear programming codes since the operations performed within an iteration vary significantly; see [3]. If all possible problem functions are determined by the same arithmetic operations (linear, quadratic, or geometric programming, linear least squares, etc.), one should omit this criterion. Whenever possible, one should try to relate efficiency to solution accuracy, since the calculation of more precise solutions requires more calculation time and function evaluations.

b) Reliability: One of the most important criteria for the practical application of a program is the reliability, i.e. a measure of the probability that a given problem will successfully be solved by the algorithm. For a battery of test runs, the following subcriteria can be used to define reliability:

Percentage of nonsuccessful solutions with,

(i) Nonsuccessful solution being defined as failure in objective function,

(ii) Nonsuccessful solution being defined as failure in satisfying constraint functions, or

(iii) Percentage of interrupted test runs.

Since an optimal solution cannot always be achieved exactly or even at least within machine precision, a user has to define tolerances within which he will consider the result of a solution process as either successful or not. Accordingly, the test designer of a comparative study could present the percentage of nonsuccessful solutions dependent upon some predetermined tolerance as a measure for reliability. Whenever possible, he should evaluate the corresponding failures in terms of both the objective function and the constraint set. Other valuable information would include how many test runs have been interrupted by overflow and exceeding calculation time.

c) Global convergence: In many mathematical models, e.g. general nonlinear programming problems, one must be aware of the existence of local solutions. Since a user is usually interested in obtaining a global rather than a local solution, a test designer should develop a criterion for the probability that an algorithm approximates a global solution. For a given set of test runs, this could be measured by the following:

(i) Percentage of global solutions.

(ii) Difference between objective function value obtained by code and objective function value of the global solutions.

The presentation of the average objective function values of all global solutions allows a more detailed investigation of the global convergence behavior: in particular a user might first choose among codes based on the percentage of global solutions obtained and if two codes tied, then check the difference between the objective function values when global solutions were not obtained.

d) Robustness: The performance criterion robustness should give a measure of the ability of code to handle extreme situations. Such situations arise if problems are formulated with extraordinary numerical properties (e.g. ill-conditioned problems) or if a user prepares a problem in an unsuitable way (e.g. starting point far away from the solution). A measure of the robustness may be valuable in the following cases:

(i) Choice of initial values (e.g. starting point).

(ii) Ill-conditioned problem.

(iii) Degenerate problem.

(iv) Indefinite problem.

(v) Perturbations of a problem.

A precise definition of the special problem types (ii), (iii), and (iv) depends on the mathematical model, cf. [3] for general nonlinear programming algorithms. For evaluating the subcriteria, one of the main questions is whether a code is capable of solving the problem in these extreme situations. Therefore one could determine reliability scores for problems having such numerical characteristics. Furthermore one could evaluate the efficiency scores of a) and relate them to the corresponding values obtained for the less difficult situations (e.g. well-conditioned problems).

4. Test Examples

The evaluation of performance criteria requires the definition of suitable test problems. All test examples can be classified into real life, hand selected, or randomly generated test problems. To gain an impression of their structure, consider the general nonlinear programming model. Test problems of the first two categories are found in Himmelblau [1] and Hock, Schittkowski [2], for example, and instances of the last category are constructed in [3]. A test designer should weigh the possible advantages and disadvantages of each type. We list them below. However, for many mathematical models, test problems are only available for one or the other category, for instance, in large scale linear programming. Even under such circumstances, I would like to emphasize that the test designer of a comparative study should always try to gather a large number of problems. The basic intention is to make statistical statements about some performance criterion, and the test examples must be considered as a suitable sample for the problem type under consideration.

a) Real life test problems: Problems of this kind are based on practical applications of the mathematical model, such as the re-entry problem in optimal control.

Advantages:

(i) They reflect typical practical applications.

(ii) They show typical difficulties obtained for real life problems only (e.g. bad scaling, round-off errors).

Disadvantages:

(i) Only a limited number of test problems is available.

(ii) The precise solution is not known so that, in particular, the achieved final accuracy is not computable.

(iii) Special numerical properties are not known, in general (e.g. condition number).

(iv) The implementation is difficult and time consuming.

b) Hand selected problems: These problems are determined by hand -- they are set more or less arbitrarily by a test designer.

Advantages:

(i) The precise solution is known, in general.

(ii) They are usually computationally simple (well scaled, no round-off errors, simple arithmetic operations).

(iii) The implementation is easy.

Disadvantages:

(i) Only a limited number of test problems is available.

(ii) In most cases, they are too small to characterize practical situations.

(iii) They are too simple compared with practical problems and do not show typical difficulties (e.g. bad scaling, round-off errors).

(iv) Special numerical properties are unknown in general.

c) Randomly generated test problems: Proceeding from a predetermined optimal solution and a set of data such as dimension, number of constraints, and so on, a test problem generator constructs arbitrarily many problems randomly.

Advantages:

(i) The precise solution is known.

(ii) Special properties can be predetermined (e.g. condition number, degeneracy).

(iii) Arbitrarily many problems are available.

(iv) Implementation is easy.

Disadvantages:

(i) They do not show typical numerical difficulties of real life problems (e.g. bad scaling, round-off errors).

(ii) They are not available for all problem types (e.g. optimal control).

5. Evaluation of the Results

The evaluation of the results of a comparative study is handicapped by the extensive amount of computational data which must be collected.

The first step consists of determining scores for each subcriterion in some uniform fashion, mean values, for example. After choosing these scores, one should use an appropriate measure according to a statistical distribution (e.g. the arithmetic vs geometric mean). Even this composition of single results could be too complex to gain an impression of the performance of competing programs. However, the combination of different scores for the subcriteria must be related to the individual significance of these subcriteria for a decision maker. For instance, deciding whether calculation time is more or less important than number of function evaluations is dependent upon the analyst's practical problems. Since, in addition, the subcriteria possess different statistical distributions, I recommend the usage of weighted rank numbers to get a single score for each criterion. The user has to be informed about the way to evaluate the scores subject to individual weights.

On a second level, a user could be interested in obtaining one final score for each program as a basis for his decision. Since each criteria may have a different weights for different decision makers, the usage of weighted rank numbers is again recommended. In particular, they are easily calculated and give a total measure for each algorithm subject to individual priorities. Application of the proposed performance evaluation for comparing nonlinear programming codes is found in [3].

References

[1] D.M. Himmelblau, Applied Nonlinear Programming, McGraw-Hill Book Company, New York, 1972.

[2] W. Hock, K. Schittkowski, Test examples for nonlinear programming codes, to appear: Lecture Notes in Economics and Mathematical Systems, Springer-Verlag, Berlin, Heidelberg, New York.

[3] K. Schittkowski, Nonlinear programming codes - Information, tests, performance, Lecture Notes in Economics and Mathematical Systems, No. 183, Springer-Verlag, Berlin, Heidelberg, New York, 1980.

REMARKS ON THE COMPARATIVE EVALUATION OF ALGORITHMS FOR MATHEMATICAL PROGRAMMING PROBLEMS*

by

A. Miele**

In this presentation, we consider the minimization of a function $f = f(x)$, where f is a scalar and x is an n-vector whose components are unconstrained. We consider the comparative evaluation of algorithms for unconstrained minimization. We are concerned with the measurement of the computational speed and examine critically the concept of equivalent number of function evaluations N_e, which is defined by

$$N_e = N_0 + nN_1 + mN_2. \tag{1}$$

Here, N_0 is the number of function evaluations; N_1 is the number of gradient evaluations; N_2 is the number of Hessian evaluations; n is the dimension of the vector x; and $m = n(n+1)/2$ is the number of elements of the Hessian matrix above and including the principal diagonal. We ask the following question: Does the use of the quantity N_e constitute a fair way of comparing different algorithms?

The answer to the above question depends strongly on whether or not analytical expressions for the components of the gradient and the elements of the Hessian matrix are available. It also depends on the relative importance of the computational effort associated with algorithmic operations vis-a-vis the computational effort associated with function evaluations.

Both theoretical considerations and extensive numerical examples carried out in conjunction with the Fletcher-Reeves algorithm, the Davidon-Fletcher-Powell algorithm, and the quasilinearization algorithm suggest the following idea: the N_e-concept, while accurate in some cases, has drawbacks in other cases; indeed, it might lead to a distorted view of the relative importance of an algorithm with respect to another.

---

*This study was supported by the National Science Foundation, Grant No. ENG-79-18667.

**Professor of Astronautics and Mathematical Sciences, Rice University, Houston, Texas.

The above distortion can be corrected through the introduction of a more general parameter, the time-equivalent number of function evaluations

$$\tilde{N}_e \doteq T/t_0, \tag{2}$$

where T denotes the time required to solve a particular problem on a particular computer and t_0 denotes the time required to evaluate the objective function once on that computer. This generalized parameter is constructed so as to reflect accurately the computational effort associated with function evaluations and algorithmic operations. It can be rewritten in the following form:

$$\tilde{N}_e = (1+C_3)(N_0+C_1N_1+C_2N_2), \tag{3}$$

where

$$C_1 = t_1/t_0, \quad C_2 = t_2/t_0, \quad C_3 = T_a/T_e. \tag{4}$$

Here, t_0 is the time required to calculate the objective function once; t_1 is the time required to calculate the gradient once; t_2 is the time required to calculate the Hessian once; T_a is the algorithmic time, and T_e is the function evaluation time. The coefficients C_1, C_2, C_3 can be computed experimentally or can be determined through a weighted operational count.

Five test functions are considered, namely: (i) Rosenbrock function, n = 2; (ii) Wood function, n = 4; (iii) Powell function, n = 4; (iv) Miele function, n = 4; and (v) generalized Rosenbrock function, n = 5 through 30. For test functions (i)-(v), the experimentally determined values of the coefficients C_1 and C_2 are in the following range:

$$0.07n < C_1 < 0.45n, \tag{5}$$

$$0.03m < C_2 < 0.34m. \tag{6}$$

Therefore, the use of the n-rule in (1) is equivalent to overestimating the effort associated with gradient evaluations by a numerical factor ranging between 2 and 14. Analogously, the use of the m-rule in (1) is equivalent to overestimating the effort associated with Hessian evaluations by a numerical factor ranging between 3 and 33.

Three minimization algorithms are considered, namely: (a) the Fletcher-Reeves algorithm, FR; (b) the Davidon-Fletcher-Powell algorithm, DFP; and (c) the quasilinearization algorith, QL. Algorithms FR and DFP exemplify first-order methods; and algorithm QL exemplifies second-order methods. For algorithms (a)-(c) and for test functions (i)-(v), the experimentally determined values of the coefficient C_3 are in the following range:

$$0.36 < C_3 < 1.65\ , \quad \text{FR algorithm}\ ; \tag{7}$$

$$1.04 < C_3 < 8.06\ , \quad \text{DFP algorithm}\ ; \tag{8}$$

$$0.57 < C_3 < 8.28\ , \quad \text{QL algorithm}\ . \tag{9}$$

Therefore, the neglect of the algorithmic time by comparison with the function evaluation time, which is implied in the use of (1), is not justified experimentally. Indeed, not only the algorithmic time is not negligible by comparison with the function evaluation time, but in some cases the algorithmic time can be much larger than the function evaluation time.

From the analyses performed and the results obtained, it is inferred that, due to the weaknesses of the N_e-concept, the use of the $\tilde{N}_e$-concept is advisable. In effect, this is the same as stating that, in spite of its obvious shortcomings, the direct measurement of the CPU time is still the more reliable way of comparing different minimization algorithms. Of course, this requires that special precautions be taken; namely, the comparison of different algorithms should be done on a single computer, with the same programming language, with the same compiler, with the same subroutines, under similar workload conditions of the computer, and by the same programmer.

Remark. Some of the material contained in this presentation can be found in Ref. 1. A condensed version has appeared in Ref. 2. A recent extension can be found in Ref. 3.

Copies of Refs. 1-3 can be obtained by writing at the following addresses:

Dr. A. Miele
Aero-Astronautics Group
230 Ryon Building
Rice University
Houston, Texas 77001

Dr. S. Gonzalez
Institute of Electrical Investigations
Division of Power Systems
Department of Simulation
Palmira, Morelos, Mexico

References

[1] Miele, A., Gonzalez, S., and Wu, A.K., On Testing Algorithms for Mathematical Programming Problems, Rice University, Aero-Astronautics Report No. 134, 1976.

[2] Miele, A., and Gonzalez, S., On the Comparative Evaluation of Algorithms for Mathematical Programming Problems, Nonlinear Programming 3, Edited by O.L. Mangasarian, R.R. Meyer, and S.M. Robinson, Academic Press, New York, New York, pp. 337-359, 1978.

[3] Gonzalez, S., Comparison of Mathematical Programming Algorithms, Based on the CPU Time (in Spanish), UNAM, Institute of Engineering, Mexico City, Mexico, Report No. 8196, 1979.

REMARKS ON A TESTING CENTER*

by
Angelo Miele

My first reaction to the idea of a test center was on the positive side. However, the more I thought about the operational details, the more I cooled to the idea. After sober reflection, my conclusions were negative, for the reasons indicated below.

(i) A test center can be located either inside a governmental organization (for example, NBS or NASA) or on the campus of some university. Regardless of the mode of operation of the center, there are two ways to staff the test center: (a) with personnel that have participated actively in the development of mathematical programming algorithms or (b) with personnel that have not participated actively in the development of mathematical programming algorithms. In Case (a), the matter of the objectivity of the test center must be raised. In Case (b), the matter of the competence of the test center must be raised. Thus, we find ourselves between Scylla and Charybdis, with no satisfactory solution in the middle.

(ii) If a test center is created, its financial support can be provided in two ways: (a) by allocating entirely new funds; or (b) by employing some of the funds presently available for algorithm research and development. Practical wisdom suggests that almost surely Case (b) would apply, with the following negative consequence: there would be a diversion of funds from the creative part of algorithm research and development to the bureaucratic part of algorithm research and development.

(iii) If the idea of a test center for mathematical programming problems is valid, by inference we must ask the following question: Why not a test center for ordinary differential equations? Why not a test center for partial differential equations. Why not a test center for practically everything? Before we realize it, the size of the overall operation might approach that of a governmental department!

(iv) Philosophically speaking, this writer believes that science is both basically honest and basically efficient, in this sense: over a long period of time, superior theories and procedures tend to prevail over inferior theories and procedures.

---

*The following remarks were written by Angelo Miele after the conference in response to some of the issues raised during the panel discussion.

Of course, it is entirely possible that, over the short range, a particular theory or a particular procedure might be given an inordinate amount of public attention. However, over the long range, superior theories and procedures tend to remain, while inferior theories and procedures tend to disappear.

There is a continuous and automatic process of checks and balances going on in science, which is achieved through the publication of papers and books all over the world, and through the subsequent discussion and criticism of their content. This process of checks and balances is ultimately responsible for weeding out inferior theories and procedures. This process of checks and balances has taken place and is taking place in the field of algorithm research and development. This is why I believe that there is no need to set up an official "test center".

SYSTEMATIC APPROACH FOR COMPARING THE COMPUTATIONAL SPEED OF UNCONSTRAINED MINIMIZATION ALGORITHMS

by

S. Gonzalez*

Abstract

The comparison of mathematical programming algorithms is inherently difficult, especially when deriving general conclusions about the relative usefulness, applicability, and efficiency of different algorithms. The problem is complicated by the variety of approaches used to compare algorithms. Most often, some approaches ignore essential aspects of the comparison or fail to provide sufficient information about the following items: (a) clarification of the objectives of the comparison; (b) clear and complete description of the algorithms being compared; (c) specification of the memory requirements; (d) use of the same experimental conditions for all of the algorithms being compared; (e) sufficient information about the experimental conditions and the numerical results, so as to make them easily reproducible; (f) use of enough performance indexes to ensure the fulfillment of the objectives of the comparison; (g) use of a reasonably large set of test problems having different characteristics; (h) clarification of the way in which the derivatives are computed; (i) measurement of the computational speed; (j) measurement of the effect of the stopping conditions; (k) measurement of the sensitivity to scaling; (l) presentation of the convergence rates; (m) use of several nominal points for each test problem; and (n) use of a standard format to present the result of the comparison.

Concerning the measurement of the computational speed, the use of the time-equivalent number of function evaluations $\tilde{N}_e$ is recommended. A method of comparing the computational speed of unconstrained minimization algorithms has been developed, that uses $\tilde{N}_e$ as a performance index. A step-by-step description of the method is given below.

First, for each test problem and each algorith, Steps 1 through 6 below must be carried out.

Step 1. Compute the basic times t_f, t_g, t_h. These are the times required to compute the function, the gradient, and the Hessian once.

*Researcher, Instituto de Investigaciones Electricas, Palmira, Morelos, Mexico.

Step 2. Compute the function evaluation time

$$T_e = N_f t_f + N_g t_g + N_h t_h \tag{1}$$

where N_f, N_g, N_h denote the number of function, gradient, and Hessian evaluations.

Step 3. Compute the basic time t_a. This is the time per iteration associated with algorithmic operations.

Step 4. Compute the algorithmic time

$$T_a = N t_a , \tag{2}$$

where N denotes the number of iterations.

Step 5. Compute the total time

$$T = T_e + T_a . \tag{3}$$

Step 6. Compute the time-equivalent number of function evaluations

$$\tilde{N}_e = T/t_f . \tag{4}$$

After Step 1 through 6 have been carried out for each test problem and each algorithm, Steps 7 and 8 below follow.

Step 7. For each algorithm, compute the total time-equivalent number of function evaluations

$$\tilde{N}_{et} = \Sigma \tilde{N}_e , \tag{5}$$

the summation being extended to all of the test problems.

Step 8. Sort the performance indexes $\tilde{N}_{et}$ in ascending order. In this way, the algorithms are ordered from best to worst in terms of computational speed.

The advantages of the method described above are as follows: (a) the method takes into account both the function evaluation time and the algorithmic time; (b) for a standard set of test problems, the basic times t_f, t_g, t_h can be evaluated once and stored for future use; (c) for some standard algorithms, the basic time t_a can be evaluated once and stored for future use; (d) because of the definition of the performance index $\tilde{N}_e$, algorithms of different order can be compared; (e) reasonably precise total times T can be obtained, without having to solve the test problems under consideration more than once for each nominal point; and (f) the results of the comparison are easily interpretable, once Steps 7 and 8 are carried out.

The disadvantages are as follows: (a) the ideal conditions for the applicability of the method require that all the experiments be done on the same computer, using the same compiler, the same programming language, and the same programming style; (b) the results are associated with the particular computer system used; and (c) in order to compute t_a, some special assumptions have to be made, having an obvious influence on the results of the comparison.

References

[1] Miele, A., and Gonzalez, S., On the Comparative Evaluation of Algorithms for Mathematical Programming Problems, Nonlinear Programming 3, Edited by O.L. Mangasarian, R.R. Meyer, and S.M. Robinson, Academic Press, New York, New York, pp. 337-359, 1978.

[2] Gonzalez, S., Comparison of Mathematical Programming Algorithms Based on the CPU Time (in Spanish), UNAM, Institute of Engineering, Mexico City, Mexico, Report No. 8196, 1979.

THE EVALUATION OF OPTIMIZATION SOFTWARE FOR ENGINEERING DESIGN

by

K.M. Ragsdell, P.E.*

Abstract

The design and implementation of two major comparative experiments is reviewed with particular emphasis on testing methodology. These studies took place at Purdue University. The first, an investigation of the merits of general purpose nonlinear programming codes with major funding from the National Science Foundation, was conducted over the period 1973 to 1977. The second, an investigation of the relative merits of various geometric programming strategies and their code implementations with funding from the Office of Naval Research, was conducted over the period 1974 to 1978. The various major decisions associated with such studies are discussed, such as the selection and collection of problems and codes, the nature of data to be collected, evaluation criteria, ranking schemes, presentation and distribution of results, and the technical design of the experiment itself. The statistical implications of the results in light of the experiment design are examined, as are the effects of various experiment parameters such as number of variables, number of constraints, degree of nonlinearity in the objective and constraints, and starting point placement. Finally, recommendations for future experiments are given.

1. Introduction

In this paper I intend to review two relatively major comparative studies with primary emphasis on testing methodology. For the sake of brevity, the two studies will be referred to as the NSF study and the ONR study. I shall make no effort to give new comparative results, since these have previously been reported [1, 2, 3, 4, 5], and results from a more recent and complete study are now available [6]. Some representative results will be given, but only as needed to demonstrate various aspects of the aforementioned studies.

2. The NSF Study

In this study we considered methods which address the nonlinear programming problem (NLP):

*Visiting Professor, Aerospace and Mechanical Engineering, The University of Arizona, Tucson, Arizona

MINIMIZE:	$f(x)$;	$x \varepsilon R^N$	(1)
subject to:	$g_j(x) \geq 0$	$j=1,2,3,\ldots,J$	(2)
	$h_k(x) \equiv 0$	$k=1,2,3,\ldots,K$	(3)
and	$x_i^{(\ell)} \leq x_i \leq x_i^{(u)}$	$i=1,2,3,\ldots,N$	(4)

given $x^{(0)}$, an initial estimate of the solution, x^*, a Kuhn-Tucker point.

Before coming to Purdue, I, as most others interested in nonlinear programming, was aware of the pioneering work of Al Colville of IBM [7]. Colville sent eight problems having from three to sixteen design variables and a standard timing routine to the developers of thirty codes. Each participant was invited to submit his "best effort" on each problem, and the time required to execute the standard timing routine on his machine. One can only marvel at the economic wisdom of Colville's decision to send the problems around rather than collect the codes and run the tests himself. Unfortunately, this approach contains at least three flaws. Eason [8] has shown that Colville's timing routine does not adequately remove the effect of compiler and computing machine selection on code performance. Accordingly, the data collected at one site in the Colville study is not comparable to data collected at another. Furthermore, each participant was allowed to attack each problem as many times as he felt necessary in order to optimize the performance of his code. Thus another investigator could not reasonably expect to produce similar results with the same code in the absence of the special insight which only its originator would possess. Finally, and possibly most importantly, no two participants reported solutions to the same accuracy, which in our experience is a major difficulty in fair comparison. These shortcomings cast a very real shadow on the validity of the Colville results and conclusions. Quite unfortunately from the scientific point of view the Colville experiment was not repeatable, so that it was essentially impossible for latter investigators to confirm his findings directly. In 1974 Eason and Fenton [9] reported on a comparative study of twenty codes on thirteen problems. Dr. Eason did his testing on one machine, and included in his test set the Colville problems plus several problems from mechanical design. The major contribution of the Eason study is the introduction of error curves, which allow convenient comparison of codes at exactly the same error criteria. The major shortcomings are the lack of difficulty in the problem set and the failure to include the more powerful algorithms. There have been other important studies reported [10, 11, 12, 13, 14,

15, 16, 17], but the work of Colville and Eason have had the most profound impact on the work reported here.

2.1 Major Objectives of Study

The major goal of this study was to discern the utility of the world's leading NLP methods for use in engineering design. A secondary objective was to design the experiment and present the results so as to enhance the utility of the results and conclusions in an industrial environment. Theoretical convergence rates were and are of little importance to me. I wanted to rate the methods in a manner that would allow a typical designer to choose the appropriate method or possibly class of methods for his particular problem. It seemed desirable to include as many current industrial problems and codes in the experiment as possible. Industry supported the work by contributing codes, problems and by direct financial support of visits in the summers. I spent two summers in industry during the study helping to formulate problems and collecting codes. The companies who contributed most to the study are Whirlpool Corporation, Honeywell Corporation, York Division of Borg Warner, and Gulf Oil Corporation. The same person, Eric Sandgren, performed all of the numerical experiments, and all calculations were performed on the same machine using the same compiler, a CDC-6500 at Purdue University. Furthermore, the student was not involved in algorithm development. The major steps in the study were:

1. assemble codes and problems.
2. qualify codes via preliminary test set of 14 problems.
3. apply 24 qualified codes to full 35 problem test set.
4. eliminate problems on which fewer than 5 codes were successful.
5. compile and tabulate results for 24 codes on 23 problems.
6. prepare individual and composite utility curves.

2.2 Codes and Problems

The complete set of codes and problems is given in Table 1 and 2. 30 problems and 35 codes were assembled. 23 of the problems had previously been used by Eason, Colville and Dembo [18], and the remaining seven came more or less from industrial applications. A more complete description including Fortran listings of the problems is given by Sandgren [1]. As can be seen from the table, the problems vary greatly in size and structure. The number of variables range from 2 to 48, while the number of constraints goes from 0 to 19. The number of variable bounds range from 3 to 72. The problems used in the preliminary testing are numbers 1 through 8, 10 through 12, 14, 15, and 16; a

TABLE 1: COMPLETE TEST SET

PROBLEM NUMBER	NAME AND/ OR SCOURCE	N	J	K	NUMBER OF VARIABLE BOUNDS
1	EASON #1	5	10	0	5
2	EASON #2	3	2	0	6
3	EASON #3	5	6	0	10
4	EASON #4	4	0	0	8
5	EASON #5	2	0	0	4
6	EASON #6	7	0	4	12
7	EASON #7	2	1	0	4
8	EASON #8	3	2	0	6
9	EASON #9	3	9	0	4
10	EASON #10	2	0	0	4
11	EASON #11	2	2	0	4
12	EASON #12	4	0	0	8
13	EASON #13	5	4	0	3
14	COLVILLE #2	15	5	0	15
15	COLVILLE #7	16	0	8	32
16	COLVILLE #8	3	14	0	6
17	DEMBO #1	12	3	0	24
18	DEMBO #3	7	14	0	14
19	DEMBO #4	8	4	0	16
20	DEMBO #5	8	6	0	16
21	DEMBO #6	13	13	0	26
22	DEMBO #7	16	19	0	32
23	DEMBO #8	7	4	0	14
24	WELDED BEAM	4	5	0	3
25	COUPLER CURVE	6	4	0	6
26	WHIRLPOOL	3	0	1	6
27	SNG (HONEYWELL)	48	1	2	72
28	FLYWHEEL	5	3	0	10
29	AUTOMATIC LATHE	10	14	1	20
30	WASTE WATER	19	1	11	38

TABLE 2: CODES IN STUDY

CODE NUMBER	NAME AND/ OR SOURCE	CLASS	UNCONSTRAINED SEARCH METHOD
1	BIAS	exterior penalty	variable metric (DFP)
2	SEEK1	interior penalty	random pattern
3	SEEK3	interior penalty	Hooke-Jeeves
4	APPROX	linear approximation	none
5	SIMPLX	interior penalty	simplex
6	DAVID	interior penalty	variable metric (DFP)
7	MEMGRD	interior penalty	memory gradient
8	GRGDFP	reduced gradient	variable metric (DFP)
9	RALP	linear approximation	none
10	GRG	reduced gradient	variable metric (BFS)
11	OPT	reduced gradient	conjugate gradient (FR)
12	GREG	reduced gradient	conjugate gradient (FR)
13	COMPUTE II (0)	exterior penalty	Hooke-Jeeves
14	COMPUTE II (1)	exterior penalty	conjugate gradient (FR)
15	COMPUTE II (2)	exterior penalty	variable metric (DFP)
16	COMPUTE II (3)	exterior penalty	simplex/Hooke-Jeeves
17	Mayne (1)	exterior penalty	pattern
18	Mayne (2)	exterior penalty	steepest descent
19	Mayne (3)	exterior penalty	conjugate direction
20	Mayne (4)	exterior penalty	conjugate gradient (FR)
21	Mayne (5)	exterior penalty	variable metric (DFP)
22	Mayne (6)	exterior penalty	Hooke-Jeeves
23	Mayne (7)	interior penalty	pattern
24	Mayne (8)	interior penalty	steepest descent
25	Mayne (9)	interior penalty	conjugate direction
26	Mayne (10)	interior penalty	conjugate gradient
27	Mayne (11)	interior penalty	variable metric (DFP)
28	SUMT IV (1)	interior penalty	Newton
29	SUMT IV (2)	interior penalty	Newton
30	SUMT IV (3)	interior penalty	steepest descent
31	SUMT IV (4)	interior penalty	variable metric (DFP)
32	MINIFUN (0)	mixed penalty	conjugate directions
33	MINIFUN (1)	mixed penalty	variable metric (BFS)
34	MINIFUN (2)	mixed penalty	Newton
35	COMET	exterior penalty	variable metric (BFS)

total of 14. The codes tested in our study are listed and briefly described in Table 2. Included in the test set are 4 generalized reduced gradient codes, a Method of Multipliers (BIAS) code, two methods based on extensions of linear programming, and a large variety of penalty methods. RALP was submitted by Spencer Schuldt of Honeywell Corporation, and COMPUTE II was contributed by Gulf Oil Corporation. All codes were modified to run on our CDC-6500 computer system. The changes varied somewhat from code to code and were usually quite small. All codes were converted to single precision, since the CDC-6500 system has a 60 bit word (48 bit mantissa and 12 bit exponent), which gives approximately 14 places of precision in single precision. Any code which required analytical gradient information was altered to accept numerical gradient approximations using forward differences. Finally, all print instructions were removed from the basic iteration loop of the algorithms in order that we might more accurately measure the time of calculation. Since we wished to study those algorithms which might have real value in an engineering design setting, and the anticipated amount of computer time required to solve the 30 test problems by each of the 35 codes was great, we decided to run each code thru preliminary tests in advance of the more complete final test phase. Each code was applied to a set of 14 problems selected from the Colville and Eason and Fenton sets, and codes which did not solve at least half these problems were eliminated from additional study. It should be noted that in this phase any code which required more than three times the average time to solve a given problem was considered to have failed on that problem. Accordingly, codes 2, 4, 5, 6, 7, 17, 18, 23, 24, 25 and 30 were not considered in the final testing, leaving 24 codes in the final tests.

2.3 Data Collection and Evaluation Criteria

We collected data so that we could rank the codes using criteria suggested by others, such as Eason and Fenton. This included:

1. number of objective evaluations to termination for a given error.
2. number of constraint evaluations to termination for a given error.
3. CPU time to termination for a given error.

Many criteria have been proposed for measuring the performance of nonlinear programming algorithms. Himmelblau [19] speaks of efficiency (speed) and robustness as desirable qualities of a code. The literature is filled with comparative results which report number of function evaluation, number of iterations and the like. We collected this data, but prefer a much different method of ranking. We feel that the ability to solve a large variety of problems in a reasonable amount of time is the paramount performance criteria for NLP codes. One can view this criteria as a combination of efficiency and robustness, or as a robustness objective with an efficiency constraint. We define the total relative error as:

$$\varepsilon_t = \varepsilon_f + \sum_{j-1}^{J} \langle g_j(x) \rangle + \sum_{k=1}^{K} |h_k(x)| \tag{5}$$

where

$$\varepsilon_f = \frac{|f(x) - f(x^*)|}{f(x^*)} \quad \text{for } f(x^*) \neq 0 \tag{6}$$

and

$$\varepsilon_f = |f(x)| \text{ for } f(x^*) = 0 \tag{7}$$

Recall the bracket operator notation, that is:

$$\langle \alpha \rangle \; \begin{cases} 0, & \text{if } \alpha \geq 0 \\ -\alpha, & \text{if } \alpha < 0 \end{cases} \tag{8}$$

or in other words, only the violated constraints appear in the first summation of (5). Plots of ε_t for each algorithm on each of the test problems were prepared. After Eason, this allowed us to compare codes at exactly the same error criteria. Since the ability to solve a large number of problems in a reasonable amount of time is the desired ranking criteria, we rank the codes on the basis of the number of problems solved within a series of specified limits on relative solution time. The limits on the solution time are based on a fraction of the average time for all codes on each problem. Each solution time for a problem was normalized by dividing by the average solution time on that problem. This produces a low normalized solution time for a code with a relatively fast solution time and high normalized solution time for a code with a relatively slow solution time. This normalization process allows us to directly compare results on problems of varying difficulty. Let us examine the performance of the codes in the final tests at a convergence criterion of $\varepsilon_t = 10^{-4}$. In Table 3 we list the number of problems solved for each code at various solution time limits. From the table it is easy to identify the very fast codes, since they have large numbers in the first column labeled .25. In the first column we list the number of problems solved by each code in 25% of the average solution time, and so on for the other columns. Only 14 of the codes in the final tests solved half or more of the rated test problems.

TABLE 3: NUMBER OF PROBLEMS SOLVED WITH $\varepsilon_t = 10^{-4}$

CODE NUMBER	NAME AND/ OR SOURCE	NUMBER OF PROBLEMS SOLVED					
		.25*	.50	.75	1.00	1.50	2.50
1	BIAS	0	9	14	17	19	20
3	SEEK3	0	1	2	3	6	10
8	GRGDFP	10	15	17	17	18	18
9	RALP	12	13	13	14	14	16
10	GRG	15	17	19	19	19	19
11	OPT	16	21	21	21	21	21
12	GREG	14	18	20	23	23	23
13	COMPUTE II (0)	2	7	9	11	13	15
14	COMPUTE II (1)	2	4	6	6	8	9
15	COMPUTE II (2)	4	9	11	15	15	15
16	COMPUTE II (3)	1	5	7	8	9	11
19	MAYNE (3)	0	0	2	3	7	11
20	MAYNE (4)	1	4	10	10	11	11
21	MAYNE (5)	7	13	14	16	16	16
22	MAYNE (6)	2	4	7	8	8	9
26	MAYNE (10)	0	1	4	6	9	10
27	MAYNE (11)	2	7	11	14	14	17
28	SUMT IV (1)	0	0	0	0	3	9
29	SUMT IV (2)	0	0	0	1	2	7
31	SUMT IV (4)	0	1	3	5	9	13
32	MINIFUN (0)	0	0	0	4	8	15
33	MINIFUN (1)	0	2	4	7	13	15
34	MINIFUN (2)	0	1	2	4	8	10
35	COMET	0	2	5	7	9	15

* times the average solution time.

The final rated test problems are those on which many codes made progress. This final rated set contains 23 problems, and excludes problems 9, 13, 21, 22, 28, 29 and 30.

2.4 Presentation of Results

We considered presenting the results as raw data in tabular form, but rejected this method because of our anticipated audience; engineering designers. Instead we chose to present the results in graphical form in the hope that the information could more efficiently be transmitted. We normalize the results as a function of problem difficulty by using fraction of average time as the abscissa. Number of problems solved is the ranking criteria and is therefore the ordinate. Average time is the average solution time to a given termination error of all successful codes on a given problem. Figure 1 gives a graphical representation of the performance of the 14 codes which solved half or more of the rated test problems. All codes in this final category have exhibited superior speed and robustness. Furthermore, the shape of an individual utility curves gives insight to a codes character. The perfect code would solve all 23 problems in no time; obviously impossible. On the other hand, codes which have a utility curve with a steep slope and a high final ordinate are both fast and robust. In Figures 2 and 3 the effect of termination criterion is given. We see that all codes have difficulty solving the problems when the convergence or termination criterion become smaller. Since we tested several coded implementations of similar algorithms, we wondered if algorithm class trends might be evident from the data. The three algorithm classes considered are generalized reduced gradient, exterior penalty, and interior penalty. We selected the best three codes of each class and prepared average or composite utility curves as given in Figures 4, 5 and 6.

2.5 Major Conclusion

We were able to confirm the major conclusion of the Colville study; that is, that the Generalized Reduced Gradient codes are superior as a class to the others tested. Sandgren [20] shows that the trends displayed in Figures 4, 5 and 6 are statistically significant.

2.6 An Alternate Comparative Strategy

In the closing stages of this study we decided to investigate the effect of varying various problem parameters on code performance. The previously described approach and problem set did not seem appropriate.

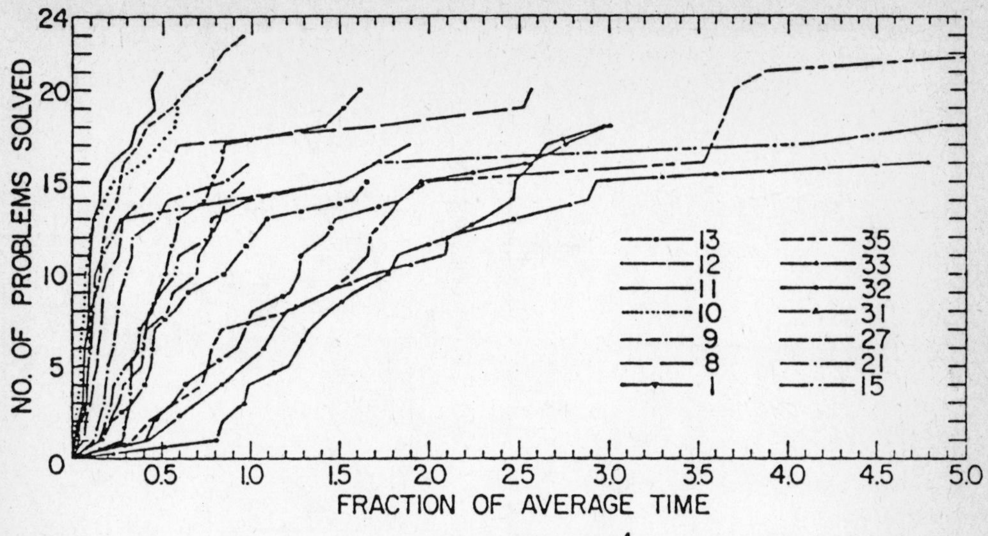

Fig. 1 Algorithm utility ($\varepsilon_t = 10^{-4}$)

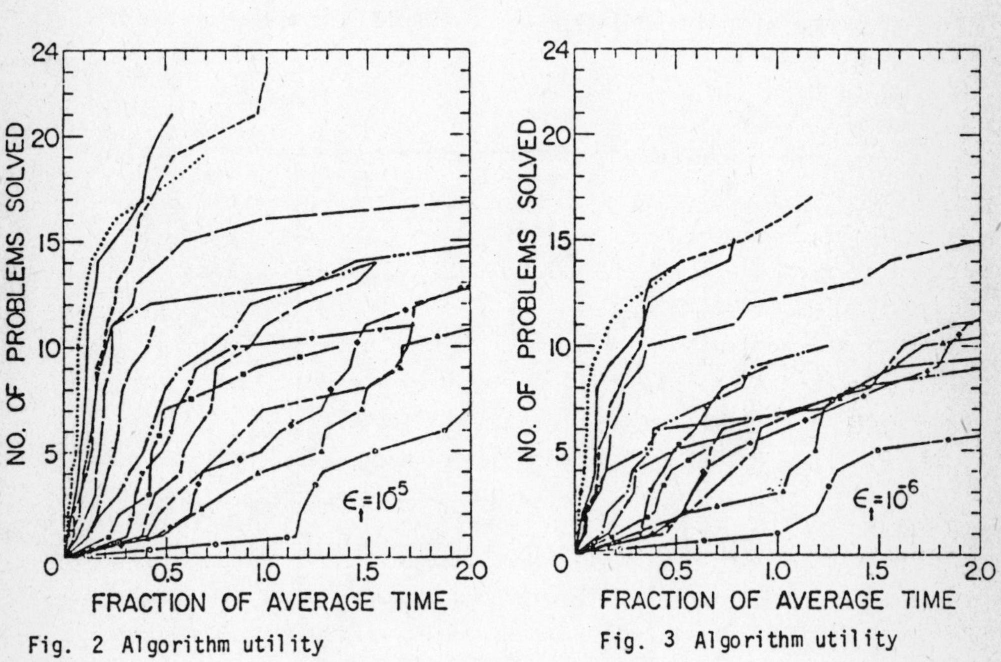

Fig. 2 Algorithm utility

Fig. 3 Algorithm utility

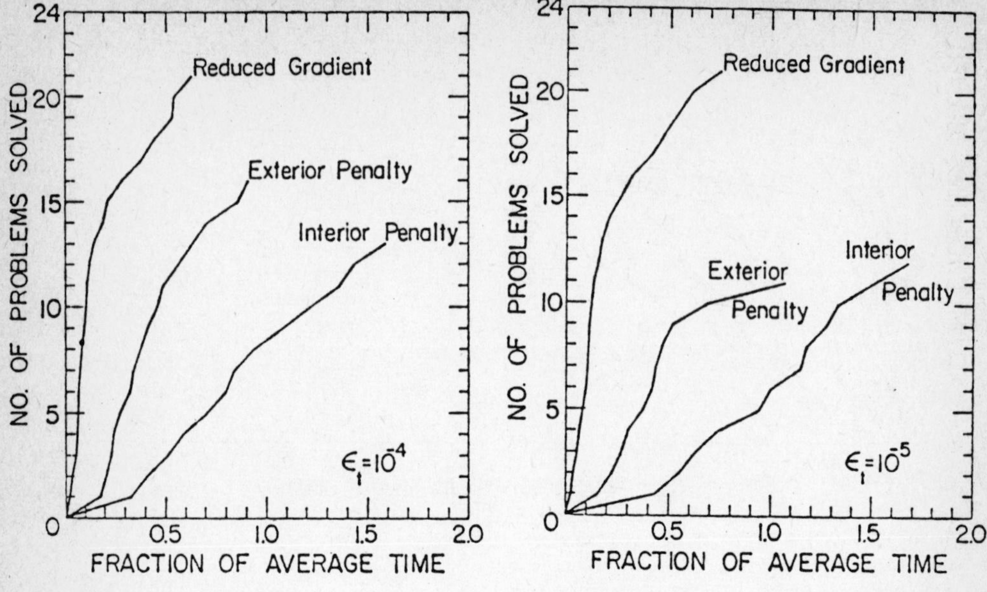

Fig. 4 Average algorithm utility

Fig. 5 Average algorithm utility

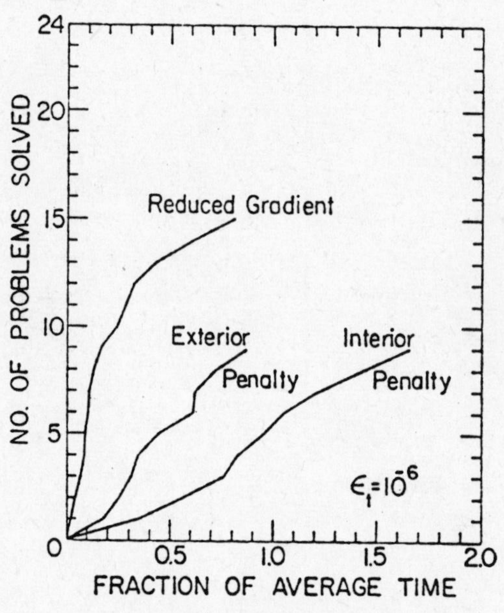

Fig. 6 Average algorithm utility

We decided to use a very special class of test problems, and simply to record the change in solution time as a function of change in problem characteristic. A problem containing five variables and ten inequality constraints with a quadratic objective function and constraints was selected as the standard problem. This standard problem was then altered by changing one problem factor such as the number of variables, the number of inequality constraints, the number of equality constraints or the degree of nonlinearity to form another problem class. The solution times reported for each problem class are the average time for each algorithm on a set of ten randomly generated problems within that problem class to an accuracy of $\varepsilon_t = 10^{-4}$. The test problems were generated following the procedure of Rosen and Suzuki [21]. The quadratic form for the objective function may be expressed as

$$f(x) = x^T Q_0 x + ax \tag{9}$$

and for each constraint as

$$g_j(x) = x^T Q_j x + b_j x + C_j \geq 0; \; j = 1, 2, \ldots, J \tag{10}$$

For these expressions the Q_0 and Q_j are randomly generated N by N matrices with Q_0 forced to be positive definite to guarantee unimodality and the Q_j forced to be negative definite to guarantee a convex feasible region. The a, b_j and C_j are all column vectors containing N elements. Not only are the Q matrices selected but the b_j vectors, and the Lagrange Multipliers are randomly generated, and the solution vector is also selected. The Lagrange Multipliers are either set to zero if a constraint is not to be active at the solution or to a random number between .5 and 10. So that the problem is not over constrained the number of constraints allowed to be active at the solution was also selected as a random integer between one and N-1. Now the a_j and C_j may be determined by the conditions required to make the selected optimal vector a Kuhn Tucker point. The procedure used to guarantee Q_0 is positive definite and the Q_j are negative definite is described by Sandgren [1]. The starting vector for all problems was the origin and the solution point for all of the five variable problems was $x_i = 2.0$. As the number of variables increased, the solution point was adjusted so that the distance from the origin to the solution vector was the same as for the five variable problem. Extending the standard problem to include additional variables or inequality constraints is elementary, and the basic procedure for the addition of equality constraints remains unchanged with the exception that the Lagrange Multipliers for the equality constraints may now be positive or negative. However the addition of nonlinear equality constraints does introduce the possibility of local minima. This problem was handled by including only the problems generated where all of the algorithms reached the selected optimal vector. For the increase in nonlinearity, additonal higher order terms were added to the basic quadratic form. Again for this case no check was made for positive and negative definiteness and problems where alternate solutions were found were not included. The average solution times on each set of ten problems for the increase in design variables, the number of inequality constraints, the number of equality constraints and the increase in nonlinearity has been given by Sandgren [3]. The codes considered were 1, 9, 10, 11, 15, 21 and 31. The demonstrated trends are of considerable interest, and are consistent with the earlier tests.

3. The ONR Study

In this study we considered methods which address the prototype posynomial geometric programming problem (GP):

$$\text{MINIMIZE:} \quad g_o(x) \quad ; \quad x \in R^N \qquad (11)$$

$$\text{subject to} \quad g_k(x) \leq 1 \quad k=1,2,3,\ldots K \qquad (12)$$

$$\text{and} \quad x_i > 0 \quad i=1,2,3,\ldots,N, \qquad (13)$$

where the posynomial functions $g_k(x)$ are defined as:

$$g_k(x) = \sum_{t=s_k}^{T_k} c_t \prod_{n=1}^{N} x_n^{a_{nt}} \qquad (14)$$

with specified positive coefficients c_t and specified real exponents a_{nt}. The term indices t are defined as:

$$s_0 = 1 \qquad (15)$$

$$s_{k+1} = T_k + 1 \qquad (16)$$

$$T_k = T \qquad (17)$$

This problem is in general a non-convex programming problem which because of the nonlinearities of the constraints can be expected to severely tax conventional nonlinear programming codes. However, despite the apparent difficulty of the primal problem, there are structural features of the generalized posynomial functions which can be exploited to facilitate direct primal solutions. In addition various transformations can be employed to give equivalent formulations. The additional formulations which we considered in this study are:

1. The Convexified Primal

This formulation results from the transformation $x_i = e^{z_i}$; $i=1,2,3,\ldots,N$.

The new functions of z become convex functions.

2. The Transformed Primal

   The additional transformation, $w = A^T z + \ell n c$ gives this formulation.

3. The Dual

   The relationship of the dual and primal is well known.

4. The Transformed Dual

   An alternate way of formulating the dual program is to eliminate the linear equality constraints by solving them for the dual variables in parametric form. Using this device the dual variable δ_t can be expressed as the sum of a particular solution and a linear combination of T-N-1 homogeneous solutions of the N+1 dual constraints.

This study was conducted in collaboration with Professor Reklaitis of the School of Chemical Engineering at Purdue, and was begun with the support of Neal Glassman of the Office of Naval Research. In this study we gathered experimental data on the performance of ten codes or code variants in solving 42 test problems from up to 20 randomly generated starting points.

Four previous comparative studies of prototype GP solution approaches have been reported in the literature. Two of these, Rijckaert and Martens [22] and Dembo [23] primarily focused on generalized GP's but did include prototype problems in their test slate. The study by Dinkel [24] was restricted to the examination of alternative cutting plane methods used for the solution of the convex primal. Sarma [25], in what may be viewed as a pilot to the present work, considered primal, dual, and transformed primal solution approaches and attempted to draw conclusions about the preferred approach. Rijckaert and Marten's tests were restricted to eight prototype problems, used single starting points, and generally employed penalized slack variables to avoid difficulties with loose constraints. Dembo included six prototype problems in his testing but these six were parameter variants of only three original problems. The codes tested consisted of several good general NLP codes applied directly to the primal as well as several specialized GP codes. The test problems were run by the code authors on their own machines; using a single set of starting points; allowing tuning of programs by the authors; but requiring the solutions to meet fixed tolerances. Solution times were compared using Colville standard times.

3.1 Major Objectives of Study

The goals of this study were to determine:

1. Whether the constructions resulting from GP theory offer any computational advantages over conventional NLP methodology.
2. Which of the various equivalent GP problem formulations are preferable and under what conditions.
3. Which GP algorithm/formulation combination is most likely to be successful for a given problem.
4. Whether a criteria can be defined by means of which GP problem difficulty can be gauged.

In designing this experiment, we attempted to rectify some of the inadequacies of previous studies. In particular we did the following:

1. Used a large number (42) of problems, with randomly generated starting points (up to 20 per problem).
2. Results were obtained at several precise error levels.
3. Execution time is measured such that starting point generation and extraneous I/O is excluded.
4. Designed the tests such that formulation effects can be separated from algorithm effects.
5. Finally appropriate statistical tests were used for the comparisons.

3.2 Codes and Problems

Ten codes or code variants were used in this study. The first four are general purpose NLP codes included because of favorable performance in the NSF study. These codes were selected because they were readily available, represent the major algorithm classes in the NSF study, and exhibited superior robustness and speed in the NSF tests. These codes are listed as 1 (BIAS), 9 (RALP), 11 (OPT) and 27 (MAYNE (11)) in Table 2. In addition Root [26] prepared a special version of his Method of Multipliers Code, BIAS-SV, which exploits the convexified GP primal formulation for use in the study. BIAS-SV employs Newton's method for the unconstrained iterations, and uses second derivatives in the line searches. This was conveniently accomplished due to the availability of first and second derivatives in the convexified primal form. The remaining five codes are special purpose GP codes. GPKTC [27], a code developed by Martens and Rijckaert, solves the convexified primal using a Kuhn-Tucker optimality condition iteration; GGP [28], a code developed by Dembo, uses Kelley's cutting plane method and also addresses the convexified primal; MCS [29] is a GP specialization of the convex simplex method of Zangwill as modified by Beck and Ecker; QUADGP is a GP code developed by Bradley [30], which solves the transformed dual as a series of quadratic programs; and finally, DAP [31] is an adaptation of the differential algorithm of Beightler and Wilde given by Reklaitis.

42 prototype GP problems were selected from those available in the literature, with about half representing engineering applications. Many of the problems have been previously used by Dembo, and Beck and Ecker; and complete source references are given by Fattler [5]. A review of the problem set reveals the following general characteristics: the number of primal variables range from 2 to 30, primal terms from 8 to 197, constraints from 1 to 73, and exponent matrix density from 3.3% to 83%. It would have been desirable to include large and sparce GP problems in the study, but they were not readily available in the literature.

3.3 Test Procedure:

We conducted this comparative experiment along the general lines of the NSF study with the following major exceptions: multiple, randomly generated starting points were used; a computer program was written which essentially automated the data collection process; formulation effects were explored; and finally the significance of the results was statistically verified. Primal and dual starting points for each problem were generated randomly by sampling points on a hypersphere whose center is the solution. Only feasible points were used. Typically, two radii were used and 10 points retained for each radius for each problem. The obvious difficulty associated with multiple starting points is the increase in computational effort. In this study, some 10,000 test runs were made, of which 6,000 produced useful results; but we feel the need for multiple starting points can not be overstated. We used the pseudo-Lagrangian in the error function for purposes of comparison:

$$\varepsilon_L = \left|(g_0^n - g_0^*)/g_0^* + \sum_k \lambda_k^* \left|(g_k^n - g_k^*)\right|/g_0^*\right| \tag{18}$$

The starred quantities in this expression are the known optimal values of the problem functions and Lagrange Multipliers. The sum includes only those constraints tight at the solution. Values of this error function were computed for the intermediate iteration points recorded during each run. The intermediate solution times (which excluded all I/O time required for recording test data) and error function values were fitted to polynomials and these used as interpolating functions to determine solution times at specified error function levels. Mean solution times and standard deviations were tabulated for each problem/code or formulation combination at relative error levels of 10^{-2}, 10^{-3},

10^{-4}, for all successful runs and at termination for all runs. Each code was run with a fix set of program parameters which were determined in preliminary tests, and usually were very close to the values suggested by the various code authors. Since adjustment of these values was not allowed during the formal tests, some codes did not achieve a 10^{-4} error level on some problems. As this did not occur often, we do not view this as a serious problem.

3.4 Results and Conclusions

The results from this study are simply too voluminous to adequately abstract briefly here, but some brief remarks can be given along with general conclusions. The reader is referred to Fattler [5] for a complete presentation and discussion. A summary of the number of problems and runs attempted with each code-formulation pair is given in Table 4. As can be seen, not all problems were attempted with some codes. This was sometimes impossible, but more often unnecessary due to well established trends. The general NLP codes RALP and MAYNE seem particularly prone to failure, as was the code OPT when used on the transformed primal formulation. Most surprising was the erratic performance of GPKTC, which at times produces extremely fast solutions but in other cases failed completely. Similarly surprising is the high number of failures of the special version of BIAS. Both solved the convexified form of the primal: the former used a modified Newton algorithm with analytic derivatives; the latter a DFP algorithm with numerical derivatives. The most reliable performance seems to have been achieved by the general NLP codes OPT and BIAS when applied to the convexified primal formulation. The next best performance was attained by the specialized codes GGP and DAP. The results were aggregated into the following classes:

1. Comparison of solution times of various algorithms for a given GP problem formulation.
2. Cross-comparison of solution times for the various formulations all solved using the same algorithm.
3. Cross-comparisons of the most successful algorithms found for each GP formulation type.
4. Examination of how solution time varies with problem characteristic dimensions for each of the various formulations.

Finally, statistical tests were administered to discern the significance of the variation in code performance. Assuming that the solution times x and y of two codes for any given problem are normally distributed variables each with their own variances σ_x^2 and σ_y^2, then code solution

Table 4. Number of Problems Attempted and Solved

Code	Problems Attempted	Runs Attempted	Runs Failed	% Unsuccessful Attempts
OPT-P	40	399	27	6.77
OPT-CP	41	616	1	0.16
OPT-TP	25	452	124	27.43
GGP	41	598	24	4.01
GPKTC	39	589	240	40.75
MAYNE	31	379	61	16.09
RALP-P	34	446	146	32.74
RALP-CP	37	552	115	20.83
BIAS-P	39	260	1	0.38
BIAS-CP	40	457	0	0.0
BIAS-SV	29	166	21	12.65
MCS	26	412	60	14.56
DAP	40	406	14	3.45
QUAD-GP	34	149	13	8.72

Table 5. CODE Ranking

Relative Error	10^{-2}		10^{-3}		10^{-4}	
Ranking	1st	2nd	1st	2nd	1st	2nd
OPT-CP	13	14	18	8	14	7
DAP	16	1	8	5	4	2
GGP	6	9	7	7	3	8
QUADGP	10	5	5	7	5	6
MCS	3	6	7	4	13	1
OPT-P	0	4	0	4	0	3

time comparison is equivalent to the problem of testing whether the true mean solution times, M_x and M_y, of the two codes for the given problem are equal. This is the Behrens-Fisher problem of statistics [27]. Appropriate expressions were used to test the difference in the means of the various solution times. For several problems, the means were not significantly different. These calculations were completed at the 10^{-2}, 10^{-3} and 10^{-4} relative error levels. The results were used to determine the number of problems for which each code achieved the best or second best solution time. A 90% significance level was required before means were considered to be different, otherwise they were considered to be the same. The result of this ranking process is given in Table 5.

The major conclusions of this study are:

1. the convex primal is inherently the most advantageous formulation for solution.
2. a general purpose GRG code applied to the convex primal will dominate even the reputedly best specialized GP codes currently available.
3. the differences between the primal and convex primal formulations lie mainly in scaling and function evaluation time.
4. transformed primal solution approaches are not likely to lead to more effecient GP solution than the convex primal.
5. the dual approaches are only likely to be competitive for small degree of difficulty, tightly constrained problems.
6. posynomial GP problem difficulty as measured in solution time is best correlated to an exponential of the number of variables in the formulation being solved and is proportional to the total number of multi-term primal constraints.

4. Recommendations for Future Work

All comparative experiments to date, including those reported here, leave much to be desired. We need to increase our efforts to make our experiments repeatable [28] and statistically significant. Furthermore, we need to come to grips effectively with the fact that software evaluation is a multi-criteria decision problem. The work of Lootsma [29] in applying Saaty's priority theory [30] is especially promising here. I recommend that we establish at once software evaluation centers with the responsibility and resources to carry on long range (5 to 10 years) testing efforts. These centers would develop significant performance data bases against which new codes could be measured as they become available [31, 32]. Finally, these software centers should be organized such that direct involvement by a wide cross section of scholars in the area is possible and in fact encouraged. These centers will provide a valuable service to algorithm developers and to users alike. Finally, as testing procedures become more widely accepted we shall be able to expand our testing scope to other important application areas, such as mixed discrete-continuous problems, unconstrained problems, large sparse problems, etc.

5. Acknowledgements

I am grateful to Morris Ojalvo, Neal Glassman, Mike Gaus, Clif Astill, Eric Sandgren, Gary Gabriele, Ron Root, Ed Fattler, A. Ravindran, F.A. Lootsma, J.N. Siddall, E.D. Eason, K. Schittkowski, K. Hoffman, R.S. Dembo and G.V. Reklaitis for the real contributions they have made to my thinking in this area, and to the work described here. I am indebted to the National Science Foundation, the Office of Naval Research, the Purdue Research Foundation, Whirlpool, Honeywell, Gulf Oil, and York Division of Borg Warner for their financial support.

References

[1] Sandgren, E., "The Utility of Nonlinear Programming Algorithms", Ph.D. Dissertation, Purdue University, Dec., 1977, available from University Microfilm, 300 North Zeeb Road, Ann Arbor, Michigan 48106, USA, document no. 7813115.

[2] Ragsdell, K.M., "On Some Experiments Which Delimit The Utility of Nonlinear Programming Algorithms", ORSA/TIMS meeting, Los Angeles, CA, November 13-15, 1978.

[3] Sandgren, E. and Ragsdell, "On Some Experiments Which Delimit the Utility of Nonlinear Programming Methods for Engineering Design", Tenth International Symposium on Mathematical Programming, Montreal Canada, August 27-31, 1979.

[4] Sandgren, E. and Ragsdell, K.M., "The Utility of Nonlinear Programming Algorithms: A Comparative Study-Part 1 and 2", ASME Journal of Mechanical Design, 102, no. 3, pp. 540-551, July 1980.

[5] Fattler, J.E., Sin, Y.T., Root, R.R., Ragsdell, K.M. and Reklaitis, G.V., "On the Computational Utility of Posynomial Geometric Programming Solution Methods", to appear in Mathematical Programming.

[6] Schittkowski, K., "Nonlinear Programming Codes: Information, Tests, Performance", Lecture Notes in Economics and Mathematical Systems, 183, Springer-Verlag, Berlin, 1980.

[7] Colville, A.R., "A Comparative Study on Nonlinear Programming Codes", Proceeding of the Princeton Symposium on Mathematical Programming, Kuhn, H.W., ed., Princeton, N.J., pp. 481-501, 1970.

[8] Eason, E.D., "Validity of Colville's Time Standardization for

Comparing Optimization Codes", ASME Design Engineering Technical Conference, Chicago, September 1977, 77-DET-116.

[9] Eason, E.D. and Fenton, R.G., "A Comparison of Numerical Optimization Methods for Engineering Design", ASME Journal of Engineering for Industry, Series B, 96, No. 1, Feb. 1974, pp. 196-200.

[10] Stocker, D.C., "A Comparative Study of Nonlinear Programming Codes", M.S. Thesis, The University of Texas, Austin, Texas, 1969.

[11] Himmelblau, D.M., Applied Nonlinear Programming, McGraw-Hill, New York, 1972, pp. 368-369.

[12] Pearson, J.D., "Variable Metric Methods of Minimization", Computer Journal, 12, 1969, pp. 171-178.

[13] Huang, H.Y., and Levy, A.V., "Numerical Experiments on Quadratically Convergent Algorithms for Functional Minimization", JOTA, 6, 1970, pp. 269-282.

[14] Murtagh, B.A., and Sargent, R.W.H., "Computational Experience with Quadratically Convergent Minimization Methods", Computer Journal, 13, 1970, pp. 185-194.

[15] Pappas, M. and Moradi, J.Y., "An Improved Direct Search Mathematical Programming Algorithm", Journal of Engineering for Industry, Trans., ASME, Series B, Vol. 97, No. 4, Nov. 1975, pp. 1305-1310.

[16] Schuldt, S.B., Gabriele, G.A., Root, R.R., Sandgren, E. and Ragsdell, K.M., "Application of a New Penalty Function Method to Design Optimization", Journal of Engineering for Industry, Trans. ASME, Series B, Vol. 99, No. 1, Feb. 1977, pp. 31-36.

[17] Gabriele, G.A. and Ragsdell, K.M., "The Generalized Reduced Gradient Method: A Reliable Tool for Optimal Design", Journal of Engineering for Industry, Trans. ASME, Series B, Vol. 99, No. 2, May 1977, pp. 394-400.

[18] Dembo, R.S., "A Set of Geometric Programming Test Problems and Their Solutions", Working paper #87, Department of Management Sciences, University of Waterloo, Waterloo, Ontario, June 1974.

[19] Himmelblau, D.M., Applied Nonlinear Programming, McGraw-Hill, New York, 1972.

[20] Sandgren, E., "A Statistical Review of the Sandgren-Ragsdell Comparative Study", Mathematical Programming Society COAL Meeting, Boulder CO, Jan. 5-6, 1981.

[21] Rosen, J.B. and Suzuki, S., "Construction of Nonlinear Programming Test Problems", ACM Communications, 8, No. 2, Feb. 1965.

[22] M.J. Rijckaert and X.M. Martens, "A Comparison of Generalized GP Algorithms", Report CE-RM-7503, Katholieke Universiteit, Leuven, Belgium, 1975.

[23] R.S. Dembo, "The Current State-of-the-Art of Algorithms and Computer Software for Geometric Programming", JOTA, 26, 149, 1973.

[24] J.J. Dinkel, W.H. Elliot, and G.A. Kochenberger, "Computational Aspects of Cutting Plane Algorithms for Geometric Programming

Problems", Math. Progr., 13, 200, 1977.

[25] P.V.L.N. Sarma, X.M. Martens, G.V. Reklaitis, and M.J. Rijckaert, "A Comparison of Computational Strategies for Geometric Programs", JOTA, 26, 2, 1978.

[26] Root, R.R., "An Investigation of the Method of Multipliers", Ph.D. Dissertation, Purdue University, December 1977.

[27] Hoel, P.G., Introduction to Mathematical Statistics, John Wiley & Sons, New York, 1962, p. 279.

[28] Crowder, H., Dembo, R.S., and Mulvey, J.M., "On Reporting Computational Experiments with Mathematical Software", ACM Trans on Mathematical Software, 5, No. 2, June 1979, pp. 193-203.

[29] Lootsma, F.A., "Ranking of Nonlinear Optimization Codes According to Efficiency and Robustness" in Konstruktive Methoden der finiten nichtlinearen Optimierung, ed. Collatz, Meinardus and Wetterling, Birkhauser, Basel Switzerland, 1980, pp. 157-158.

[30] Saaty, T.L., "A Scaling Method for Priorities in Hierarchical Structures", J. Math. Psych., 15, 1977, pp. 234-281.

[31] Han, S.P., "A Globally Convergent Method for Nonlinear Programming", JOTA, 22, No. 3, 1977, pp. 297-309.

[32] Schittkowski, K., "Nonlinear Programming Methods with Linear Least Squares Subproblems", Mathematical Programming COAL Meeting, Boulder CO, Jan. 5-6, 1981.

Vol. 101: W. M. Wonham, Linear Multivariable Control. A Geometric Approach. X, 344 pages. 1974.

Vol. 102: Analyse Convexe et Ses Applications. Comptes Rendus, Janvier 1974. Edited by J.-P. Aubin. IV, 244 pages. 1974.

Vol. 103: D. E. Boyce, A. Farhi, R. Weischedel, Optimal Subset Selection. Multiple Regression, Interdependence and Optimal Network Algorithms. XIII, 187 pages. 1974.

Vol. 104: S. Fujino, A Neo-Keynesian Theory of Inflation and Economic Growth. V, 96 pages. 1974.

Vol. 105: Optimal Control Theory and its Applications. Part I. Proceedings 1973. Edited by B. J. Kirby. VI, 425 pages. 1974.

Vol. 106: Optimal Control Theory and its Applications. Part II. Proceedings 1973. Edited by B. J. Kirby. VI, 403 pages. 1974.

Vol. 107: Control Theory, Numerical Methods and Computer Systems Modeling. International Symposium, Rocquencourt, June 17–21, 1974. Edited by A. Bensoussan and J. L. Lions. VIII, 757 pages. 1975.

Vol. 108: F. Bauer et al., Supercritical Wing Sections II. A Handbook. V, 296 pages. 1975.

Vol. 109: R. von Randow, Introduction to the Theory of Matroids. IX, 102 pages. 1975.

Vol. 110: C. Striebel, Optimal Control of Discrete Time Stochastic Systems. III. 208 pages. 1975.

Vol. 111: Variable Structure Systems with Application to Economics and Biology. Proceedings 1974. Edited by A. Ruberti and R. R. Mohler. VI, 321 pages. 1975.

Vol. 112: J. Wilhelm, Objectives and Multi-Objective Decision Making Under Uncertainty. IV, 111 pages. 1975.

Vol. 113: G. A. Aschinger, Stabilitätsaussagen über Klassen von Matrizen mit verschwindenden Zeilensummen. V, 102 Seiten. 1975.

Vol. 114: G. Uebe, Produktionstheorie. XVII, 301 Seiten. 1976.

Vol. 115: Anderson et al., Foundations of System Theory: Finitary and Infinitary Conditions. VII, 93 pages. 1976

Vol. 116: K. Miyazawa, Input-Output Analysis and the Structure of Income Distribution. IX, 135 pages. 1976.

Vol. 117: Optimization and Operations Research. Proceedings 1975. Edited by W. Oettli and K. Ritter. IV, 316 pages. 1976.

Vol. 118: Traffic Equilibrium Methods, Proceedings 1974. Edited by M. A. Florian. XXIII, 432 pages. 1976.

Vol. 119: Inflation in Small Countries. Proceedings 1974. Edited by H. Frisch. VI, 356 pages. 1976.

Vol. 120: G. Hasenkamp, Specification and Estimation of Multiple-Output Production Functions. VII, 151 pages. 1976.

Vol. 121: J. W. Cohen, On Regenerative Processes in Queueing Theory. IX, 93 pages. 1976.

Vol. 122: M. S. Bazaraa, and C. M. Shetty, Foundations of Optimization VI. 193 pages. 1976

Vol. 123: Multiple Criteria Decision Making. Kyoto 1975. Edited by M. Zeleny. XXVII, 345 pages. 1976.

ol. 124: M. J. Todd. The Computation of Fixed Points and Applications. VII, 129 pages. 1976.

l. 125: Karl C. Mosler. Optimale Transportnetze. Zur Bestimmung es kostengünstigsten Standorts bei gegebener Nachfrage. VI, 2 Seiten. 1976.

l. 126: Energy, Regional Science and Public Policy. Energy and vironment I. Proceedings 1975. Edited by M. Chatterji and P. Van mpuy. VIII, 316 pages. 1976.

l. 127: Environment, Regional Science and Interregional Modeling. ergy and Environment II. Proceedings 1975. Edited by M. Chatterji d P. Van Rompuy. IX, 211 pages. 1976.

l. 128: Integer Programming and Related Areas. A Classified bliography. Edited by C. Kastning. XII, 495 pages. 1976.

. 129: H.-J. Lüthi, Komplementaritäts- und Fixpunktalgorithmen in mathematischen Programmierung. Spieltheorie und Ökonomie. 145 Seiten. 1976.

Vol. 130: Multiple Criteria Decision Making, Jouy-en-Josas, France. Proceedings 1975. Edited by H. Thiriez and S. Zionts. VI, 409 pages. 1976.

Vol. 131: Mathematical Systems Theory. Proceedings 1975. Edited by G. Marchesini and S. K. Mitter. X, 408 pages. 1976.

Vol. 132: U. H. Funke, Mathematical Models in Marketing. A Collection of Abstracts. XX, 514 pages. 1976.

Vol. 133: Warsaw Fall Seminars in Mathematical Economics 1975. Edited by M. W. Loś, J. Loś, and A. Wieczorek. V. 159 pages. 1976.

Vol. 134: Computing Methods in Applied Sciences and Engineering. Proceedings 1975. VIII, 390 pages. 1976.

Vol. 135: H. Haga, A Disequilibrium – Equilibrium Model with Money and Bonds. A Keynesian – Walrasian Synthesis. VI, 119 pages. 1976.

Vol. 136: E. Kofler und G. Menges, Entscheidungen bei unvollständiger Information. XII, 357 Seiten. 1976.

Vol. 137: R. Wets, Grundlagen Konvexer Optimierung. VI, 146 Seiten. 1976.

Vol. 138: K. Okuguchi, Expectations and Stability in Oligopoly Models. VI, 103 pages. 1976.

Vol. 139: Production Theory and Its Applications. Proceedings. Edited by H. Albach and G. Bergendahl. VIII, 193 pages. 1977.

Vol. 140: W. Eichhorn and J. Voeller, Theory of the Price Index. Fisher's Test Approach and Generalizations. VII, 95 pages. 1976.

Vol. 141: Mathematical Economics and Game Theory. Essays in Honor of Oskar Morgenstern. Edited by R. Henn and O. Moeschlin. XIV, 703 pages. 1977.

Vol. 142: J. S. Lane, On Optimal Population Paths. V, 123 pages. 1977.

Vol. 143: B. Näslund, An Analysis of Economic Size Distributions. XV, 100 pages. 1977.

Vol. 144: Convex Analysis and Its Applications. Proceedings 1976. Edited by A. Auslender. VI, 219 pages. 1977.

Vol. 145: J. Rosenmüller, Extreme Games and Their Solutions. IV, 126 pages. 1977.

Vol. 146: In Search of Economic Indicators. Edited by W. H. Strigel. XVI, 198 pages. 1977.

Vol. 147: Resource Allocation and Division of Space. Proceedings. Edited by T. Fujii and R. Sato. VIII, 184 pages. 1977.

Vol. 148: C. E. Mandl, Simulationstechnik und Simulationsmodelle in den Sozial- und Wirtschaftswissenschaften. IX, 173 Seiten. 1977.

Vol. 149: Stationäre und schrumpfende Bevölkerungen: Demographisches Null- und Negativwachstum in Österreich. Herausgegeben von G. Feichtinger. VI, 262 Seiten. 1977.

Vol. 150: Bauer et al., Supercritical Wing Sections III. VI, 179 pages. 1977.

Vol. 151: C. A. Schneeweiß, Inventory-Production Theory. VI, 116 pages. 1977.

Vol. 152: Kirsch et al., Notwendige Optimalitätsbedingungen und ihre Anwendung. VI, 157 Seiten. 1978.

Vol. 153: Kombinatorische Entscheidungsprobleme: Methoden und Anwendungen. Herausgegeben von T. M. Liebling und M. Rössler. VIII, 206 Seiten. 1978.

Vol. 154: Problems and Instruments of Business Cycle Analysis. Proceedings 1977. Edited by W. H. Strigel. VI, 442 pages. 1978.

Vol. 155: Multiple Criteria Problem Solving. Proceedings 1977. Edited by S. Zionts. VIII, 567 pages. 1978.

Vol. 156: B. Näslund and B. Sellstedt, Neo-Ricardian Theory. With Applications to Some Current Economic Problems. VI, 165 pages. 1978.

Vol. 157: Optimization and Operations Research. Proceedings 1977. Edited by R. Henn, B. Korte, and W. Oettli. VI, 270 pages. 1978.

Vol. 158: L. J. Cherene, Set Valued Dynamical Systems and Economic Flow. VIII, 83 pages. 1978.

Vol. 159: Some Aspects of the Foundations of General Equilibrium Theory: The Posthumous Papers of Peter J. Kalman. Edited by J. Green. VI, 167 pages. 1978.

Vol. 160: Integer Programming and Related Areas. A Classified Bibliography. Edited by D. Hausmann. XIV, 314 pages. 1978.

Vol. 161: M. J. Beckmann, Rank in Organizations. VIII, 164 pages. 1978.

Vol. 162: Recent Developments in Variable Structure Systems, Economics and Biology. Proceedings 1977. Edited by R. R. Mohler and A. Ruberti. VI, 326 pages. 1978.

Vol. 163: G. Fandel, Optimale Entscheidungen in Organisationen. VI, 143 Seiten. 1979.

Vol. 164: C. L. Hwang and A. S. M. Masud, Multiple Objective Decision Making – Methods and Applications. A State-of-the-Art Survey. XII, 351 pages. 1979.

Vol. 165: A. Maravall, Identification in Dynamic Shock-Error Models. VIII, 158 pages. 1979.

Vol. 166: R. Cuninghame-Green, Minimax Algebra. XI, 258 pages. 1979.

Vol. 167: M. Faber, Introduction to Modern Austrian Capital Theory. X, 196 pages. 1979.

Vol. 168: Convex Analysis and Mathematical Economics. Proceedings 1978. Edited by J. Kriens. V, 136 pages. 1979.

Vol. 169: A. Rapoport et al., Coalition Formation by Sophisticated Players. VII, 170 pages. 1979.

Vol. 170: A. E. Roth, Axiomatic Models of Bargaining. V, 121 pages. 1979.

Vol. 171: G. F. Newell, Approximate Behavior of Tandem Queues. XI, 410 pages. 1979.

Vol. 172: K. Neumann and U. Steinhardt, GERT Networks and the Time-Oriented Evaluation of Projects. 268 pages. 1979.

Vol. 173: S. Erlander, Optimal Spatial Interaction and the Gravity Model. VII, 107 pages. 1980.

Vol. 174: Extremal Methods and Systems Analysis. Edited by A. V. Fiacco and K. O. Kortanek. XI, 545 pages. 1980.

Vol. 175: S. K. Srinivasan and R. Subramanian, Probabilistic Analysis of Redundant Systems. VII, 356 pages. 1980.

Vol. 176: R. Färe, Laws of Diminishing Returns. VIII, 97 pages. 1980.

Vol. 177: Multiple Criteria Decision Making-Theory and Application. Proceedings, 1979. Edited by G. Fandel and T. Gal. XVI, 570 pages. 1980.

Vol. 178: M. N. Bhattacharyya, Comparison of Box-Jenkins and Bonn Monetary Model Prediction Performance. VII, 146 pages. 1980.

Vol. 179: Recent Results in Stochastic Programming. Proceedings, 1979. Edited by P. Kall and A. Prékopa..IX, 237 pages. 1980.

Vol. 180: J. F. Brotchie, J. W. Dickey and R. Sharpe, TOPAZ – General Planning Technique and its Applications at the Regional, Urban, and Facility Planning Levels. VII, 356 pages. 1980.

Vol. 181: H. D. Sherali and C. M. Shetty, Optimization with Disjunctive Constraints. VIII, 156 pages. 1980.

Vol. 182: J. Wolters, Stochastic Dynamic Properties of Linear Econometric Models. VIII, 154 pages. 1980.

Vol. 183: K. Schittkowski, Nonlinear Programming Codes. VIII, 242 pages. 1980.

Vol. 184: R. E. Burkard and U. Derigs, Assignment and Matching Problems: Solution Methods with FORTRAN-Programs. VIII, 148 pages. 1980.

Vol. 185: C. C. von Weizsäcker, Barriers to Entry. VI, 220 pages. 1980.

Vol. 186: Ch.-L. Hwang and K. Yoon, Multiple Attribute Decision Making – Methods and Applications. A State-of-the-Art-Survey. XI, 259 pages. 1981.

Vol. 187: W. Hock, K. Schittkowski, Test Examples for Nonlinear Programming Codes. V. 178 pages. 1981.

Vol. 188: D. Bös, Economic Theory of Public Enterprise. VII, 142 pages. 1981.

Vol. 189: A. P. Lüthi, Messung wirtschaftlicher Ungleichheit. IX, 287 pages. 1981.

Vol. 190: J. N. Morse, Organizations: Multiple Agents with Multiple Criteria. Proceedings, 1980. VI, 509 pages. 1981.

Vol. 191: H. R. Sneessens, Theory and Estimation of Macroeconomic Rationing Models. VII, 138 pages. 1981.

Vol. 192: H. J. Bierens: Robust Methods and Asymptotic Theory in Nonlinear Econometrics. IX, 198 pages. 1981.

Vol. 193: J. K. Sengupta, Optimal Decisions under Uncertainty. VII, 156 pages. 1981.

Vol. 194: R. W. Shephard, Cost and Production Functions. XI, 104 pages. 1981.

Vol. 195: H. W. Ursprung, Die elementare Katastrophentheorie: Eine Darstellung aus der Sicht der Ökonomie. VII, 332 pages. 1982.

Vol. 196: M. Nermuth, Information Structures in Economics. VIII, 236 pages. 1982.

Vol. 197: Integer Programming and Related Areas. A Classified Bibliography. 1978 – 1981. Edited by R. von Randow. XIV, 338 pages. 1982.

Vol. 198: P. Zweifel, Ein ökonomisches Modell des Arztverhaltens. XIX, 392 Seiten. 1982.

Vol. 199: Evaluating Mathematical Programming Techniques. Proceedings, 1981. Edited by J. M. Mulvey. XI, 379 pages. 1982.